World Scientific Publishing Co Pte Ltd
P.O. Box 128
Farrer Road
Singapore 9128

ISBN 9971–950–00–6
Printed by Singapore National Printers (Pte) Ltd

Partons in Soft-Hadronic Processes

Proceedings of the
Europhysics Study Conference
Erice, Italy
March 8-14, 1981

Edited by R. T. VAN DE WALLE
University of Nijmegen — Netherlands

World Scientific Singapore
1981

FOREWORD

From March 8-14, 1981, a meeting was held in the Ettore Majorana Centre for Scientific Culture - Erice (Sicily) under the auspices of the European Physical Society, the Dutch National Institute for Nuclear and High-Energy Physics and the Italian National Research Council. The aim of this conference was to further the understanding of low-p_t hadron reactions in terms of fundamental concepts. Apart from its intrinsic merits such an understanding could also shed light on the long-range interaction component of Quantum-Chromo-Dynamics i.e. on the region of QCD related to the confinement problem. To this end a comparison was made of the present phenomenological models. During the last years substantial progress had been achieved with the continuing developments of the recombination and fragmentation models. A study conference on the limited topic of the phenomenology of low-p_t interactions in terms of partons seemed timely.

The format of the meeting was chosen so as to make its 'proceedings' result directly in a systematic and coherent book on the topic. To this end a very limited number of speakers was invited to present a series of lectures on their topic of expertise within well defined boundaries. The following pages contain the fruits of their efforts. I would like to thank them all for their efforts before, during and after the conference.

The value of a topical book such as the present one depends heavily on the speed with which it is published. To this end (and also to keep costs at a minimum) the solution of printing from camera-ready typing was adopted. We hope that the reader will find partial compensation in those factors of speed and cost for the typographical and esthetic imperfection inherent in such a reproduction.

I would like to thank Prof. S. Ratti, my co-organizer, for his patient support, Mrs. Dikmans-de Koning for her assistence in all phases of the organization and the present proceedings, Prof. A. Zichichi and Dr. A. Gabriele of the Ettore Majorana Centre for their encouragement and help with the conference. Apart from the sponsors mentioned above, we also gratefully acknowledge support from the National Institute for Nuclear Physics - Pavia, from the Italian Ministries of Public Education and of Scientific and Technological Research and from the Faculty of Science of the University of Nijmegen.

Nijmegen, 13 July 1981
R.T. Van de Walle

CONTENTS

Chapter I

SOFT HADRONIC INTERACTIONS - REVIEW OF THREE BASIC OBSERVATIONS

(W. Kittel - U. of Nijmegen)

Chapter II

LEPTON-HADRON INTERACTIONS AND NUCLEON STRUCTURE

(H. Lubatti - CERN)

Chapter III

MODELS FOR COLOR JETS AND FRAGMENTATION FUNCTIONS

(U.P. Sukhatme - U. of Illinois at Chicago Circle)

Chapter IV

HADRON STRUCTURE AND SOFT PROCESSES

(R. Hwa - U. of Oregon)

Chapter V

DUAL FRAGMENTATION MODELS OF SOFT HADRON-HADRON AND HADRON-NUCLEUS INTERACTIONS

(A. Capella - L.Ph.TPE, Orsay)

Chapter VI

THE QUARK-PARTON MODEL AND PRODUCTION OF LOW p_t PARTICLES IN NUCLEI

(A. Dar - Technion-Haifa)

Chapter VII

QCD-CONFINEMENT, DTU-MODEL AND HADRON-NUCLEUS COLLISIONS

(Ch.B. Chiu - U. of Texas, Austin)

Chapter VIII

QCD AND LOW-p_t PHYSICS

(J.F. Gunion - U. of California, Davis)

Chapter IX

CONCLUDING REMARKS

Chapter I: SOFT HADRONIC INTERACTIONS - REVIEW OF THREE BASIC OBSERVATIONS

W. KITTEL
University of Nijmegen
Nijmegen, The Netherlands

Abstract:

Soft hadron-hadron collisions are compared to each other and to deep inelastic lepton-hadron and e^+e^- collisions in the light of quark-parton pictures. Three basic observations can be made. They are the success of quark statistics to explain different quantum number hadron yields, the reflection of the valence quark distribution in hadronic particle production and the similarity of jets in hadron-hadron, deep inelastic lepton-hadron and e^+e^- collisions.

Three observations have recently been drawing increased attention to low p_t hadronic collisions

1) Cross sections for resonance and particle production in central and fragmentation regions can be understood from quark statistics.
2) Pion production in the nucleon fragmentation region of soft hadron-hadron collisions seems to reflect the valence quark distribution in the nucleon as observed in moderately deep inelastic lepton-nucleon collisions.
3) Quark fragmentation jets from e^+e^- annihilation and deep inelastic collisions seem to resemble soft hadronic production in longitudinal, transverse and multiplicity behaviour of the produced hadrons.

These observations lead to the expectation that the parton structure of hadrons also governs hadron-hadron collisions at low p_t. To test this unifying concept and to use its far-reaching consequences to illuminate not only the complicated hadron-hadron collisions, but the (soft) quark fragmentation in general, is the basis of present experimental and theoretical effort in this field. Before discussing the status of this effort in the later sections, we shall first review the three basic observations.

1. Quark Additivity

Early data [Bö77,Ki77,Zi77] give evidence for a two domain picture of inclusive particle and resonance production. The fragmentation domain depends on the quantum numbers of the produced particle or resonance and of the incoming fragmenting particle. The momentum shape in the central domain is universal, i.e. does neither depend on the quantum numbers of the incoming particles nor on those of the produced particle or resonance. In particular, the following experimental results have been obtained:

- suppression of mesons in the proton fragmentation region (see fig.1) [Bö78];
- a ratio $\rho^+:\rho^0:\rho^- \approx 2:1:0$ for $\pi^+p \to \rho$ forward (fig.1) [Bö78];
- suppression of $\bar{K}^{*0}$ in π^- fragmentation, suppression of K^{*-} in π^+ fragmentation (fig.2a) [Ev77];
- suppression of ϕ in proton or pion fragmentation (fig.2b) [Gh77];
- more $\pi \to \rho^0$ than $K \to \rho^0$ (fig.2c) [Sc78];
- a ratio of $(3 + 3\lambda):(3\lambda) \gtrsim 4\text{-}8$ for K^{*+} produced from K^+ and π^+ beams (or charge conjugate, see fig.2d) [Bö80].

Furthermore, it has been noted that only of the order of 10% of the pions are produced directly. The largest pion sources are $\rho^{\pm 0}$ and ω [Bö77].

An interesting approach to understand these observations is an extension of the Satz additive quark model for multiparticle production [Sa67] to inclusive resonance production by Anisovich and Shekhter [An73] and Bjorken and Farrar [Bj74]. According to this probably too simple model, one quark of the beam particle reacts with one quark of the target to produce a large quark-antiquark ($q\bar{q}$) sea in the center as drawn below:

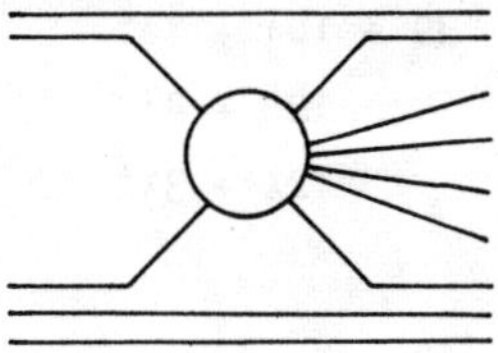

One or more quarks (or antiquarks) from the center join respectively the beam spectator and the target spectator to produce hadrons in the fragmentation regions. The remaining quarks (and antiquarks) in the sea join to form central hadrons.

SU(6) symmetry is assumed to be broken only by a relative suppression λ of strange quark production. The probability of quark production in the sea does, therefore, not depend on spin projection and charge. Furthermore, due to the large quark multiplicity at high energy, the probability of quark production will not depend any more on the nature of the incoming quarks. Thus, the distribution of particles produced in the $x \ll 1$ range at high energy should not depend on the nature of incident particles, i.e.:

$$\frac{d^2 \sigma_i}{dx\, dp_t^2} = C_i F(x, p_t^2) \tag{1}$$

where $F(x,p_t^2)$ is universal (as indeed observed). The coefficient C_i for production of hadron i can be calculated in relative scale from the model [An73]. For the production of the two $q\bar{q}$ S-wave nonets $J = 0^-, 1^-$ the coefficients C_i are given in table 1. A factor 2J+1 is included in C_i to account for the number of possible spin states.

For $J=0^-, 1^-$ production in the fragmentation region, one spectator is joined by one quark from the sea. Obviously, the distribution depends on the nature of the spectator. If the incoming particle is a proton, two spectators have to be joined by anti-quarks from the sea, which leads to an extra suppression factor δ. Various meson yields for the fragmentation region [An73] are given in table 2.

Table 1

Relative probabilities of $J = 0^-, 1^-$ meson production in the small longitudinal momentum region.

	C_i	Direct + Decay (1^- Mesons + Baryons)
π^+, π^-	3	$36 + 16\lambda + 6\lambda^2$
π^0	3	$39 + 16\lambda + 7\lambda^2$
K^+, K^-	3λ	$12\lambda + 4\lambda^2$
$K^0, \bar{K}^0$	3λ	$12\lambda + 3\lambda^2$
η	$1 + 2\lambda^2$	
X	$2 + \lambda^2$	
$\rho^+, \rho^0, \rho^-, \omega$	9	
ϕ	$9\lambda^2$	
$K^{*\pm}, K^{*0}, \bar{K}^{*0}$	9λ	

Table 2

Relative probabilities of vector meson production in the p, π^+ and K^+ fragmentation region

	p	π^+	K^+
ρ^+	12δ	6	3
ρ^0	9δ	3	3/2
ρ^-	6δ	–	–
ω	9δ	3	3/2
ϕ	–	–	3λ
K^{*+}	$12\delta\lambda$	3λ	$3 + 3\lambda$
K^{*0}	$6\delta\lambda$	–	3
$\bar{K}^{*0}$	–	3λ	–
K^{*-}	–	–	–

As can be seen from table 1, the strange quark suppression factor λ can be obtained from the ratio of K/π, K^*/ρ or ϕ/ρ production, or more generally in a simultaneous fit of all available resonances as in fig.3a [Dr81,Sp80]. Results are given in table 3. There seems to be a slight increase of λ with energy.

Table 3. Strange quark suppression factor

	from	experiment	reference
0.15±0.02	K^{*+}/ρ^o	pp 12 GeV/c	[Bö80]
0.18±0.02	K^{*+}/ρ^o	pp 24 GeV/c	[Bö80]
0.21±0.03	ϕ/ρ^o	pp 24 GeV/c	[Bl75]
0.28±0.15	K^*/ρ^o	pp 69 GeV/c	[Am76]
0.22	K^+/π^+	pp 240 GeV/c	[Bo80]
0.32±0.10		pp 405 GeV/c	[Su79]
0.35±0.06	all	pp $\sqrt{s}$=52.2 GeV	[Dr81,Sp80]
0.20±0.04	K^{*+}/ρ^o	π^+p 16 GeV/c	[Bö80]
0.17±0.02	ϕ/ρ^o	π^-p 16 GeV/c	[Gh77]

The model may be oversimplified and its success in explaining the results in figs.1 and 2 not expected at the relatively low energies, but it is certainly good for serving as a first guideline.

One of the main assumptions of the model is that a "gas" of quarks and antiquarks with non-correlated spin projections is formed. As a consequence, the number of $q\bar{q}$ pairs with total spin $s_{q\bar{q}}$ is proportional to the statistical weight $2s_{q\bar{q}}+1$. This gave the expected ratio 3:1 for ρ/π or $K^*(890)/K$, more generally for mesons containing the same quarks and belonging to the same SU(6) multiplet.

The actually observed ratio depends on the contamination from decays of unobserved resonances. In ref. [An80], the ratio 3:1 is checked on K, K^* and K^{**} mesons (which appear in decay processes to a smaller extent and in a better controlled way than π and ρ) from pp [Ki79] and K^-p [Co79] data.

The meson L=1 SU(6) multiplet contains the four SU(3) nonets $J^P= 0^+$, 1^+, 1^+ and 2^+, with statistical weigths 2J+1. The total amount of particles in the SU(6) multiplet is therefore 12/5 times the amount T of tensor particles. Again 3/4 of the particles have to have $s_{q\bar{q}}=1$. Neglecting higher SU(6) multiplets, the s=1 resonance decay contribution to K production comes from vector decay V and from 9/5 of the tensor decay T.

From the data in table 4, the experimental value of the above mentioned quantity is in each case near to 75% of the total kaon production. This observation speaks for the hypothesis of the "gas" of non-correlated quarks and antiquarks in multihadron production processes. A point which needs further clarification, I feel, is the role of SU(6) breaking due to the different masses of the produced particles.

Table 4

Production of kaons in K^-p collisions (32 GeV/c)
and pp collisions (405 GeV/c)

	$K^-p \to K^-$		$K^-p \to \bar{K}^0$		$K^-p \to K^+$		$K^-p \to K^0$		$pp \to K_s^0$	
inclusive	mb	%	mb	%	mb	%	mb	%	mb	%
	8.3±1.5	100	8±0.5	100	1.6±0.5	100	1.6±0.5	100	7.4±0.5	100
Total contribution of vector mesons (V)	4±0.4	48	4.2±0.3	53	1.1±0.3	69	0.9±0.2	56	3.4±1	46
Total contribution of tensor mesons (T)	0.9±0.2	11	0.8±0.2	10	0.1±0.01	5	0.1±0.01	5	1.7±0.8	23
Mesons with $s_{q\bar{q}}=1$ (V+ 9/5 · T)	5.6±0.4	68 ±5	5.6±0.3	71 ±4	1.2±0.3	78 ±19	1.0±0.2	65 ±13	6.5±1.9	88 ±26

The value of the inclusive cross section $K^-p \to K^-$ is decreased by the value of the diffraction dissociation cross section. In the vextor meson cross section, production from tensor meson decays are not taken into account.

Gaisser and Halzen [Ga81] have used the data of [Bo80] to extract the strange quark mass. They assume that particles are produced by drawing quarks out of the large soup of quarks and antiquarks produced in the collision. From statistics, the particle cross sections are

$$\sigma_i = \eta(2I_i+1)(2J_i+1)\frac{1}{m_i^2}F(\frac{Q}{m_i}) \tag{2}$$

where $\eta = 1$ ($\frac{1}{6}$) for mesons (baryons) and Q is the Q-value of the process. The assumption is that heavy particle cross sections scale in Q/m. The factor $1/m_i^2$ then results from dimensions. The main consequence of the scaling assumption is that the difference in quark masses leads to the suppression of $\bar{\Lambda}$ versus $\bar{p}$ and $\bar{\Omega}$ versus $\bar{\Lambda}$ beyond the statistical counting factors. Fig.3b shows a fit to data with

$$m_i = m_o + (n_i - 1)m_s \tag{3}$$

where n_i is the number of strange quarks and $m_o=1/2\ (M_\Lambda + M_\Sigma)$. The fit gives a strange quark mass $m_s = 0.3 \pm 0.05$ GeV.

As described above, in a collision of a meson with a nucleon one constituent quark of the projectile meson interacts with the target and breaks into partons. Then the newly made sea quarks and antiquarks recombine with each other to hadrons in the central region, or recombine with a constituent quark-spectator to fragmentation hadrons. In the original version [An73], the probabilities for a quark-spectator to bind with a sea quark or a sea antiquark were assumed to be equal. As a result, the fragmentation mesons M_i and baryons B_i where produced in a ratio 2:1,

$$q_i \rightarrow \frac{1}{3} B_i + \frac{2}{3} M_i \ . \tag{4}$$

Experimentally, the baryon multiplicity in the meson fragmentation region is considerably smaller [Ed78,Cu79] while it seems to work in the case of e^+e^- annihilation [An79]. Such a difference can be understood if there is a reason for a quark-spectator in meson fragmentation to bind more preferably with a sea-antiquark (or an antiquark spectator with a sea-quark).

For the meson fragmentation region Shabelski [Sh80] assumes that baryon exchange in the t-channel is suppressed in comparison to meson exchange. As a consequence, the spectator quark (antiquark) more often picks up a sea antiquark (quark). Let us suppose that the constituent quark-spectator q_i picks up the sea-antiquark with probability β and the sea-quark with probability $1-\beta$. Because of symmetry, the antiquark-spectator $\bar{q}_i$ picks up the sea-quark with probability β and the sea-antiquark with probability $1-\beta$. Consequently q_i turns into $\beta M_i + (1-\beta)q_i q$ in the first iteration and $(1-\beta)^2 B_i$ in a second. Accounting for the whole chain,

$$q_i \rightarrow \frac{\beta}{1-\beta+\beta^2} M_i + \frac{(1-\beta)^2}{1-\beta+\beta^2} B_i \ . \tag{5}$$

The value of β can be determined from the meson and baryon multiplicities at intermediate x values ($x \simeq 1/2$). The experimental value of $\bar{\Lambda}/K^o_s$ at $0.4 \lesssim x \lesssim 0.7$ ($p_t \simeq 0.07$ GeV/c) in π^- fragmentation at 200 GeV/c is 0.11 ± 0.01

[Ed78]. With $2(1-\beta)^2/\beta$, this leads to $\beta \approx 0.8$. For Λ/K^o_s from K^- fragmentation the prediction is $\approx 2.3\ (1-\beta)^2/\beta$. From an experimental value of 0.135 ± 0.025 [Ed78] one again finds $\beta \approx 0.8$. For these predictions a strange quark suppression $\lambda = 1/3$ has been used, in good agreement with $(\Lambda+\bar{\Lambda})/(n+\bar{n}) \approx 0.35$ from π^- fragmentation compared to the experimental value of 0.31 ± 0.04 [Ed78], and with the value obtained from fig.3a.

A detailed application of quark additivity is the formulation of Kalinowski, Pokorski and Van Hove [Ka79]. The authors investigate the overall fate of the valence quarks of an incident hadron when this fragments in a low p_t hadronic collision. In this formulation there are six parameters A_1, A_2, A_3, η, η', and ξ: The parameter A_1 is the probability that the three incident quarks of a proton emerge in three distinct outgoing hadrons; A_2 is the probability that the three incident quarks are shared between two outgoing hadrons; A_3 is the probability that the three quarks emerge in one outgoing baryon. A single incident quark may recombine with non-incident antiquarks or quarks to form a meson with probability η, or a baryon with probability η'. With these definitions one has the constraints

$$A_1 + A_2 + A_3 = 1 \tag{6}$$

and

$$\eta + \eta' = 1 \tag{7}$$

so that only four of the six parameters are independent.

Finally, the probabilities for any non-incoming quark (or antiquark) to be strange, up, or down are ξ, $(1-\xi)/2$, and $(1-\xi)/2$, respectively (neglecting the production of heavier flavours). It is connected to the strange quark suppression λ by $\lambda=2\xi(1-\xi)$ and therefore of the order of $\xi=0.1$ (see table 3).

From a careful analysis of existing data on meson, hyperon and antibaryon production, the authors determine the following values

$$\begin{aligned} A_1 &= 0.05 - 0.1 \\ A_2 &= 0.6 - 0.5 \\ A_3 &= 0.35 - 0.4 \end{aligned} \tag{8}$$

Furthermore, when the outgoing hadron contains one and only one incident valence quark, it is most of the time a meson or meson resonance and only in 1-2% of cases a baryon or baryon resonance.

Crucial tests of the model are provided by a comparison of the predictions with the experimental results for strange baryon and antibaryon production [Bo80a].

For antibaryon production in the fragmentation region the main diagram is shown in Fig.4a. It is proportional to $A_2\,\eta'$. To first order in η' the mean multiplicities $\langle n_{\bar{B}}\rangle$ of the antibaryons produced in the fragmentation of the incoming proton (x > 0) are given by

$$\langle n_{\bar{p}}\rangle = 1/2\ \eta' A_2 (1 - \xi)^3 \tag{9}$$
$$\langle n_{\bar{\Xi}}\rangle = 3/2\ \eta' A_2 (1 - \xi)\xi^2 \tag{10}$$
$$\langle n_{\bar{\Omega}}\rangle = \eta' A_2 \xi^3 \tag{11}$$

The mean multiplicity of strangeness S = +1 antibaryons is

$$\langle n_{S=+1}\rangle = 3\eta' A_2 (1 - \xi)^2 \xi,$$

with $\langle n_{\bar{\Sigma}^-}\rangle + \langle n_{\bar{\Sigma}^+}\rangle = \langle n_{\bar{\Sigma}^o}\rangle + \langle n_{\bar{\Lambda}}\rangle$, so that

$$\langle n_{\bar{\Sigma}_{ch}}\rangle = \langle n_{\bar{\Sigma}^-}\rangle + \langle n_{\bar{\Sigma}^+}\rangle = 3/2\ \eta' A_2\ (1 - \xi)^2 \xi\ . \tag{12}$$

Since all the fragmentation antibaryons are produced by the same diagram (Fig.4a), the x and p_t dependence of their invariant cross-sections are expected to be the same. One can therefore make a direct comparison with the predictions of the model at a given average (x, p_t) value. The observed and expected $\bar{B}/\pi^-$ ratios at x=0.48 and p_t=600 MeV/c are listed in table 5. For the expected ratios, the measured $\bar{p}/\pi^-$ value is used as input, the other $\bar{B}/\pi^-$ ratios are calculated from $\bar{p}/\pi^-$ using eqs. (9) to (12) and ξ =0.10.

The table shows a clear disagreement between the observed and the expected $\bar{B}/\pi^-$ ratios, and the disagreement increases with the antibaryon strangeness. Agreement would be reached by using a ξ value of 0.15 instead of 0.10. However, such a large value of ξ would generate difficulties in the baryon sector (see below). Assuming that, with ξ=0.10, the model gives a good description of the fragmentation contribution to the antibaryon production, the observed disagreement suggests that there is additional contribution from non-fragmentation antibaryon-baryon pairs, also in the non-central region. An unambiguous distinction between the pair production and the fragmentation contributions to antibaryon production can only be achieved with correlation measurements. For example, in the case of Ω^- production the fragmentation contribution would correspond to an $[\bar{\Omega}^-\Xi^-\Lambda(\Sigma)]$ system, whereas the pair production would give mainly $[\bar{\Omega}^-\Omega^-]$. The pair production is expected to depend mainly upon the mass of the

Table 5

Observed and expected $\bar{B}/\pi^-$ ratios at $\sqrt{s}$ = 21.2 GeV, x = 0.48, and p_T = 600 MeV/c

$\bar{B}/\pi^-$	$\bar{p}/\pi^-$	$\overline{\Sigma_{ch}}/\pi^-$	$\overline{\Xi^-}/\pi^-$	$\overline{\Omega^-}/\pi^-$
Observed	$(1.21 \pm 0.12) \times 10^{-2}$	$(6.2 \pm 1.0) \times 10^{-3}$	$(1.10 \pm 0.12) \times 10^{-3}$	$(1.6 \pm 0.5) \times 10^{-4}$
Expected	$(1.21 \pm 0.12) \times 10^{-2}$ input	$(4.1 \pm 0.4) \times 10^{-3}$	$(0.47 \pm 0.05) \times 10^{-3}$	$(0.36 \pm 0.04) \times 10^{-4}$
$\frac{\text{Observed}}{\text{Expected}}$	1	1.50 ± 0.28	2.33 ± 0.34	4.5 ± 1.5

Table 6

Average particle multiplicities in the forward hemisphere at $\sqrt{s}$ = 21.2 GeV

	p	Λ	Σ^+	Σ^-	Ξ^-	Ω^-
$\langle n_B \rangle$	0.68	6.3×10^{-2}	3.0×10^{-2}	1.4×10^{-2}	-	-
$\langle n_{\bar{B}} \rangle$	3.0×10^{-2}	1.3×10^{-2}	$\langle n_{\overline{\Sigma^+}} \rangle + \langle n_{\overline{\Sigma^-}} \rangle = 1.0 \times 10^{-2}$		1.9×10^{-3}	3.0×10^{-4}

system, rather than upon the strangeness of the antibaryon, which could explain the general trend shown by table 5.

The production of protons, $\Lambda\,(\Sigma^o)$, Σ^+, and Σ^-: In the model of Kalinowski et al. [Ka79] the particle multiplicities in the forward hemisphere are given by

$$\langle n_p\rangle = A_3 + 2/3\, A_2(1-\xi) + 3/2\,(A_1+1/3\,A_2)\eta'(1-\xi)^2 + 1/2\,A_1\eta^3(1-\xi)^3 \quad (13)$$

$$\langle n_\Lambda\rangle = \xi[\,2/3\,A_2 + 3\,(A_1+1/3\,A_2)\eta'(1-\xi) + 3/2\,A_1\eta^3(1-\xi)^2] \quad (14)$$

$$\langle n_{\Sigma^+}\rangle = \xi[\,1/3\,A_2 + 2\,(A_1+1/3\,A_2)\eta'(1-\xi) + 3/4\,A_1\eta^3(1-\xi)^2] \quad (15)$$

$$\langle n_{\Sigma^-}\rangle = \xi\,[\,(A_1+1/3\,A_2)\eta'(1-\xi) + 3/4\,A_1\eta^3(1-\xi)^2]. \quad (16)$$

They correspond to the diagrams of fig.4a-d and, for the term A_3 of $\langle n_p\rangle$, to the non-fragmented proton. Using the condition (6) and the forward particle multiplicities of table 6, a graphical representation of (13)-(16) is given in fig.5 for $\eta'=0$, $\eta=1$, and $\xi=0.10$ (To retain the fragmentation contribution only, the corresponding antiparticle multiplicity is subtracted from each particle multiplicity). Note that, for $\eta'=0$, (9)-(12) do not contribute to the antiparticle multiplicities which are in that case entirely due to the B-$\bar{B}$ pair production. The bands shown in fig.5 correspond to the uncertainties on the multiplicities. The Σ^+ and the $(\Lambda+\Sigma^o)$ bands overlap exactly. The intersection of the three bands gives $A_2 \simeq 0.6$ and $A_1 \simeq 0.15$. There is no overlap between the three bands when $\xi \simeq 0.15$ is assumed and therefore no acceptable solution for this high ξ value.

The Ξ^- and Ω^-multiplicities are

$$\langle n_{\Xi^-}\rangle = \xi^2[3/2\,(1-\xi)A_1\eta^3 + (A_1 + 1/3\,A_2)\eta'] \quad (17)$$

$$\langle n_{\Omega^-}\rangle = \xi^3 A_1\eta^3 \quad (18)$$

The ratio $\bar{\Omega}^-/\bar{\Xi}^-$ at x=0.48 and p_t=600 MeV/c is $(1.9\pm0.5)\cdot10^{-2}$. The terms which contain $A_1\eta^3$ in $\langle n_{\Xi^-}\rangle$ and $\langle n_{\Omega^-}\rangle$ both correspond to the diagram of fig.4c and are therefore expected to have identical x and p_t dependences. The second term of (17) corresponds to the diagram of fig.4a, which also produces the antibaryons, and to the diagram of fig.4d. The measured ratio $\bar{\Xi}^-/\Xi^-$ at x=0.48 and p_t=600 MeV/c is $(3.75\pm0.45)\cdot10^{-2}$. Using the values of A_1 and A_2 given above to determine the coefficients of η' in $\langle n_{\Xi^-}\rangle$ and $\langle n_{\bar{\Xi}^-}\rangle$, one can deduce from the measured $\bar{\Xi}^-/\Xi^-$ ratio that the η' term would contribute at most 1.5 x 10^{-2} of the observed Ξ^- p_t=600 MeV/c and x=0.48, if the p_t and x dependence of the η' terms were the same for Ξ and $\overline{\Xi^-}$. With the $(1-x)^7$ dependence of the antiparticle distributions and the observed $(1-x)^3$ dependence for Ξ^- above x=0.48, we may reasonably assume an enhancement factor of about $4/8(1-0.48)^{-4}$= 6.8 for the Ξ^- contribution. Assuming no contribution to the enhancement factor from the p_t dependence, the contribution of the η' term amounts to about 10% of the observed Ξ^- at p_t=600 MeV/c and x=0.48. Neglecting this small contribution, (17) and (18) give:

$$\Omega^-/\Xi^- = 2/3\xi\ /(1-\xi).$$

This is 7.4 x 10^{-2} for ξ=0.10 and 12 x 10^{-2} for ξ=0.15, in strong disagreement with the measured ratio of (1.9 ± 0.5) x 10^{-2}. Including a $B\bar{B}$ pair contribution, as shown to be needed above, the disagreement becomes even worse.

Kalinowski et al. have considered one fragmentation of the proton per collision. Assuming that <u>m</u>ulti <u>f</u>ragmentation may occur within the same interaction, the A_2^{MF} probability [Bo80] can be expressed as a function of the elementary probability a_2:

$$A_2^{MF} = a_2 + a_2^2 + \ldots = a_2/(1 - a_2). \qquad (19)$$

From (19) one gets

$$a_2 = A_2^{MF}\ /(1 + A_2^{MF}).$$

The main effect of the multifragmentation scheme for the s=-1 hyperon probabilities is to introduce a new A_2 term in $\langle n_{\Sigma^-}\rangle$. The effect of this term is shown in fig.5 where the two axes now correspond to the variables A_2^{MF} and A_1^{MF}.

The a_2 multifragmentation now becomes the dominant contribution to Ξ^- and Ω^- production at large x. The lowest-order diagrams are shown in figs.4e and 4f for Ξ^- and Ω^-, respectively. Neglecting the other terms, the Ω^-/Ξ^- ratio at x=0.48 is given:

$$\Omega^-/\Xi^- = a_2\xi\varepsilon\ , \qquad (20)$$

where the factor ε represents the effect of the different x dependences of the Ξ^- and Ω^- cross-sections. Since the observed dependence on x for Σ^- is $(1-x)^{1.5}$ and for Ξ^- is $(1-x)^3$, on can assume a $(1-x)^{4.5}$ dependence for Ω^- production, compatible with the measured invariant cross-sections at x=0.48 and x=0.55. Using, in addition, the value A_2^{MF}=0.6 one obtains Ω^-/Ξ^- =1.8 x 10^{-2} for ξ=0.10 -- in agreement with the observed ratio (and Ω^-/Ξ^- =2.8 x 10^{-2} for ξ=0.15).

Including the multifragmentation contributions, the final set of parameters for ξ=0.10 is:

$$A_1^{MF} \approx 0.1,\ A_2^{MF} \approx 0.6,\ A_3^{MF} \approx 0.3$$
$$\eta' < 0.1 \quad \text{and} \quad \eta > 0.9, \tag{21}$$

in good agreement with the original values.

The three-quark system of the incident proton appears to fragment into two quarks and one quark about 60% of the time. It is fully broken into three separate quarks in only 10% of the cases, while in the remaining 30% the three quarks stay together. Such a model gives a comprehensive representation of data on the production of charged baryons and antibaryons for strangeness S=-3 to +3.

In the light of the above data [Bo80a], Van Hove [Ho80] suggests a quark fragmentation-recombination picture, simpler than that reached in [Ka79]: each valence quark of the incident proton either recombines with an antiquark to give an outgoing meson or recombines with other quarks to give an outgoing baryon. These two possibilities turn out to occur in an uncorrelated way for each valence quark of the incident proton, with the respective probabilities of about 40% and 60%.

This picture differs from [Ka79] in that recombinations are neglected where valence quarks emerge in two or three distinct baryons. This assumption is based on the result of [Bo80] that also in the proton fragmentation region, antibaryon production can be attributed dominantly to the non-fragmentation type $B\bar{B}$ pair production with local compensation of quantum numbers.

Consequently, the mean multiplicity $n_f(B)$ of the fragmentation baryon B can be expressed as

$$n_f(B) = n(B) - n(\bar{B}) \tag{22}$$

with $n(B)$ and $n(\bar{B})$ being the mean multiplicities of B and $\bar{B}$ in the corresponding c.m. hemisphere.

The relevant diagrams are given in fig.6a with i= 0,1,2 or 3 valence quark in the emerging baryon. The three subclasses with i=1 and 2 are assumed to contribute with equal probabilities $a_1/3$ and $a_2/3$, respectively, i.e. symmetrically in u and d valence quarks of the protons. So, here the parameters are the probabilities a_i for the baryon to contain i valence quarks, and the probabilities τ and ξ for the picked-up quarks to be non-strange or strange, respectively. With

$$\sum_i a_i = 1 \tag{23}$$

and
$$\tau = \tfrac{1}{2}(1-\xi) \tag{24}$$

so that again, four of the parameters are independent.

In terms of these parameters the multiplicities can be expressed as

$$n_f(\Omega^-) = \xi^3 a_o, \quad n_f(\Xi^-) = \xi^2 (3\tau a_o + \frac{1}{3} a_1) \tag{25}$$

$$\left.\begin{aligned} n_f(\Sigma^-) &= \xi\tau(3\tau a_o + \frac{2}{3} a_1) \\ n_f(\Lambda^o) &= n_f(\Sigma^o) = \xi(3\tau^2 a_o + \tau a_1 + \frac{1}{3} a_2) \\ n_f(\Sigma^+) &= \xi(3\tau^2 a_o + \frac{4}{3}\tau a_1 + \frac{1}{3} a_2) \end{aligned}\right\} \tag{26}$$

$$n_f(p) = 4\tau^3 a_o + \frac{62}{27}\tau^2 a_1 + \frac{35}{27}\tau a_2 + w a_3, \quad w = \frac{8}{9} \tag{27}$$

where $n_f(p)$ contains the effect of Δ production as the lowest lying three-quark state with I=3/2. No spin factors are included in this analysis. From the positivity of the a_i it follows for the hyperons that

$$n_f(\Sigma^+)+n_f(\Sigma^-) \lesssim n_f(\Lambda+\Sigma^o) \lesssim 2n_f(\Sigma^+) \lesssim n_f(\Lambda^o+\Sigma^o)+n_f(\Sigma^-) \tag{28}$$

Fig.7 shows this inequality in terms of the ratios

$$\begin{aligned} V_o &= n_f(\Lambda^o + \Sigma^o)/\, n_f(\Sigma^+), & 1.5 \lesssim V_o \lesssim 2 \\ V_- &= n_f(\Sigma^-)/\, n_f(\Sigma^+) & 0 \lesssim V_- \lesssim 1 \end{aligned} \tag{29}$$

Positivity is fulfilled in the shaded triangle and the experimental point [Bo80a] indeed falls into this region, although more accurate data would help. The line AD within the same triangle corresponds to the experimental estimate of the ratio $n_f(\Omega^-)/n_f(\Xi^-)$.

For the fraction of events that is non-diffractive ($\bar{a}$), uncorrelated behaviour of the valence quarks means that each has a probability P of fragmenting and emerging as a meson, and a probability 1−P of emerging in the baryon B.

For the a_i this gives e.g.

$$a_o = \bar{a}P^3, \quad a_1 = 3\bar{a}P^2(1-P), \quad a_2 = 3\bar{a}P(1-P)^2 \quad . \tag{30}$$

This leads to the further constraint

$$a_1^2 = 3a_o a_2 \tag{31}$$

shown as solid line in fig.7, and to

$$P = a_1/(a_1+a_2) \quad . \tag{32}$$

With the help of the constraint (31) better accuracy can be obtained on the estimation of a_i than from the experimental point directly. The new results are

$$a_o = 0.05 \pm 0.03 \qquad a_1 = 0.20 \pm 0.04$$
$$a_2 = 0.30 \pm 0.16 \qquad a_3 = 0.45 \pm 0.17 \quad . \tag{33}$$

These values give

$$\bar{a} = 0.69 \pm 0.30 \text{ and } \quad P = 0.41 \pm 0.14 \quad . \tag{34}$$

What is the dynamical interpretation of these observations? The simplest may be the quark cascade with recombination of Fukuda et al [Fu77,Fu80] reached by completing the diagrams of fig.6a to those of fig.6b. In contrast to the quark fragmentation to be discussed later, the fate of the quark resulting from such fragmentation is here specified.

The basic assumption is no correlation between valence quarks. It has to be noted that next to a_3, the probability a_2 is quite large, however. Before the constraint (31) is applied, a_2 is even larger than a_3 (as can be deduced from the experimental point in fig.7). If this is due to diquark substructure, (ud) valence may be prefered over (uu) in the final baryon as the former can also be in isospin I=0. This would lead to different weights for the subclasses for i=2 and reduce Σ^+ (and Δ^{++}) production and lead to an upper limit for V_o larger than two. It would be great if better results on $n_f(\Lambda^0+\Sigma^0)/n_f(\Sigma^+)$ could be obtained.

We can conclude that pure quark statistics can give a consistent description of relative particle yields in central and fragmentation regions. This description is very global, but it can be taken as a first hint that quarks may play a role, also in soft hadronic reactions. If we want to know how, we have to go to more differential distributions.

2. Reflection of the Proton Valence Quark Distribution

The inclusive particle distribution of a hadron h in the fragmentation region of the proton at small transverse momenta p_t is known to obey simple scaling laws and (at high energies) is independent of the nature of the other incident hadron. Therefore one can define the structure functions of the form

$$F^{p\to h}(x)=\frac{1}{\sigma_{tot}}\int E\frac{d^3\sigma^{p\to h}}{d^3p}d^2p_t. \tag{35}$$

In the quark-parton model the electromagnetic structure functions are related to the probability distributions u(x), d(x) and s(x) to find a quark with momentum fraction x of the initial proton momentum by

$$\frac{1}{x}F_2^{ep}(x)=\frac{4}{9}(u(x)+\bar{u}(x))+\frac{1}{9}(d(x)+\bar{d}(x))+\frac{1}{9}(s(x)+\bar{s}(x)). \tag{36}$$

The analysis of lepton induced reactions has shown that only about half of the momentum of the proton is carried by quarks (q) and antiquarks ($\bar{q}$) and a small fraction by antiquarks alone. As $q(x)=\bar{q}(x)$ at $x=0$, the antiquark distribution is concentrated at small x (say $x \lesssim 0.2$, the "sea" region) and the same is expected for gluons which could dissociate into a $q\bar{q}$ pair. The presence of a $\bar{q}$ component in the proton structure function implies that the proton, which primarily consists of three valence quarks, has dissociated into a state with $q\bar{q}$ pairs if hit by the electron.

It has been suggested by W. Ochs [Oc77] that the same process of dissociation into partons occurs if the proton fragments into hadrons in the field of another hadron at high energies. The longitudinal momenta of partons will be approximately conserved during the interaction and there is a correspondence between the hadron state and the parton state in the following way: A quark in the parton state outside the sea-region ($x \gtrsim 0.3$) is a valence quark of a hadron of the final hadron state, and the momentum of the quark contributes to the momentum of the hadron.

In the fragmentation region of the proton, the π^+ can be assumed to be composed of a u valence and a $\bar{d}$ sea quark. Since the latter carries very little momentum, we expect to find a $\pi^+=|u\bar{d}\rangle$ with momentum similar to that of the u-quark . The same holds for a $\pi^-=|d\bar{u}\rangle$ and the d-quark. This leads to the relation

$$\frac{u(x)}{d(x)} \approx \frac{F^{p\to\pi^+}(x)}{F^{p\to\pi^-}(x)} \tag{37}$$

Outside the sea region, where those arguments are supposed to hold, one can express (37) directly by the electromagnetic structure functions for the proton and the neutron using (36) and $F_2^{ep}(u,d)=F_2^{ep}(d,u)$ as

$$\frac{F_2^{en}(x)}{F_2^{ep}(x)} = \frac{4+R(x)}{4R(x)+1} \quad , \qquad R(x)= \frac{F^{p\to\pi^+}(x)}{F^{p\to\pi^-}(x)}. \tag{38}$$

In fig.8 this relation is compared with data using the π^+/π^- ratios from ISR [Ca74,Al75,Si78] and SLAC data on σ_n/σ_p [At75]. There is a good agreement with (38) apart from a discrepancy at large x, but there the neutron data are less reliable due to deuteron smearing effects.

A particular significance may be attributed to the value of the ratio (38) for $x\to 1$. In a model in which quarks interact by the exchange of vector gluons, Farrar and Jackson [Fa75] have derived that the helicity of a fast quark (x~1) is the same as the helicity of the proton. Therefore by examining the wave function of the proton

$$|p(\uparrow)\rangle=2|u(\uparrow)u(\uparrow)d(\downarrow)\rangle - |u(\uparrow)d(\uparrow)u(\downarrow)\rangle + \text{perm.},$$

they find $u/d \to 5$ for $x \to 1$. This corresponds to $F_2^{en}/F_2^{ep} \to 3/7$ or using (37) to $\pi^+/\pi^- \to 5$. Indeed the pion data are in agreement with this prediction in fig.8.

Furthermore, the x-distribution of a pion in the fragmentation region of an incident proton is indeed similar to that of the valence quark it shares with the proton [Si78]. Fig.9 shows that the x-distributions of π^+ and K^+ are similar to the u-quark distribution u(x) derived from SLAC data on electron-nucleon deep inelastic scattering [Mi72,Bo73]. The π^- distribution agrees with d(x) up to x ~ 0.7, and is only slightly above d(x) for larger x-values.

The similarity of the ratio R of positive and negative pions produced in pp collisions at ISR, FNAL and PS energies at small p_t [Jo78,Si78,Ka71] to the u/d quark ratio as extracted from deep inelastic lepton-nucleon collisions suggests that also soft hadronic interactions can be used to probe the structure of the hadrons. The scaling observed in fig.10a [Bu79] in the case of pp reactions from p_{lab}= 19 to 2000 GeV/c may mean that the parton structure of the incoming particles is responsible for this effect, and that this is true already at low energies. The ratio increases significantly with Feynman x towards R=5, thus further supporting the idea that the constituents producing these pions reflect the quantum numbers of the fragmenting proton. So, if constituents are responsible for pion production in the proton fragmentation region of hadron-hadron collisions, these are valence quarks rather than gluons.

We conclude that the quantum numbers and the distribution of the target proton valence quarks can be found back in the particles produced in the target fragmentation region. What is the role of the beam quantum numbers?

Fig.10 gives a comparison [Bu79,Sc81] of the ratio of the inclusive invariant cross sections for π^+ and π^- production as a function of x in the target fragmentation region for the reactions $\pi^+ p \rightarrow \pi^\pm X$ at 16 GeV/c to that for $p \rightarrow \pi^\pm X$ (scaling between 19 and 200 GeV/c). R is considerably larger for 16 GeV/c π^+p than for pp reactions. On the other hand, R falls below the proton curve for $x \lesssim -0.5$ for π^-p collisions. The tendency of R being higher in π^+p reations than in π^-p reactions persists up to $p_{lab} \approx 147$ GeV/c (fig.10b). It can be explained by a $q\bar{q}$-annihilation mechanism [Bu79].

The approach of R to a high energy limit is given in fig.11 for $\pi^\pm$ p collisions, as a function of $s^{-1/2}$, for several x intervals. The energy dependence agrees with an $s^{-1/2}$ term, the asymptotic limit is consistent with the ratio R for pp reactions. It is, however, also consistent with a flat (rather than increasing) x dependence (fig.12), above $x \approx 0.2$.

What is the influence of resonances and diffraction dissociation on the ratio R? As shown for 32 GeV/c [Aj80] in fig.13, the high R values observed for $0.2<|x|<0.4$ are related to the fact that a large fraction of π^+'s in this x-range are decay products of the Δ^{++}. The influence of diffraction has been studied in [Sc81] by removing events with a single forward pion. The effect of diffraction is found to be negligible.

Similar observations as for $\pi^\pm$ p reactions have been made for K^+p reactions [Bu79,Sc81 ,Aj80 ,Ba81a]. In fig.14a, the difference of the π^+/π^- ratios for K^+ and K^- beams is still very considerable at 32 GeV/c, from each other as well as from that for the proton beam. At 147 GeV/c K^+p and pp results agree within the large errors with the lower energy K^-p data (fig.14b), indicating that the mechanism causing a difference between K^+ and p beams at lower energies has lost its importance at ≈ 150 GeV/c. As is the case for the ratios from πp reactions, there seems to be a trend towards a constant or even decreasing ratio R in the large x region.

Furthermore, a strong energy dependence is observed for the ratio R^{-1} for inclusive $\pi^{\mp}$ production in $\pi^{-}n$ reactions between 21 and 360 GeV/c [Ab81]. In fig.15 is shown that there is - in addition to the s dependence - a p_t dependence of the ratio R^{-1} in the large x region, in particular for 21 GeV/c, not observed in earlier pn, $\pi^{+}n$ and pp data at ISR and FNAL [Go72,Jo78,Si78,Ei79,Bi79].

A p_t dependence for x<0 ratios is also reported for 32 GeV/c in ref. [Aj80] (fig.15b). Averaged over all x, the ratio tends to a plateau value of about 4.5. The shape of the distribution is remarkably similar to the x-dependence of R. This effect cannot be blamed on large p_t jet production. Since it increases towards larger x values, the p_t dependence seems to be a property of the fragmentation jet itself. A careful study of the p_t dependence would be needed from pion and proton beams at the higher energies, since conclusions can be drawn on the diquark structure of the proton.

The second basic observation remains that the quantum numbers and the distribution of the target proton valence quarks can be found back in the particles produced in the target fragmentation region. At beam momenta above ~200 GeV/c, this statement is independent of the quantum numbers of the beam.

3. "Jets" of Low p_t Particles

a) Are there high p_t jets decaying into low p_t particles?

A topic deserving special interest is the production of intermediate p_t groups of low p_t particles that have the same sign of the transverse momentum component along a principal axis in the transverse plane. This axis is chosen to maximize the squares of transverse momentum components along it. The question is whether those groups of particles could be consistent with high p_t jets.

An FNAL experiment with pp and πp at 200 GeV/c [Br78] had shown that for p_t>3 GeV/c,

1. the cross section for jets is considerably larger than that for single particle production and

2. the slope of the p_t distribution for jets is essentially the same as that for single particle production.

Fig.16a shows their results [Br78] compared to lower p_t curves obtained for the groups defined above from K^-p at 110 GeV/c [De79a], simulating the same acceptance. The 110 GeV/c data extrapolated to p_t values above 3 GeV/c, join on smoothly to the results at higher p_t both in slope and absolute normalization. Similar results come from 147 GeV/c $\pi^+/K^+/pp$ reactions [Gr79].

The momentum distribution of particles within the multiparticle system M can be studied in terms of the fraction z of momentum of the system M carried by a particle i. Fig.16b compares the data from hadron-hadron, νp and e^+e^- interactions [De79a,Gr79]. All distributions fall exponentially with similar slopes at large values of z.

Consistency between intermediate p_t hadron groups and "high p_t jets" is further found in the charged particle multiplicity of the jet and in the remainder of the event for the two experiments [Br78 ,De79a].

We may conclude that multiparticle systems found at intermediate p_t in hadron-hadron collisions around 100 GeV/c are very similar or extrapolate smoothly to "jets" at p_t>3 GeV/c in pp and πp collisions at 200 GeV/c and to νp or e^+e^- jets.

Are these jets realy jets, e.g. as predicted by several models [Fi77,Fe77,El76]? These models predict features consistent with those described above. However, the basic features, large cross section, shape of the z distribution, sphericity, thrust, multiplicity, p_t and $p_{//}$ behaviour are also found in hadron-hadron collisions at low p_t and rather low energy.

To answer this question for 16, 110 GeV/c, 32 and 147 GeV/c, the shape of the events in momentum space has been studied [De79a,Gr79,Fr79]. In fig.17a, variables s are defined [Co75] as

$$s_i = 3\lambda_i \,/\, (\lambda_1+\lambda_2+\lambda_3\,), \qquad s_1 \geq s_2 \geq s_3 \tag{39}$$

in terms of the eigen-values λ_i of the matrix $M_{ij}= \Sigma p_i p_j$, where the sum runs over the number of (charged) particles in the event and p_1, p_2, p_3 are the momentum components in the cms.

Three distinct regions are shown in fig.17a:

1. A two-jet-like events will be "cigar" shaped in momentum space ($s_1 \simeq 3, s_2 \simeq s_3 \simeq 0$).
2. An event with high p_t jet(s) as well as beam and target residues will be more like "disc" shaped ($s_1 \simeq s_2 \simeq 1.5$, $s_3 \simeq 0$).

3. An event without jets will look like a "sphere" ($s_1 \approx s_2 \approx s_3 \approx 1$).

Inspection of figs.17b,c tells us that both at 16 and at 110 GeV/c the bulk of the events is cigar shaped. There is a long tail of the maximum stretching from the upper corner roughly through the middle line of the populated triangle. The tail falls faster for 110 than for 16 GeV/c. There is no evidence for a small second maximum in the "discy" corner, so that the events there could be tail events. In any case, they are less than 30, in violent disagreement with 167 events having multiparticle groups with $p_t>2$ GeV/c [De79a]. So only few "jets" with $p_t>2$ GeV/c are high p_t jets, - they don't exist, but look like jets. Are we observing just phase space or are the low p_t beam and target fragmentation jets the ones which are similar in all the observed features to jets at high p_t, e^+e^- annihilation and deep inelastic collisions?

Fig.18a shows a similar plot for $\pi^+/K^+/pp$ data at 147 GeV/c [He80a,Br81] (Instead of s_i as in fig.17 the linear combinations $X=3/2(s_2+s_3)$ and $Y=\sqrt{3}/2\ (s_2-s_3)$ are used as axes, so that the cigar-shaped corner is now on the lower left, the disc-shaped corner on top. Again, most of the events are in the two-jet corner, but there may be a little more than just a tail into the discy corner. If we arbitrarily define as disc-shaped events those with $X>0.25$ and $s_3<0.04$ one retains 8% of the events. These have $\langle p_t \rangle$ by 15% larger than normal events and $\langle p_{/\!/} \rangle$ 30% smaller than normal events.

The shape of the events in momentum space can also be described by the polar angle Θ_B and the azimuthal angle ϕ_B of the beam direction seen in the principal axes reference frame [Co75]. At 147 GeV/c the cos Θ_B distribution is peaked at +1 with $\langle \Theta_B \rangle \approx 12^o$. The ϕ_B distribution is shown in fig.18b. It exhibits a deviation from isotropy and shows that the beam lies preferentially in the plane defined by the eigenvectors corresponding to λ_1 and λ_2.

If the effect is real, it should become clearer at ISR energies. Indeed, evidence has been claimed for a two-jet structure in the same hemisphere as the "leading proton" [Ba80b]. In "planar" events the average thrust $\langle T^* \rangle$ (eq.41) in the jet cms increases as the thrust T in the overall pp cm system decreases (fig.18c). Monte Carlo events with low T do not show this correlation with high T^* events, so the effect may be genuine: the wider the jet in the pp cm system, the narrower it is in the jet rest system. Is this evidence for the onset of high p_t jets? A similar behaviour is observed for the three-jet events (two jets in one hemisphere) in e^+e^- production [Ba80b]. The interpretation may be quite different, however, since in the case of pp interactions we expect fragmentation jets from at least two proton constituents (diquark and bachelor quark). So the reason may due to the multiple quark structure of hadrons rather than the onset of high p_t jets or glue jets.

In any case we can conclude that high p_t jets (if they exist) will not influence conclusions drawn from the bulk of low p_t hadronic collisions.

b) Low p_t fragmentation jets.

The decrease in average sphericity <S>, thrust <1-T> and spherocity <S'> with increasing energy, generally interpreted as evidence for jet production, is not only a feature of reactions in which single quark effects are expected to dominate [Be78 ,Bo79 ,Ba79], but also of low p_t hadron-hadron collisions. In refs. [Gö81,He80a] the same variables have been used to estimate the jet-like nature of K^-p events as used in the e^+e^- and νN study [Be78 ,Ba79] namely sphericity S, thrust T and sphericity S' defined as follows:

$$S = 3/2 \min \frac{\Sigma p_t^2}{\Sigma p^2} \quad (= \quad X \text{ from above}), \quad 0<S<1 \tag{40}$$

with the summation over all charged tracks. The axis which minimises the transverse momentum squared is determined by the diagonalisation of the momentum-tensor of each event as described in [De79a]. Thrust is defined by:

$$T = \max \frac{\Sigma \, |p_{//}|}{\Sigma \, |p|}, \quad 1>T>1/2 \tag{41}$$

The direction of $p_{//}$, the thrust axis, is that single axis through the event vertex which maximizes the sum of magnitudes of longitudinal momentum components of tracks (principal axis [Br64]). Finally, spherocity is defined by

$$S' = \min \left(\frac{4 \, \Sigma |p_t|}{\pi \Sigma |p|} \right)^2, \quad 0<S'<1 \tag{42}$$

In all cases, the variables are defined in the overall centre-of-mass of the hadron-proton system.

The method used to determine the thrust axis was that employed in the e^+e^- experiments [Br79a]. This method involves trying as thrust axis all directions defined by the resultant momentum of all multibody systems in the events.

One important difference between e^+e^- annihilations and hadronic reactions is that in the latter case a preferred direction for the jet axis already exists, namely the beam direction. Fig.19, shows the distribution in the angle Θ_{BT} between the beam direction and the thrust axis in the overall centre-of-mass at 110 GeV (E_{cm}=14.8 GeV). It can be seen that the thrust axis is close to the beam direction. In fact the distribution in Θ_{BT} is not much wider than the distribution in the angle Θ_{ST} between the sphericity and thrust axis. The latter angle is a measure of the error in determining the "jet" direction. The fact that K^-p distribution in Θ_{ST} at E_{cm} =14.8 GeV is narrower than that reported by the PLUTO collaboration at 9.4 GeV may be evidence that differences in <S> between e^+e^- and K^-p at low E_{cm} are most likely due to problems with determining the sphericity axis in the e^+e^- data.

As further evidence on how well the jet axis can be defined, fig.20 shows for K^+p collisions at 70 GeV/c [Ba81b] the distributions of the relative angles between the axes defined by thrust, sphericity, spherocity, the direction of the beam and the direction of the largest momentum particle. The average values of the distributions in fig.20a-c give an estimate of the uncertainty in defining the jet direction; the averages of those in fig.20d-f demonstrate its correlation to the beam direction.

In fig.21, <S>, <1-T> and <S'> are shown as functions of the hp centre-of mass energy E_{cm}, 2 charged prongs removed to be consistent with the e^+e^- and $\bar{\nu}N$ data. At 70, 110 and 147 GeV/c, <S> for hp interactions is consistent with that of the e^+e^- data. At 10 and 16 GeV/c, <S> for K^-p is not as high as in the e^+e^- case. However, one should note that at low energies the DORIS and SPEAR e^+e^- data differ by as much as the K^-p data differ from the SPEAR results. It could be that the differences in the two sets of e^+e^- data, and between the e^+e^- and K^-p data, are due to the difficulty in determining the sphericity axis for high sphericity events.

The values of <1-T> in hp interactions in fig.21b are in reasonable agreement with the e^+e^- data. Fig.21b also shows the values obtained by the νNe experiment [Ba79] plotted at E_{cm} values equal to the corrected hadronic centre-of-mass energy W of the νNe reactions. The shape of the E_{cm} variation of the hp data is similar to that of the leptonic reactions. The values of <1-T> for the hp data are in particularly good agreement with the νNe results.

The differential sphericity distribution for h^+p interactions at 147 GeV/c ($\sqrt{s}$=16.6 GeV) is compared to TASSO results at the same energy in fig.22a, separately for the diffractive and non-diffractive components [He80]. Although the statistics and resolution problems in e^+e^- data do not allow a very detailed comparison, the agreement between hadron-hadron and e^+e^- data is good. The diffractive events, as expected, have a tendency to be peaked steeper towards small S than non-diffractive events.

The energy dependence of the shape of the sphericity distribution is shown in fig.22b [Go81]. As for e^+e^- collisions, one observes a change from a dip at S=0 at low energy to a sharp peak at S=0 at higher energies. The shape of the 16 GeV/c K^-p S distribution (E_{cm}=5.6 GeV) is peaked at rather lower S values than the SPEAR data at 6.2 GeV. Very good agreement is observed between the S distribution in K^-p at 110 GeV/c (E_{cm}=14.8 GeV) with the PLUTO result at 17 GeV. The TASSO S distribution at 17 GeV [Br79b] is similar to the PLUTO data. There is also good agreement in fig.22c between the T distribution at 110 GeV/c and the νN values for 8 GeV<W<12 GeV. In all cases it can be seen that diffractive events contribute only at the most "jet-like" values i.e. low S and high T. The agreement between the shape of the K^-p and the leptonic distributions is improved when the diffractive events are removed. The only major difference in the shapes of S and T distributions in K^-p and in published leptonic data remains the peaking at lower S values in the K^-p data at 16 GeV/c in fig.22b, when compared with the SPEAR data. This probably results from the problems in defining the sphericity axis at low E_{cm} discussed earlier.

The normalized 70 GeV/c K^+p rapidity distribution, evaluated with respect to the thrust axis [Ba81b] is compared to e^+e^- results at 13 GeV/c (TASSO) [Br80] in fig.23a. At similar energies, the two rapidity distributions agree remarkably well and it can be expected from the insert (and from old results on hadronic rapidity plateaus) that hadron-hadron and e^+e^- data show a similar $\ln\sqrt{s}$ increase of the plateau hight.

Also the normalized p_t^2 distributions relative to the sphericity axis (fig.23b) show agreement at 13 GeV/c, even though there may be indication of a larger cross section in the high p_t tail for the e^+e^- results.

The conclusion of jet universality is further supported from a comparison of the energy dependence of $\langle p_t \rangle$ and $\langle p_{/\!/} \rangle$ relative to the thrust axis: in fig.23c again, the hp and e^+e^- data have essentially the same behaviour, the rise in $\langle p_t \rangle$ and $\langle p_{/\!/} \rangle$ with E_{cm} felt to be a characteristic feature of single quark jets is in fact also a feature of hadronic low p_t

particle production.

More differentially, the $\langle p_t \rangle$ vs x distribution contains the famous "sea-gull", well known [Ba63] from the golden (Polish) ages of hadron-hadron multiparticle production. Fig.24a shows agreement between h^+p data at 147 GeV/c [He80] and e^+e^- (TASSO) data at the same cms energy [Fl79]. From earlier analysis it is known that the dip at $x \approx 0$ persists after removal of kinematical factors. The energy dependence of the "sea-gull" effect (now with kinematical factors removed) is shown in fig.24b-c for $K^-p \rightarrow \pi^-$ between 12.7 and 70 GeV/c [Ba81a,La80]. The sea-gull is lifting its wings at higher energy, while $\langle p_t \rangle$ stays constant at $x \approx 0$. This is expected for jet-like particle production in the fragmentation regions [Sa77] and has been qualitatively interpreted by QCD [Ma79].

In fig.25a, the effect of lifting the wings is compared to the rapid increase of $\langle p_t^2 \rangle$ (here defined in respect to the direction of the hadronic shower) known in ν and μ interactions [Bo79 ,De80,Au80], for all hadrons with $z>0.2$. While the increase with s is larger in the ν and μ interactions for p_{lab} <70 GeV/c, there is indication that there may be a steeper increase at higher energies also for hadron-hadron collisions.

If we assume that the hadron data are dominated by quark-quark interactions, the equivalent to the energy of the hadronic shower W in hadron-hadron interactions is the forward and backward energy $\sqrt{s}/2$ (see diagrams fig.25c). At the lower energies, collective effects like quark-meson or antiquark-baryon collisions may still be at work. Then the quark-hadron total energy should be relevant with an average of more like $\sqrt{s/3}$. Fig.25b, therefore shows a comparison of the average p_t^2 as a function of W^2 and s/3, respectively [La80]. In this scale (as well as in $\sqrt{s}/2$ [La80]) the hadron data agree with the ν and μ rise!

The observation of W^2 and s/4 similarity throws up a number of questions. Firstly, what is the consequence for the other varibles like z, S, T where similarity could be found at equal W^2 and s? Secondly, where are there trivial differences to be expected between e^+e^-, lh and hh interactions and how can they be eliminated.

At this stage we may ask whether the agreement between the data is accidental, (e.g. due to longitudinal phase space) or has any more fundamental dynamical origin. In ref. [Ba81b] the sphericity, thrust, spherocity and other distributions are compared to mere longitudinal phase space and the Field and Feynman parametrization of quark-parton jets [Fi78].

In all cases, the FF parametrization as well as LPS more or less describe the data (fig.26). One may conclude that FF is a somewhat complicated way to parametrize longitudinal phase space. Jet universality reduces to the equal p_t distributions and the equal average multiplicities. The rest follows from independent emission + conservation laws, equally for hadronic, deeply inelastic and e^+e^- hadron production. In all cases the same mechanism is at work, governing multiplicities and transverse momenta in an identical way.

If quark fragmentation is indeed governed by longitudinal phase space, it will be interesting to look for the trivial differences expected from the quantum numbers of excitation modes of the fragmenting quarks.

One difference of the hadron system produced in e^+e^- annihilation and pp collisions is that the baryon number is zero in the first case and two in the second. In other words, there are more interacting valence quarks in pp collisions, than in e^+e^- hadron production. A CERN-Bologna-Frascati Collaboration [Ba80a] using the CERN Split-Field Magnet at $\sqrt{s}$=62 GeV has attempted to remove the expected effects of baryon-number conservation by selecting events with a leading proton and redefining the fractional variables of the particle after removal of that proton (criterium: fastest particle should be positive and should have $0.4 < x < 0.8$). All other particles in the same hemisphere were assumed to be the hadronic system associated with that proton. A fractional variable

$$x_R^* = p/E_{had} \tag{43}$$

was defined by dividing the cms momentum p of a given particle by the energy of the hadronic system

$$E_{had} = \frac{\sqrt{s}}{2}\ (1-|x_{proton}|) \tag{44}$$

rather than by the energy of the beams.

In fig.27, the pp data (normalized to the mean charge multiplicity are compared to (normalized) e^+e^- data from the Tasso Collaboration [Br79b] at three different energies E_{had}. The agreement between the two distributions at all three E_{had} values, both in shape and absolute value (mean charged multiplicity), suggests that the mechanism for transforming energy into particles in these two processes must be the same. This analysis is extendend to protons with small values of x (large values of E_{had}). The results compare well with a Monte Carlo simulation for high $\sqrt{s}$ e^+e^- x_R

distributions. Furthermore, the ratio $\alpha=\langle E_{ch}\rangle/\langle E_{tot}\rangle$ of charged to total available energy is compared in fig.27d for e^+e^- data and pp after removing the fast protons [Ba81c]. In both cases, almost half of the energy goes into neutrals. This fact used to be known as "energy crisis" for e^+e^- collisions. From the pp reactions it can be shown that up to $\alpha^0_{pp}=0.10\pm0.10$ the neutral energy can be explained by π^0, K^0, $\bar{K}^0$, n and $\bar{n}$ production. It is therefore not necessary to speak of an energy crisis.

However, in fig.28 a comparison [Ng80] is shown to a parameter-free model which includes the leading proton as essential ingredient and further extracts all necessary ingredients from hard scattering. There is striking agreement of the model with the data for various values of x_p and the normalization is correctly reproduced. The model also reproduces the "cross over" (fig.28a) seen in the dN/dx^*_p, but can show that this behaviour is an artifact caused by the variable x^*_R since dN/dx versus x shows no such funny behaviour. The variable x^*_R is not natural in the model. It has been used only to compare to the (distorted) data. In fact, the model is based on two chains between a spectator diquark and an interacting valence quark each. This approach is very different from the suggestion of [Ba80a] to compare to e^+e^- after exclusion of the baryons.

The 147 GeV/c h^+p collaboration studied the effect on ⟨S⟩ and ⟨T⟩ when removing the leading particle [Br81]. As expected, ⟨S⟩ and 1- ⟨T⟩ increase, but the available hadron energy E_{had} decreases. Both values are compared to the e^+e^- and N data at the corresponding values of E_{had} in fig.29. It is not obvious to me that removing the proton improves the agreement. I would suggest to look for expected differences rather than to exclude them before they are shown to exist.

The disadvantage of removing the leading particle is that also the leading parton (and its companion) is removed. The better approach seems to me to select certain events where the jet is expected to derive from qq separation as in e^+e^- jet production. A possible laboratory for this study is meson or photon diffraction dissociation [Mi80a,Ra81]. The diffractive dissociation can be assumed to be a two-step process. First, the meson dissociates into a $q\bar{q}$ pair, then this $q\bar{q}$ system hadronizes in the same manner as hadronization of neutral quark-antiquark pairs during e^+e^- annihilation to two jets of hadrons. Diffraction dissociation has the enormous advantage over e^+e^- annihilation that the $q\bar{q}$ pair is definite and known and not a mixture of $u\bar{u}$, $d\bar{d}$, $s\bar{s}$, $c\bar{c}$ or $b\bar{b}$. This provides the unique opportunity to investigate jets of known flavour!

In fig.30a, a comparison [Mi80a,b] is shown of the charge multiplicity for the system X in $\pi^- p \rightarrow Xp$ at 205 GeV/c [Wi74] and the charge multiplicity in e^+e^- annihilation [Wo79]. From M_x=2 GeV onwards, the agreement is quite good. Below 2 GeV, there may be a contamination of the e^+e^- annihilation from ρ'. Diffraction dissociation, however, dominates X only in the region M_x < 6 GeV (fig.30b), so that at least the agreement above this value needs another explanation.

More differentially, the cross section $d\sigma/dM_x^2$ can be predicted for diffraction dissociation from $d\bar{u}$ quark fragmentation including isotopic spin dependence [Mi80b]. Two very different primordial fragmentation functions [Fi78] and [Mi80b] were used, but both seem to reproduce the trend in the still crude data.

Differences to be expected for diffractive and e^+e^- jets will be the angular distribution of the jets relative to the beam axis and heavy flavor multiplicities. I think this point deserves further study, as well as application to quark-diquark separation jets in proton diffraction dissociation.

We conclude that in general very good agreement is observed between low p_t hadronic, e^+e^- and νN multiparticle production. The latter two reactions are now commonly interpreted as being due to jet-like parton fragmentation. The agreement with low p_t hadronic production therefore leads to the conclusion that parton fragmentation plays an important role also there. For the future it will be necessary to look for (trivial) differences to understand better at which energy particle production has to be compared (as in the case of $\langle p_t^2 \rangle$) or how to correctly treat the difference in quark content (as for the ISR data). Further differences have to be expected from the different parton excitation reactions in e^+e^-, νN and low p_t hadronic collisions and from the differences in the cms of the interacting quarks.

4. Conclusions

We can conclude that pure quark statistics can give a consistent description of relative particle yields in central and fragmentation regions. This description is very global, but it can be taken as a first hint that quarks may play a role, also in soft hadronic reactions. If we want to know how, we have to go to more differential distributions.

The second basic observation remains that the quantum numbers and the distribution of the target proton valence quarks can be found back in the particles produced in the target fragmentation region. At beam momenta above ~200 GeV/c, this statement is independent of the quantum numbers of the beam. This second basic observation supports the quark recombination picture of hadron production in the fragmentation region, to be reviewed by Hwa [Hw81] at this workshop.

A third basic observation is the generally very good agreement between low p_t hadronic, e^+e^- and νN multiparticle production. The latter two reactions are now commonly interpreted as being due to jet-like parton fragmentation. The agreement with low p_t hadronic production therefore leads to the conclusion that parton fragmentation plays an important role also there. For the future it will be necessary to look for (trivial) differences to understand better at which energy particle production has to be compared (as in the case of $\langle p_t^2 \rangle$) or how to correctly treat the difference in quark content (as for the ISR data). Further differences have to be expected from the different parton excitation reactions in e^+e^-, νN and low p_t hadronic collisions and from the differences in the cms of the interacting quarks. The third observation, that jets produced in e^+e^- annihilation and deep inelastic lepto-production are very similar to these produced along the beam direction in ordinary hadron-hadron collisions has led to the assumption of parton fragmentation as a common underlying mechanism. This picture will be reviewed by Capella [Ca81] at this workshop.

REFERENCES

[Ab81] H. Abramowicz et al., Z. Physik C7 (1981) 199.

[Aj80] I.V. Ajinenko et al., Z. Physik C4 (1980) 181.

[Al75] E. Albini et al., CERN NP 75-2 (1975).

[Am76] V.V. Ammosov et al., Yad. Fiz. 24 (1976) 59.

[An73] V.V. Anisovich and V.M. Shekhter, Nucl.Phys. B55 (1973) 455.

[An79] V.V. Anisovich, P.E. Volkovitsky and M.N. Kobrinsky, Yad. Fiz. 29 (1979) 1054.

[An80] V.V. Anisovich, M.N. Kobrinsky and J. Nyiri, "A Proof of the Rules of Quark Statistics in Multiparticle Processes", Leningrad 1980.

[At75] W.B. Atwood, thesis, SLAC-185 (1975).

[Au80] J.J. Aubert et al., Phys. Letters 95B (1980) 306;
B. Saitta, Erice June 1980.

[Ba63] M. Bardadin et al., Proc. Sienna. Int. Conf. on Elementary

Particles 1963, p628.

[Ba79] K.W.J. Barnham et al., Phys. Letters 85B (1979) 300.

[Ba80a] M. Basile et al., Phys. Letters 92B (1980) 367, Phys. Letters 95B (1980) 311, Nuovo Cim. 58A (1980) 193 and Nuovo Cim. Letters 30 (1981) 389.

[Ba80b] W. Bartel et al., Phys. Letters 91B (1980) 142.

[Ba81a] M. Barth et al., Z. Physik C7 (1981) 187.

[Ba81b] M. Barth et al., "Jet Like Properties of Multiparticle Systems Produced in K^+p Interactions at 70 GeV/c", CERN/EP 81-44.

[Ba81c] M. Basile et al., Phys. Letters 99 B (1981) 247.

[Be78] Ch. Berger et al., Phys.Lett. 78B (1978) 176 and 81B (1979) 410.

[Bi79] N.N. Biswas et al., Phys.Rev. D19 (1979) 1960.

[Bj74] J.D. Bjorken and G.R. Farrar, Phys.Rev. D9 (1974) 1449.

[Bl75] V. Blobel et al., Phys.Letters 48B (1974) 73, and Phys.Letters 59B (1975) 88.

[Bo73] A. Bodek et al., Phys. Rev. Letters 30 (1973) 1087 and Phys. Letters 51B (1974) 419.

[Bo79] P.C. Bossetti et al., Nucl. Phys. B149 (1979) 13.

[Bo80] M. Bourquin et al., Z. Physik C5 (1980) 275.

[Bö77] K. Böckmann, "Inclusive Vector Meson Production and Hadron Structure", Symp. on Hadron Structure and Multiparticle Production, Kazimierz 1977.

[Bö78] K. Böckmann et al., Nucl. Phys. B140 (1978) 235.

[Bö80] K. Böckmann et al., Nucl. Phys. B166 (1980) 284.

[Br64] S. Brandt, Ch. Peyrou, R. Sosnowski and A. Wroblewski, Phys.Letters 12 (1964) 57.

[Br78] C. Bromberg et al., Nucl. Phys. B134 (1978) 189.

[Br79a] S. Brandt and H.D. Dahmen, Z. Physik C1 (1979) 61.

[Br79b] R. Brandelik et al., Phys.Letters 83B (1979) 261, 86B (1979) 243 and 89B (1980) 418.

[Br80] R. Brandelik et al., Z. Physik C4 (1980) 87.

[Br81] D. Brick et al., "Hadron Production in π^+p, K^+p and pp collisions at 147 GeV/c and Properties of Jet-like Multiparticle Systems," IFNUP/AE 01/81.

[Bu79] B. Buschbeck, H. Dibon, H.R. Gerhold and W. Kittel, Z. Physik C3 (1979) 97.

[Ca74] P. Capiluppi et al., Nucl. Phys. B79 (1974) 189.

[Ca81] A. Capella, EPS Study Conference on Partons in Soft Hadronic

Processes, Erice 1981.

[Co75] M.J. Counihan, Phys.Letters 59B (1975) 367.

[Co79] C. Cochet et al., Nucl.Phys. B155 (1979) 333.

[Cu79] D. Cutts et al., Phys. Rev. Letters 43 (1979) 319.

[De79a] M. Deutschmann et al., Nucl.Phys. B155 (1979) 307.

[De79b] M. Derrick et al., Phys. Letters 88B (1979) 177.

[De80] H. Deden et al., "Transverse Momentum of Hadrons Produced in ν and $\bar{\nu}$ Interactions on an Isoscalar Target in BEBC", submitted to Physics Letters B.

[Dr81] D. Drijard et al., "Production of Vector and Tensor Mesons in Proton-Proton Collisions at $\sqrt{s}$=52.5 GeV", CERN/EP 81-12.

[Ed78] R.T. Edwards et al., Phys. Rev. D18 (1978) 76.

[Ei79] Y. Eisenberg et al., Phys.Lett. 83B (1979) 127.

[El76] S.D. Ellis, M. Jacob and P.V. Landshoff, Nucl.Phys. B108 (1976) 93.

[Ev77] C. Evangelista et al., Phys. Letters 70B (1977) 373.

[Fa75] G.R. Farrar and D.R. Jackson, Phys. Rev. Letters 35 (1975) 1416.

[Fe77] R.P. Feynman, R.D. Field and G.C. Fox, Nucl.Phys. B128 (1977) 1.

[Fi77] R.D. Field and R.P. Feynman, Phys.Rev. D15 (1977) 2590.

[Fi78] R.D. Field and R.P. Feynman, Nucl.Phys. B136 (1978) 1.

[Fl79] G. Flügge, DESY 79/26 and Tasso Collaboration DESY 79/14.

[Fr79] A. Fridman et al., "Collective Variables applied to $p\bar{p}$ Interactions at 32 GeV/c Incident Momentum", Xth Int. Symp. on Multiparticle Dynamics, Goa 1979.

[Fu77] H. Fukuda and C. Iso, Progr. Theor. Phys. 57 (1977) 483 and 57(1977) 1663.

[Fu80] H. Fukuda, Y. Ishihera and C. Iso, Tokyo Institute of Technology preprint TIT/HEP-55 (1980).

[Ga81] T.K. Gaisser and F. Halzen, Phys. Rev. D23 (1981) 1211.

[Gh77] B. Ghidini et al., Phys.Rev.Letters 68B (1977) 186.

[Go72] H. Goldberg, Nucl.Phys. B44 (1972) 149.

[Gö81] R. Göttgens et al., Nucl. Phys. B178 (1981) 392.

[Gr79] F. Grard et al., "Jet Analysis in ($\pi^+/K^+/p$)-Proton Interactions at 147 GeV/c", Xth Int. Symp. on Multiparticle Dynamics, Goa 1979.

[He80] Ph. Herquet et al., "Properties of Jet-Like Systems Observed in π^+p, K^+p and pp Interactions" Proc. XVth Renc. de Moriond 1980, Vol.I, edited by J. Tran Than Van, R.M.I.E.M., Orsay, p215.

[Ho80] L. Van Hove, "The Role of Valence Quarks in Proton Fragmentation"

Ref. TH 2997-CERN, 21 Nov. 1980.

[Hw81] R.C. Hwa, EPS Study COnference on Partons in Soft-Hadronic Processes, Erice 1981.

[Jo78] J.R. Johnson et al., Phys.Rev. D17 (1978) 1292.

[Ka71] J. Kasman and D.P. Sidhu, SUNY Nov. 1971.

[Ka79] J. Kalinowski, S. Pokorski and L. Van Hove, Z. Physik C2 (1979) 85.

[Ki77] W. Kittel, Proc. VIIIth Int. Symp. on Multiparticle Dynamics, Kaysersberg 1977.

[Ki79] H. Kichimi et al., Nuovo Cim. Letters 24 (1979) 129.

[La80] J.M. Laffaille et al., Saclay preprint DPh DE 80-16.

[Ma79] J.P. Mazzanti et al., Phys. Letters 81B, (1979) 219.

[Mi72] G. Miller et al., Phys.Rev. D5 (1972) 528.

[Mi80a] S.P. Misra, A.R. Panda and B.K. Parida, Phys. Rev. Letters 45 (1980) 322.

[Mi80b] S.P. Misra and A.R. Panda, Phys. Rev. D21 (1980) 3094.

[Ng80] A. Ng, K.E. Lasilla and U.P. Sukhatme, "Understanding Inclusive $pp \to ph^{\pm}X$ Data with Parton Fragmentation and Structure Functions", FERMILAB-Pub-80/90-THY.

[Oc77] W. Ochs, Nucl.Phys. B118 (1977) 397.

[Ra81] J. Randa, Phys. Rev. D23 (1981) 1662.

[Sa67] H. Satz, Phys.Lett. 25B (1967) 220 and Phys.Rev.Letters 19 (1967) 1453.

[Sa77] H. Satz, Proc. Europ. Conf. on Particle Physics, Budapest 1977, p 379.

[Sc78] P. Schmitz et al., Nucl. Phys. B137 (1978) 13.

[Sc81] M.M. Schouten et al., "Approach to Scaling of π^+/π^- Ratios in Target Fragmentation from $\pi^+/K^+/pp$ Interactions at 147 GeV/c", Nijmegen preprint HEN-188.

[Sh80] Yu.M. Shabelski, "Production of Projectile Fragments in Additive Quark Model; Inelasticity Coefficients and Probabilities of Inelastic Charge Exchange", Leningrad 1980.

[Si78] J. Singh et al., Nucl.Phys. B140 (1978) 189.

[Sp80] J. Spengler, Thesis Düsseldorf 1980.

[Su79] A. Suzuki et al., Nuovo Cim. Letters 24 (1979) 449.

[Wi74] F.C. Winkelman et al., Phys. Rev. Letters 32 (1974) 121.

[Wo79] G. Wolf, Proc. Int. Symp. on Lepton and Photon Interactions at High Energies, Batavia 1979.

[Zi77] A. Zieminski, Proc. EPS Conf. on Particle Physics, Budapest 1977 .

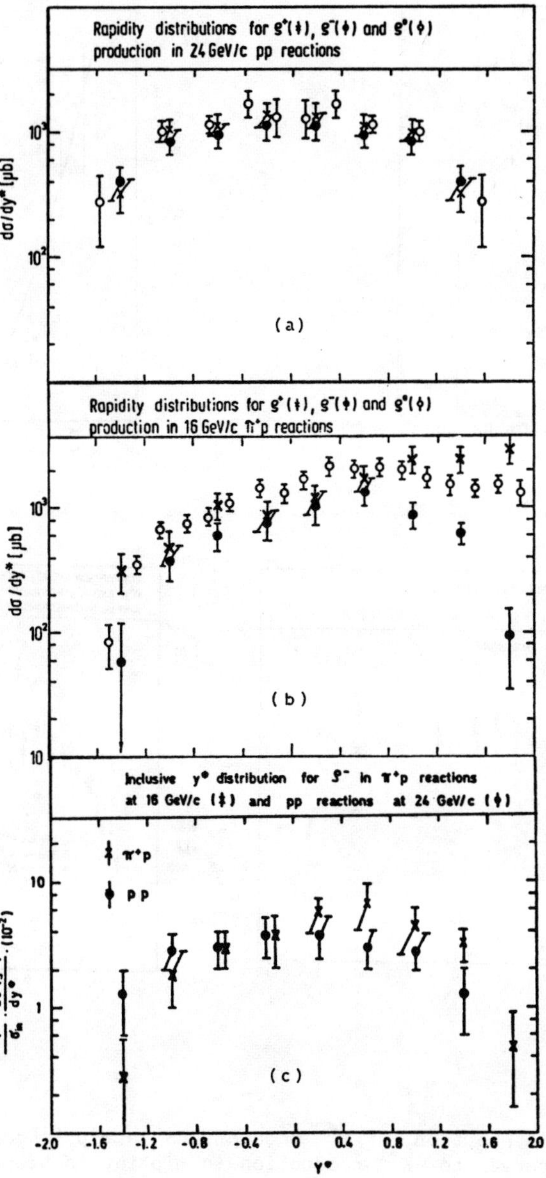

Fig.1a) Differential cross sections $d\sigma/dy^*$ for the total inclusive ρ^+ (crosses), ρ^- (full circles) and ρ^0 (open circles) production as functions of y* in pp reactions at 24 GeV/c.

b) The same as in sub-fig.a, but for π^+p reactions at 16 GeV/c.

c) Comparison of the total inclusive ρ^- density in 16 GeV/c π^+p interactions (crosses) and 24 GeV/c pp interactions (full circles) as function of y* [Bö78].

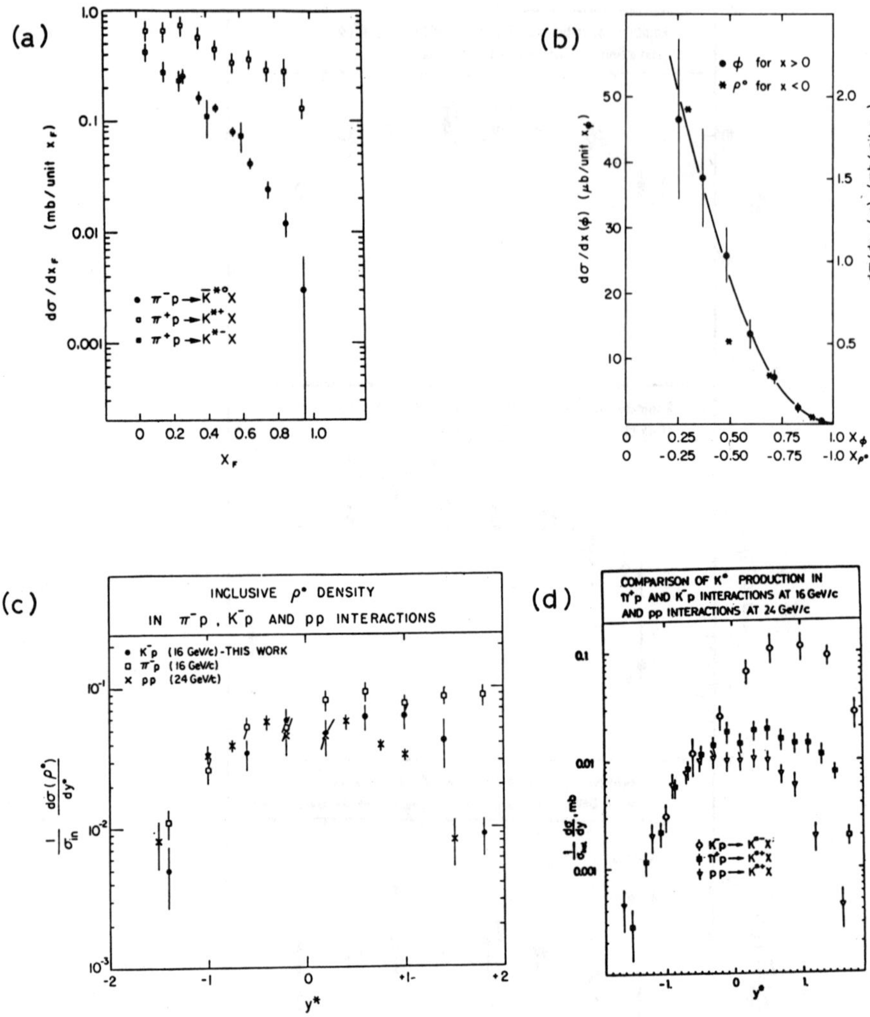

Fig.2a) ($d\sigma/dx_F$) distribution for $\bar{K}^{*0}$ productions in π^-p interactions at 16 GeV/c compared to $K^{*\pm}$ production in π^+p interactions at the same energy [Ev77].

b) Differential cross section ($d\sigma/dx_\phi$) for forward ϕ production at 16 GeV/c compared to data for backward ρ^0 production in π^-p interactions at the same energy. The curve is a fit to the ϕ data of the form $(1 - |x_\phi|)^n$ where n = (2.0 ± 0.2). The errors are statistical only [Gh77].

c) Comparison of the c.m. rapidity distribution of ρ^0 in 16 GeV/c K^-p interactions with the same in 16 GeV/c π^-p and 24 GeV/c pp interactions [Sc78].

d) Comparison of the c.m. rapidity distributions of inclusive K* production in π^+p and K^-p interactions at 16 GeV/c and in 24 GeV/c pp interactions [Bö80].

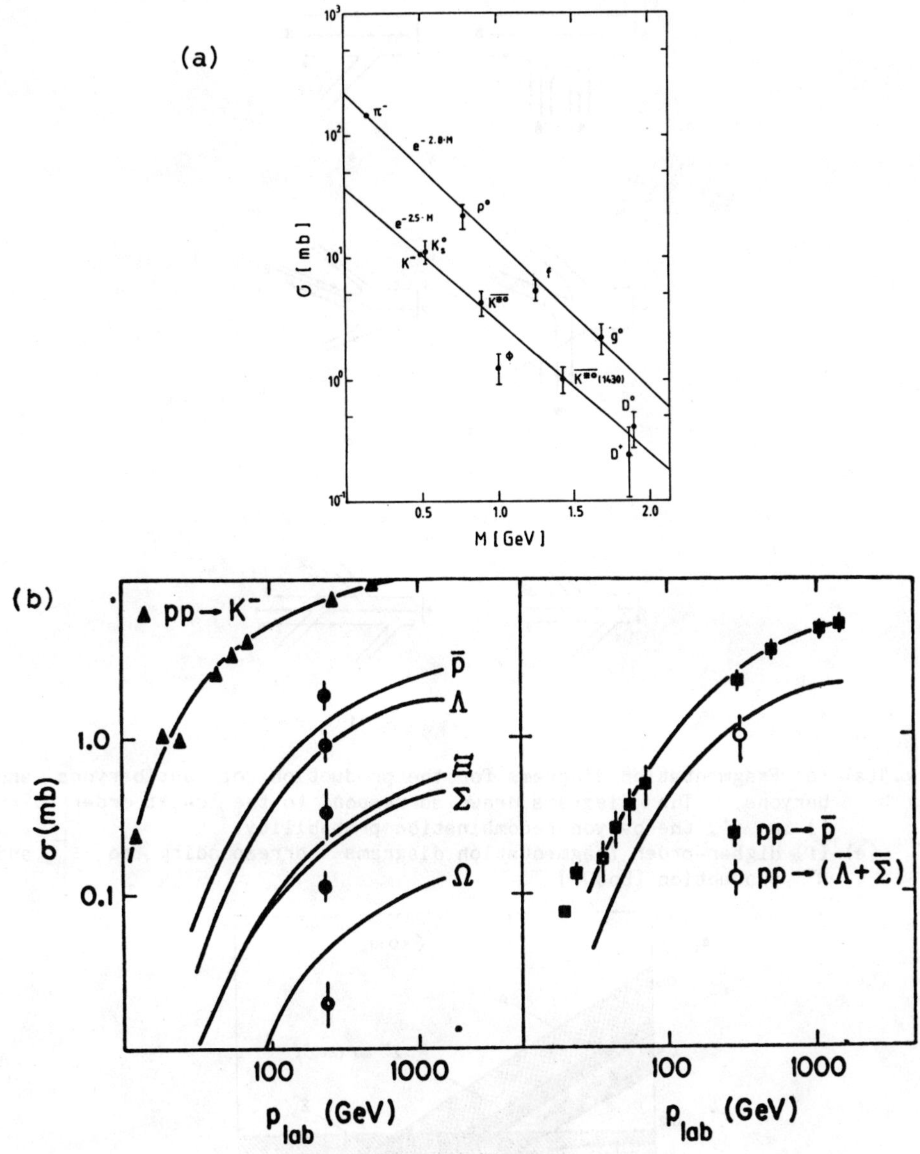

Fig.3a) Inclusive resonance cross sections for pp collisions at $\sqrt{s}$=52.5 GeV/c as a function of mass [Sp80].

b) Yields of heavy particles in pp interaction computed using quark counting according to [Ga81].

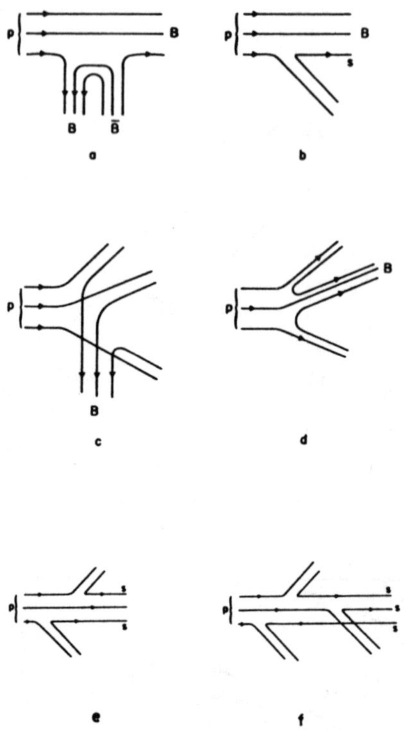

Fig.4(a)-(d) Fragmentation diagrams for the production of antibaryons and baryons. The diagrams drawn correspond to the lowest order (0 or 1) in η', the baryon recombination probability.

(e)-(f) Higher-order fragmentation diagrams corresponding to Ξ^- and Ω^- production [Bo80].

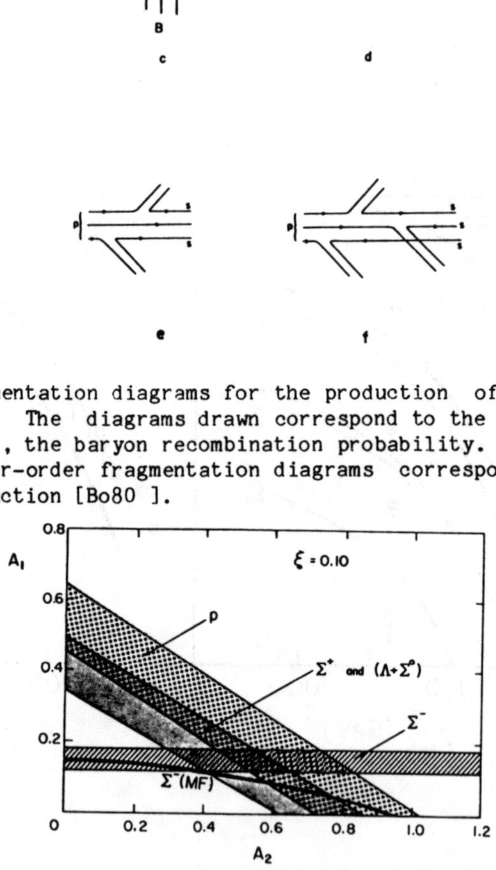

Fig.5 Graphical representation of the proton and S = -1 hyperon average multiplicities in the A_2A_1 plane for $\xi = 0.10$. The line labelled Σ^- (MF) corresponds to multifragmentation as described in the text [Bo80].

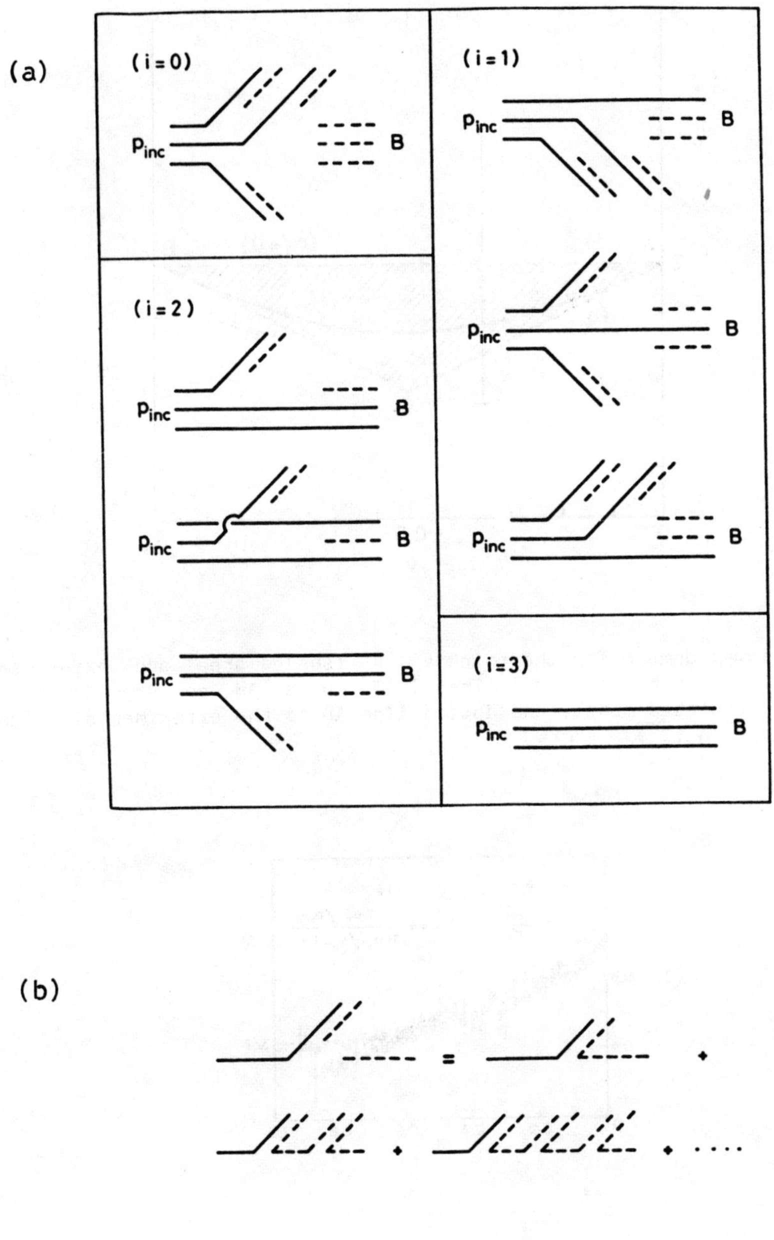

Fig.6a) Diagrams for production of baryon B by fragmentation of incident proton p_{inc}. Full lines denote valence quarks of p_{inc}, dashed lines "pick-up" quarks or antiquarks [Ho80].

b) Quark fragmentation diagrams following ref. [Fu77].

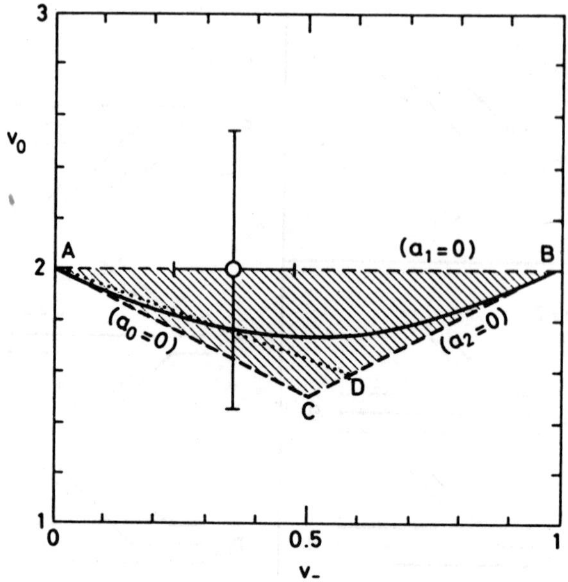

Fig.7 Allowed domain for the point v_o, v_- (shaded area) and experimental point with errors. The full curve AB corresponds to the uncorrelated model, the dotted line AD to the experimental value of $n_f(\bar{\Omega})/n_f(\bar{\Xi})$. [Ho80]

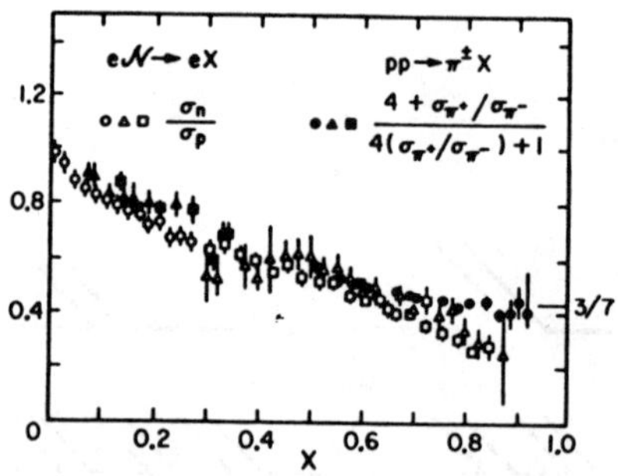

Fig.8 Test of eq. (38) in the text which relates the electromagnetic structure functions of the proton and neutron to the π^+ and π^- yields in pp collisions assuming that the dissociation of the proton into quarks is the common underlying mechanisms. The limit $\sigma_n/\sigma_p \geq 3/7$ is obtained of the helicities of the proton and leading quark are the same [Oc77].

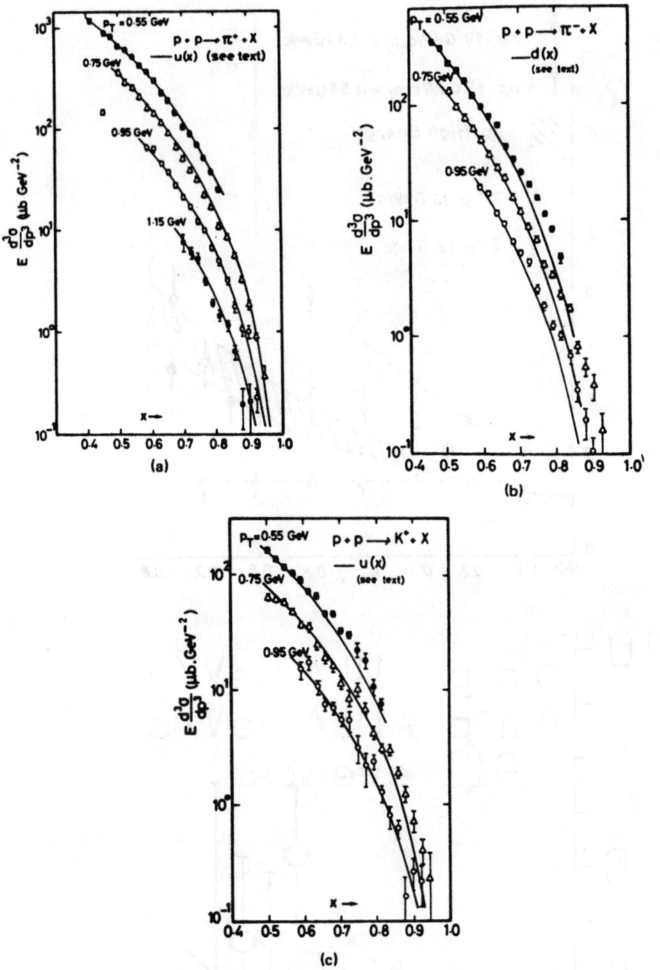

Fig.9 Comparison of the inclusive cross section (a) π^+, (b) π^-, (c) K^+ with the parton model quark distribution functions u(x) and d(x) obtained from deep inelastic collisions [Si78].

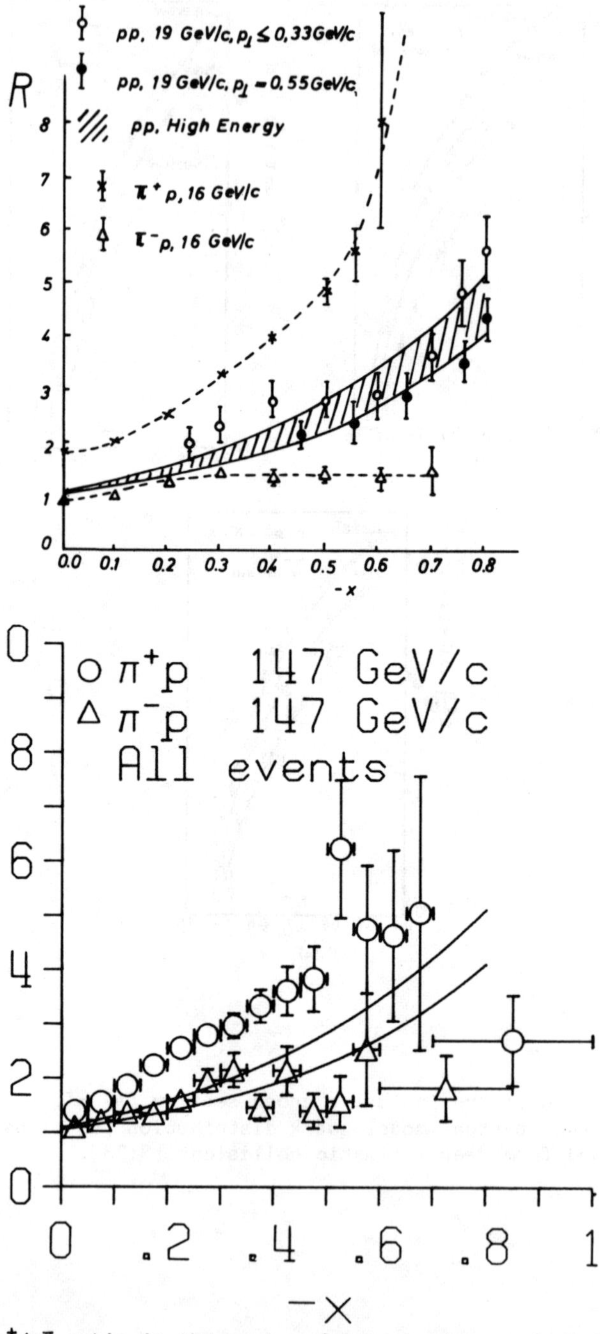

Fig.10a) The π^+/π^- ratio in the proton fragmentation region of 16 GeV/c π^+p and 19 GeV/c pp collisions as indicated. The shaded regions corresponds to pp reactions at FNAL and ISR energies [Bu79].

b) The π^+/π^- ratio in the proton fragmentation region of 147 GeV/c π^+p collisions compared to the π^+/π^- ratio of pp reactions at FNAL and ISR energies [Sc81].

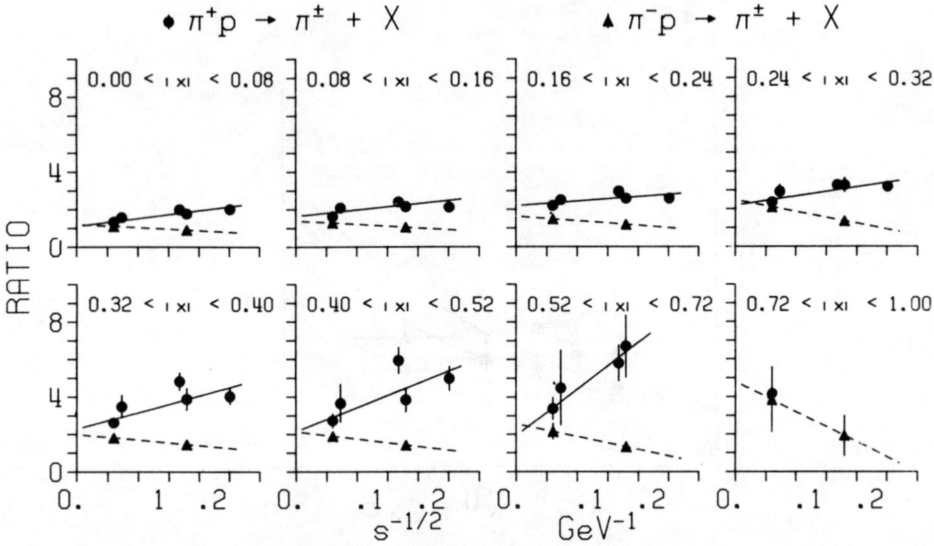

Fig.11 The π^+/π^- ratio in the proton fragmentation region for π^+p collisions as functions of $s^{-1/2}$, for the x intervals indicated [Sc81].

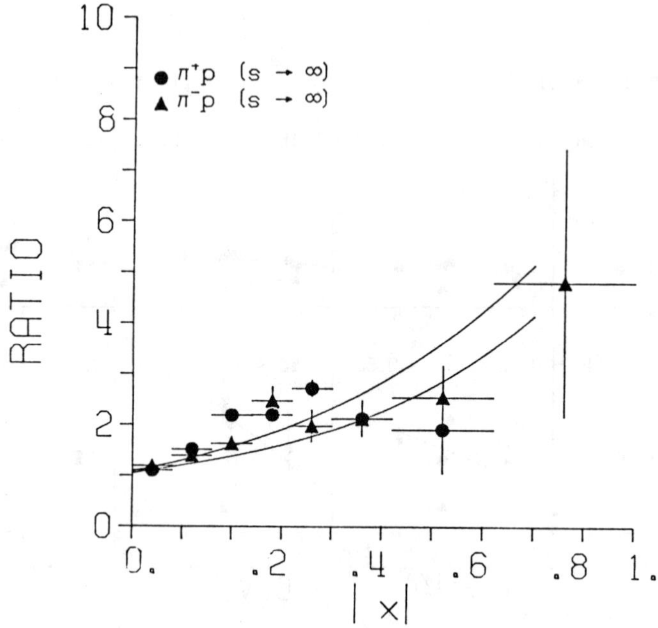

Fig.12 Extrapolation to $s^{-1/2}=0$ of the π^+/π^- ratio in the fragmentation region of π^+p and π^-p collisions, compared to the π^+/π^- ratio of pp collisions at FNAL and ISR energies (solid lines) [Sc81].

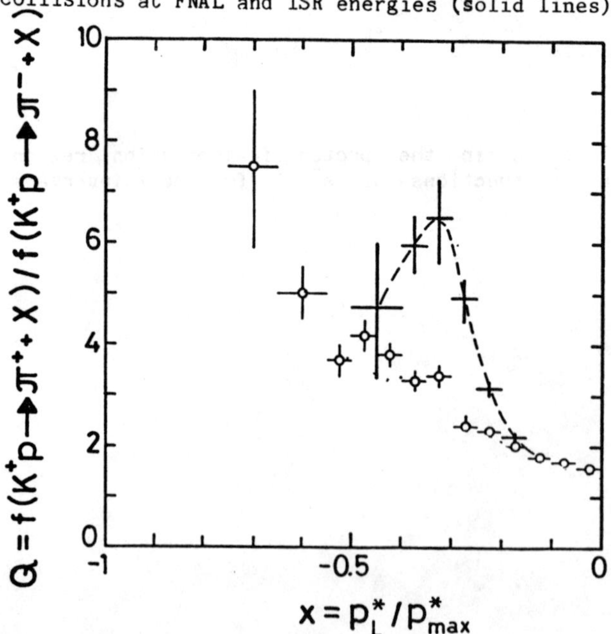

Fig.13 The π^+/π^- ratio for the proton fragmentation region of K^+p collisions at 32 GeV/c. The points connected by the dashed line correspond to the π^+/π^- ratio in a strongly Δ^{++} enriched sample (ratio for pions produced in association with protons with $|t_{pp}|<0.6$ GeV^2) [Aj80].

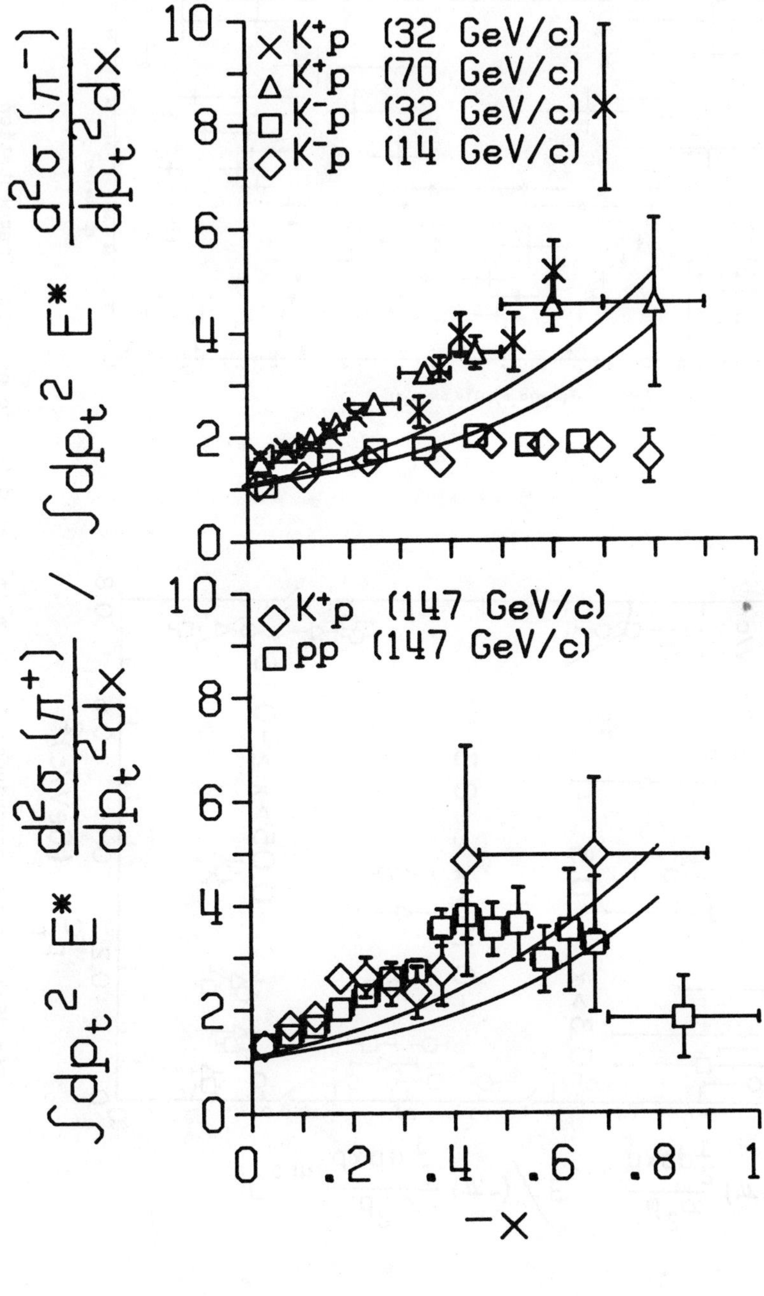

Fig.14 The π^+/π^- ratio in the proton fragmentation region of K^+p and K^-p collisions as indicated, compared to the π^+/π^- ratio in pp collisions at FNAL and ISR energies (solid lines) [Bu79,Sc81].

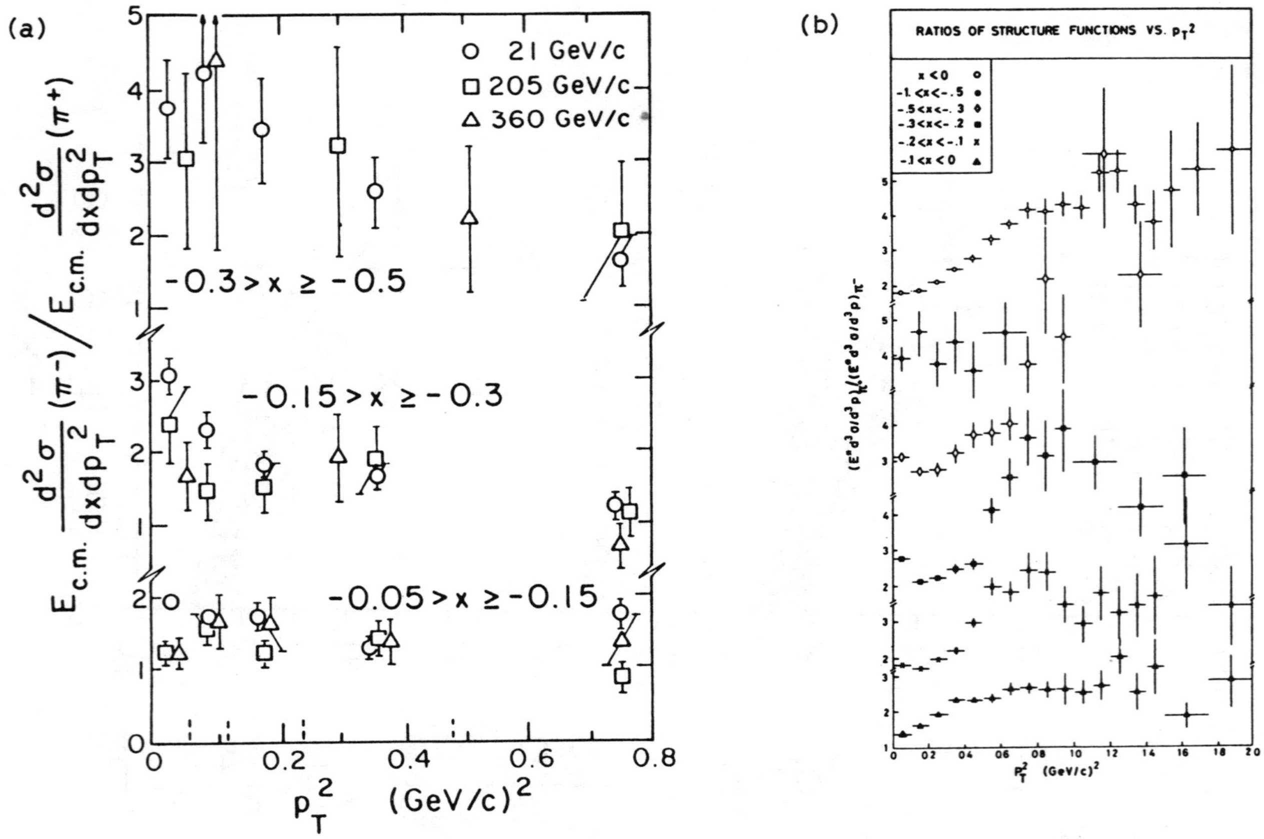

Fig.15a) The p_t^2 dependence of the π^-/π^+ ratio in the neutron fragmentation region of $\pi^- n$ collisions at 21, 205 and 360 GeV/c, for the three x intervals indicated [Ab81].

b) The p_t^2 dependence of the π^+/π^- ratio in the proton fragmentation region of K^+p collisions at 32 GeV/c, for the x intervals indicated [Aj80].

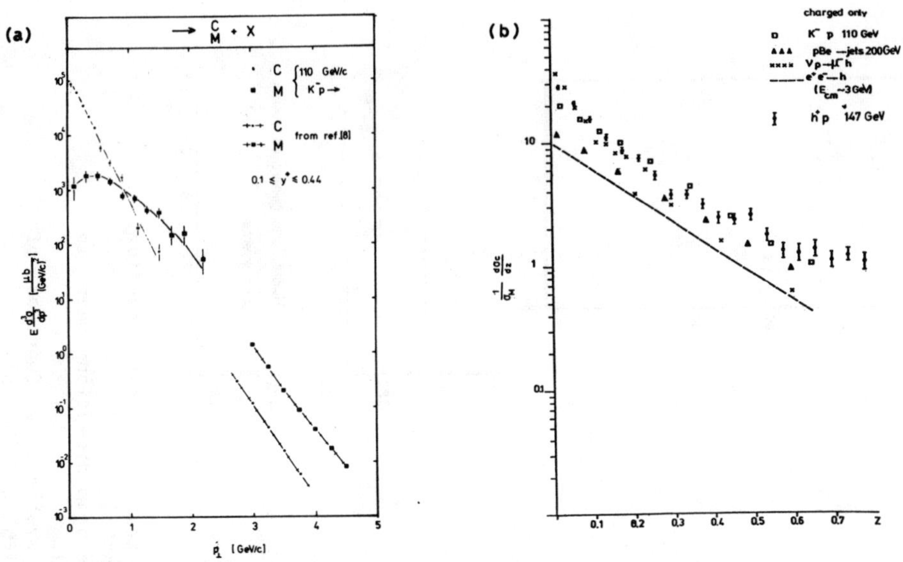

Fig.16a) Transverse momentum distribution of single charged particles C and multiparticle systems M in the rapidity interval $0.1<y^*<0.44$. All constituents of the multiparticle systems have $y^*>-0.5$. The dashed lines are to guide the eye [De79a].

b) The distribution of momentum fraction z for hadron and lepton induced multiparticle production [Gr79].

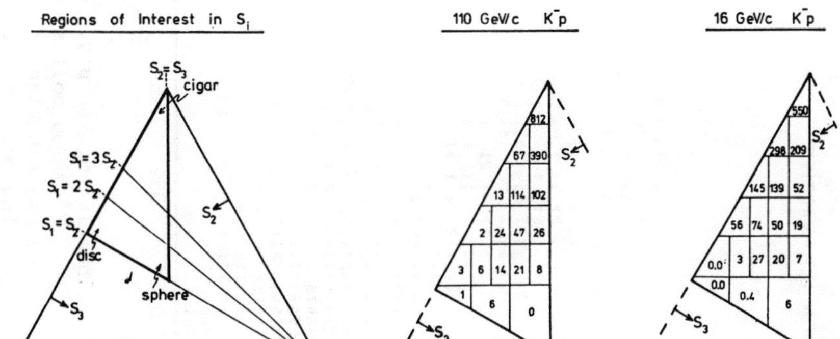

Fig.17a) Regions of interest in the variables S defined in the text.

b) The population (number of events with more than two charged particles) for K^-p data at 110 GeV/c.

c) The population for K^-p data at 16 GeV/c, normalized to that at 100 GeV/c [De79a].

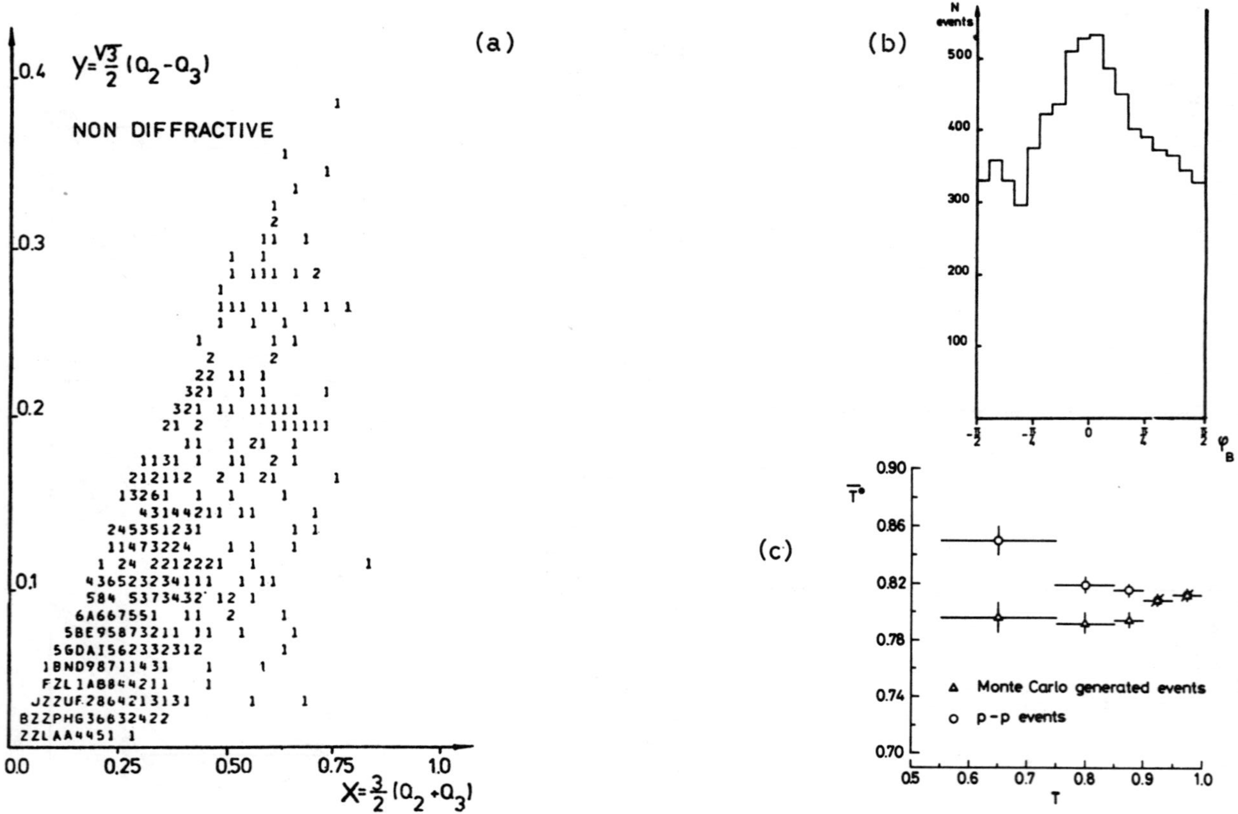

Fig.18a) Triangular plot in X and Y defined in the text for non-diffractive $K^+/\pi^+/pp$ collisions at 147 GeV/c [He80 ,Br80].

b) Distribution of the azimuthal angle of the beam direction seen in the principle axis reference frame [He80a].

c) Average value of thrust T* in the jet cms system as a function of thrust T in the overall pp cm system at $\sqrt{s}$=62 GeV [Ba80a].

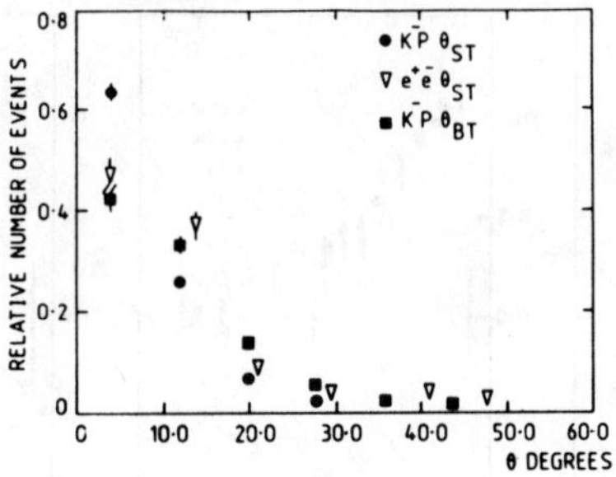

Fig.19 Angle Θ_{BT} between beam and thrust axis at 110 GeV/c in K^-p cms and angle Θ_{ST} between sphericity and thrust axis. Values of Θ_{ST} in e^+e^- annihilations at $\sqrt{s}$=9.4 GeV are also shown [Gö81,Be78].

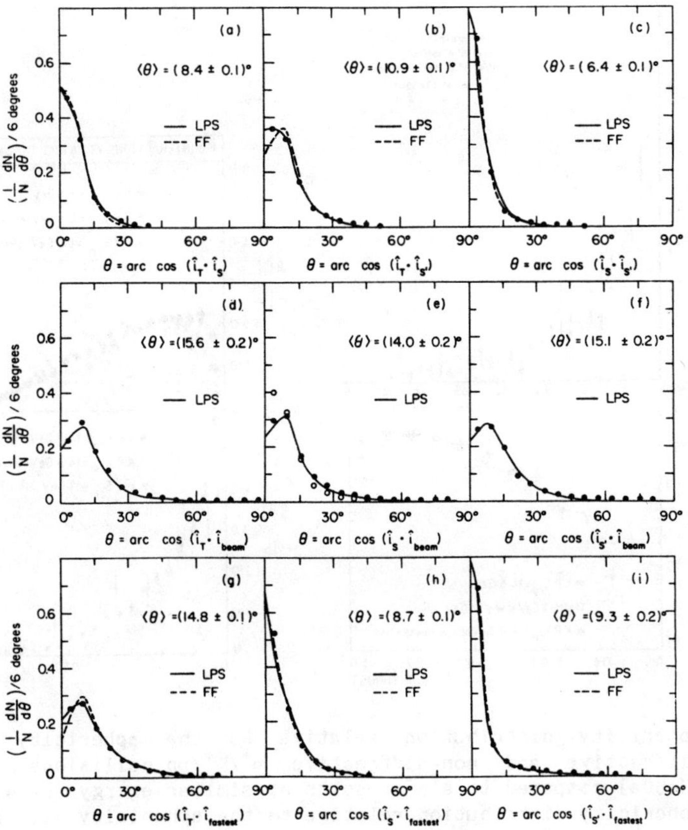

Fig.20 Distributions of angles between the axis defined by thrust, sphericity, spherocity, the direction of the beam, and the direction of the fastest particle in K^+p collisions at 70 GeV/c [Ba81b]. The lines are as in fig.26.

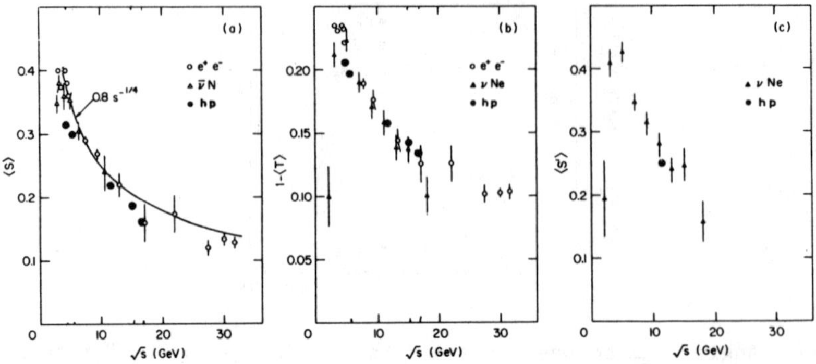

Fig.21 Average sphericity, thrust and spherocity as a function of energy for hadron-proton, e^+e^- and neutrino reactions [Gö81,Ba81a,Br81].

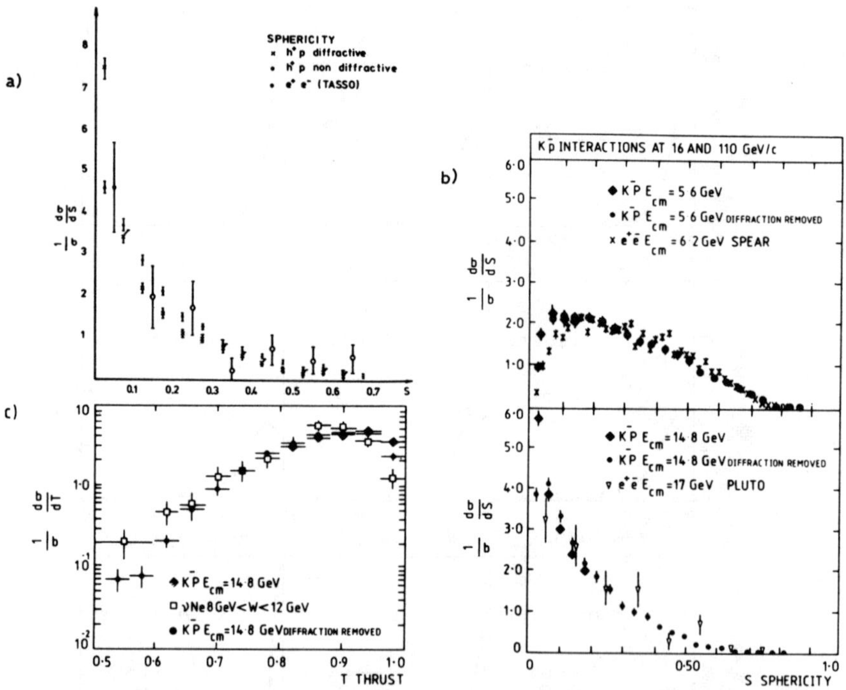

Fig.22a) Sphericity distribution relative to the sphericity axis for diffractive and non-diffractive $\pi^+/K^+/pp$ collisions at 147 GeV/c [He80a] compared to e^+e^- results at similar energy [Fl79].

b) Sphericity distribution relative to the sphericity axis for 16 and 110 GeV/c K^-p collisions (diffractive events removed and included, respectively) compared to e^+e^- data at similar energies [Gö81,Be78].

c) Thrust distribution relative to the thrust axis for 110 GeV/c K^-p collisions (diffractive events removed and included, respectively) compared to νNe interactions for $8<W<12$ GeV [Gö81,Bo79 ,Ba79].

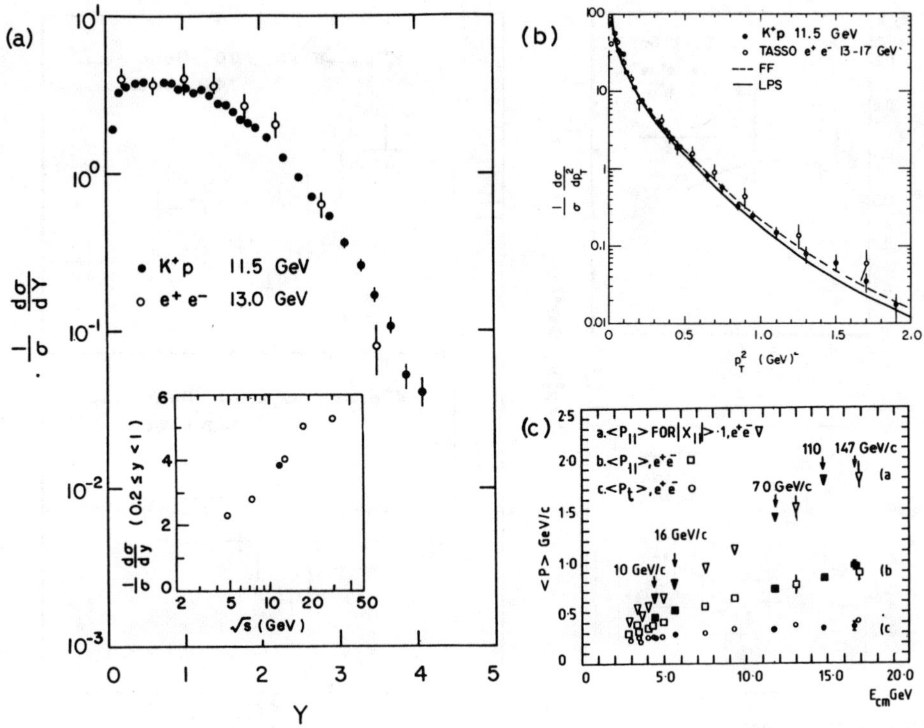

Fig.23a) Rapidity distribution $1/\sigma \cdot d\sigma/dy$ for non diffractive K^+p events at 70 GeV/c (using thrust axis). Only charged particles produced in the c.m. forward hemisphere are used. In the insert, rapidity distribution $1/\sigma \cdot d\sigma/dy$ averaged over $0.2<y<1$ as a function of the c.m. energy [Ba81b,Br80].

b) $1/\sigma \cdot d\sigma/dp_t^2$ as a function of p_t^2 relative to the sphericity axis for K^+p collisions at 70 GeV/c. The e^+e^- data are from ref. [Br79b]. The superimposed curves are the p_t^2 distributions for FF (dashed line) and for LPS (continuous line) [Ba81b].

c) Energy dependence of $\langle p_t \rangle$ and $\langle p_{/\!/} \rangle$ relative to the thrust axis for hadron proton and e^+e^- reactions [Gö81,He80 ,Fl79].

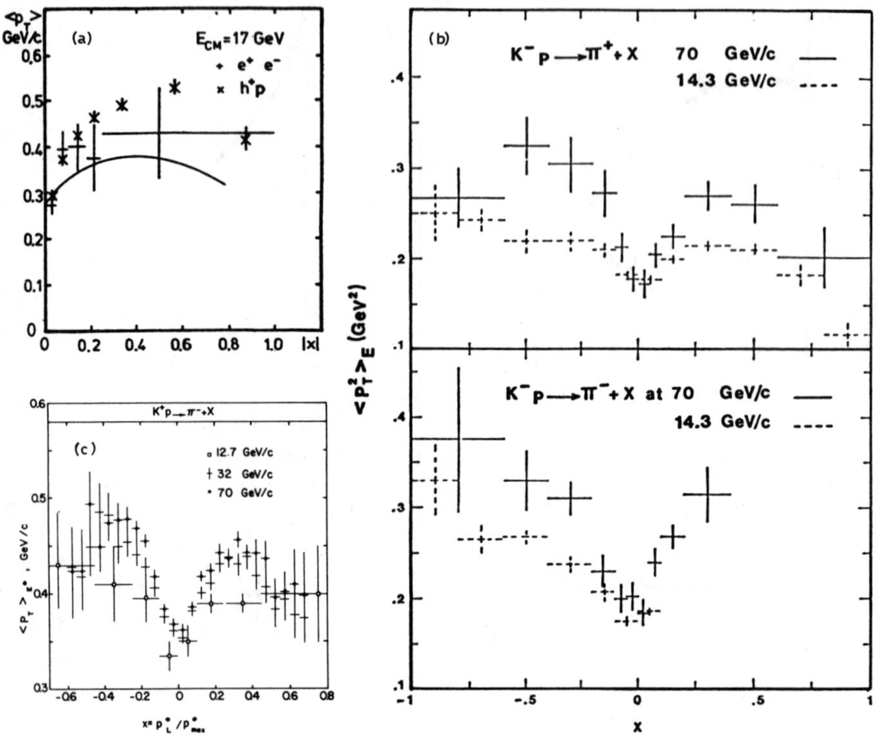

Fig.24a) "Seagull" effect observed in e^+e^- and $\pi^+/K^+/pp$ collisions at $\sqrt{s}\approx 17$ GeV/c [He80 ,Fl79].

b–c) Energy dependence of the "seagull" effect in K^+p and K^-p collisions [Ba81a,La80].

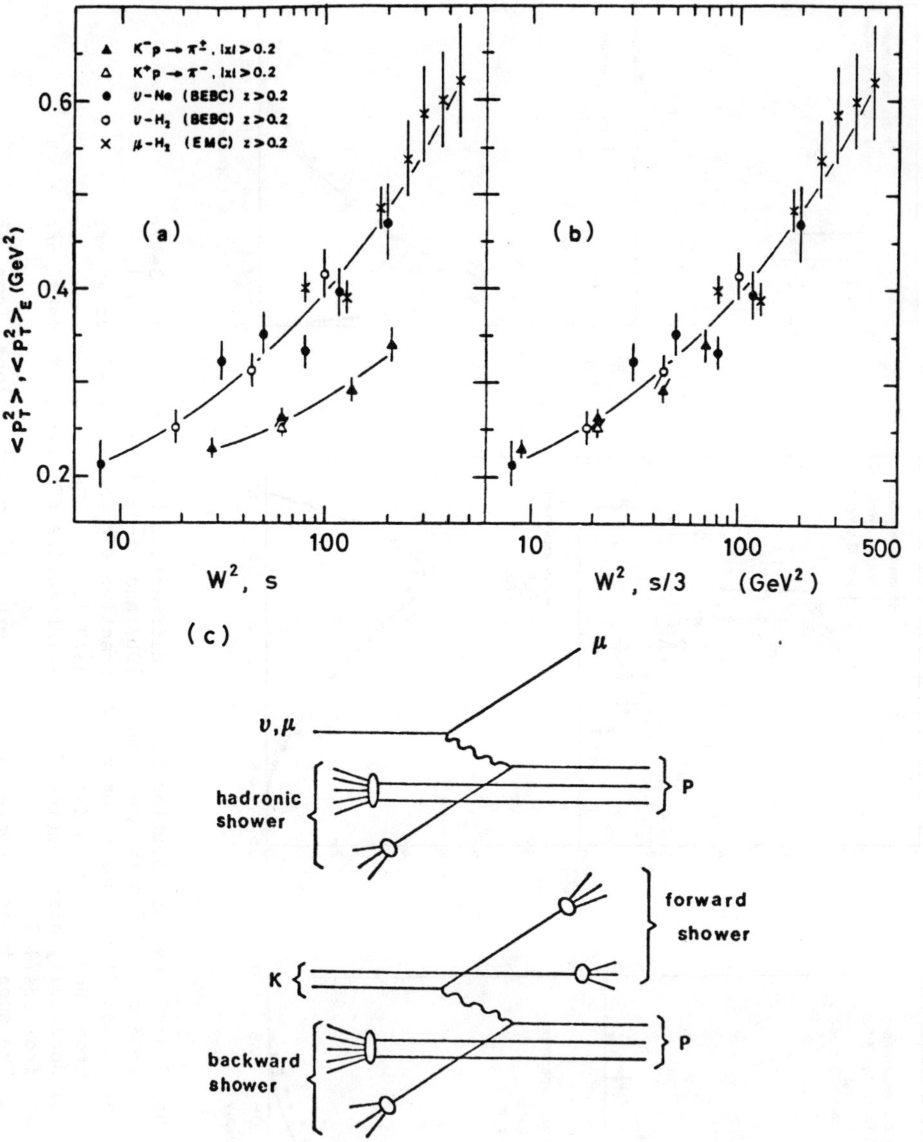

Fig.25a) Energy dependence of $\langle p_t^2 \rangle_E$ for $\pi^{\pm}$ with $|x|>0.2$ in $K^{\pm}$ interactions, and of $\langle p_t^2 \rangle$ for the hadrons with $z>0.2$ in the hadronic shower produced in ν and μ experiments [La80,Bo79 ,De80,Au80].

b) same as a) where the energy $\sqrt{s}$ in $K^{\pm}$ interactions is replaced by the quark-hadron energy $\approx \sqrt{s/3}$.

c) Corresponding diagrams for the study of the $\langle p_t^2 \rangle$ behaviour of hadronic showers produced in ν, μ and Kp interactions [La80].

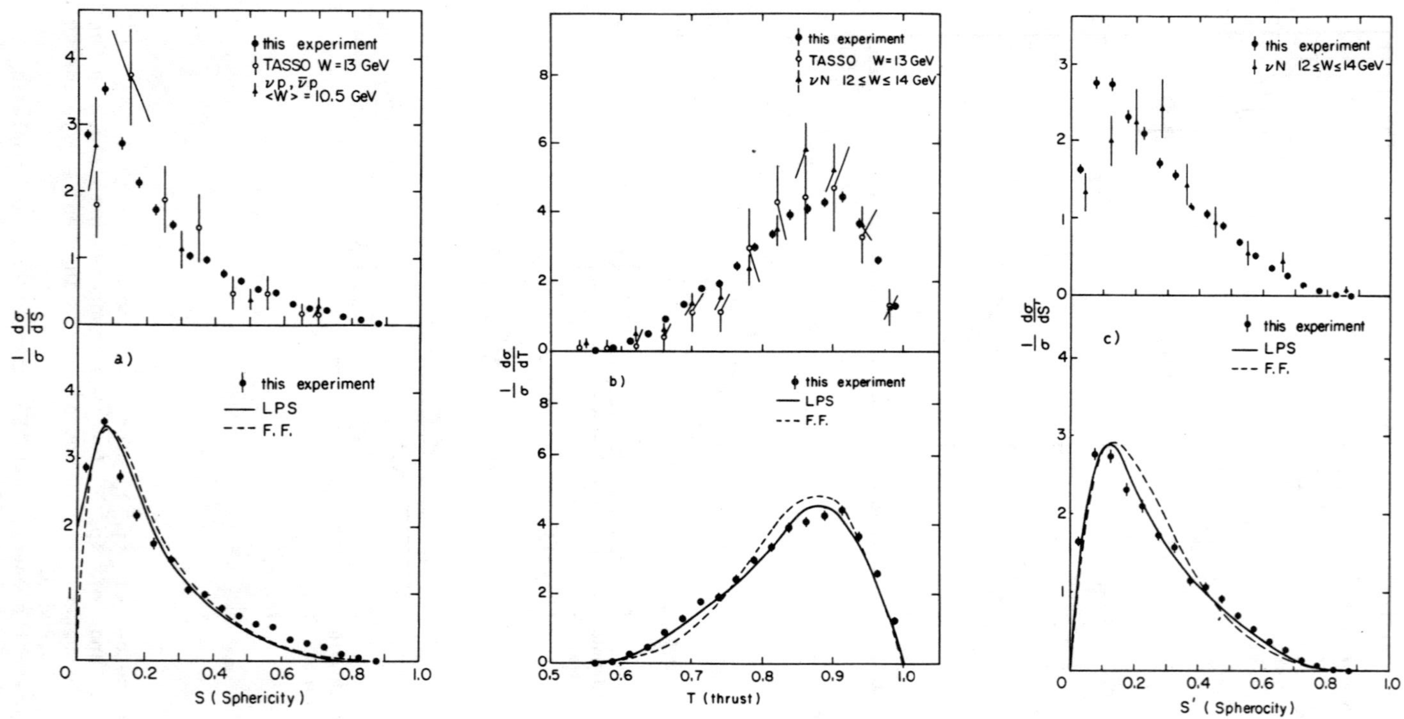

Fig.26a) Sphericity distribution for non-diffractive K^+p data at 70 GeV/c [Ba81b], e^+e^- data are from [Br80b] and $\nu(\bar{\nu})p$ data from [De79b].
b) Thrust distribution for non-diffractive K^+p data, e^+e^- data are from [Br80] and νN data from [Ba79].
c) Spherocity distribution for non-diffractive K^+p data, νN data are from [Ba79].
The superimposed curves are the expected distributions from FF (dashed line) and from LPS (continuous line), as in the text.

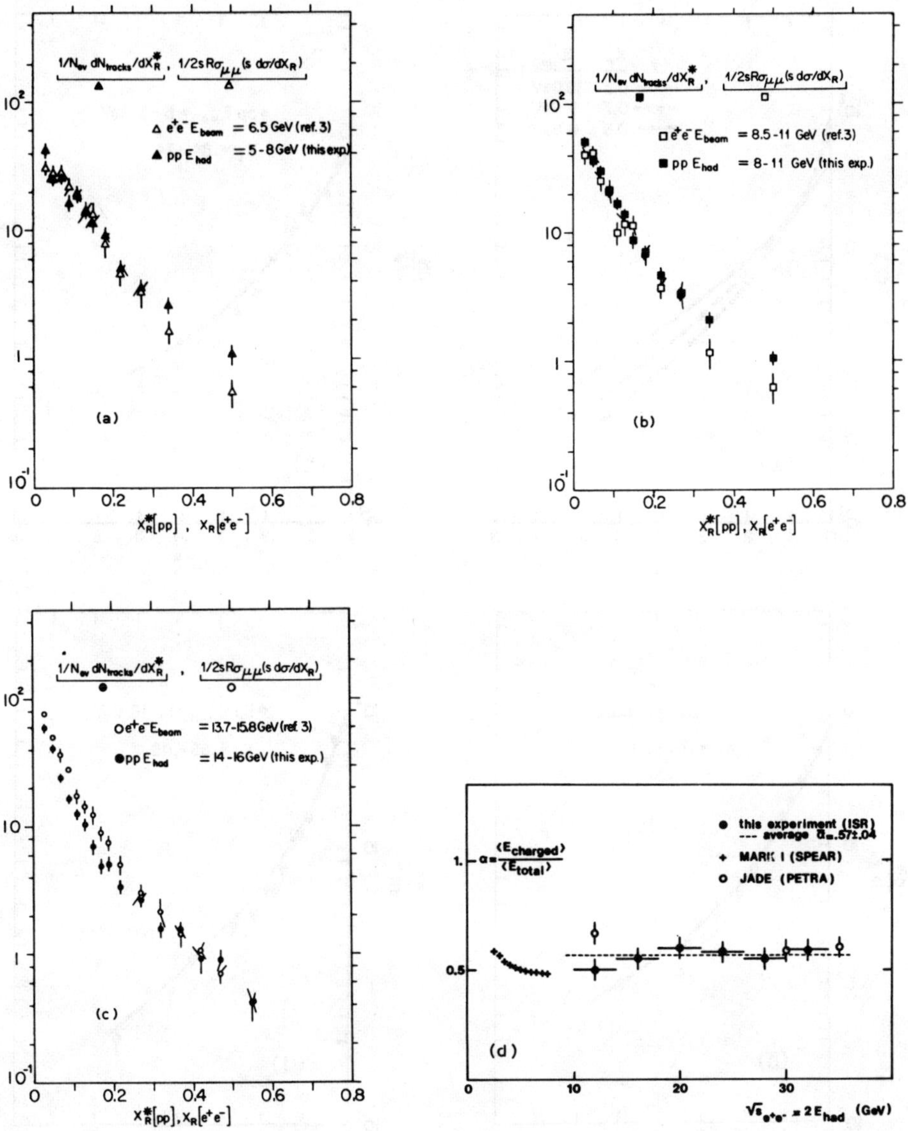

Fig.27a–c) The single-particle inclusive distributions in terms of the variable x^*_R for pp collisions at $\sqrt{s}$=62 GeV, compared to single particle inclusive distributions at three different energies as indicated [Ba80a,Br79b].

d) The ratio of charged to total energy α_{pp} as a function of $2E_{had}$ in one-proton events. The dotted line shows the mean value $\langle\alpha_{pp}\rangle$ [Ba81c].

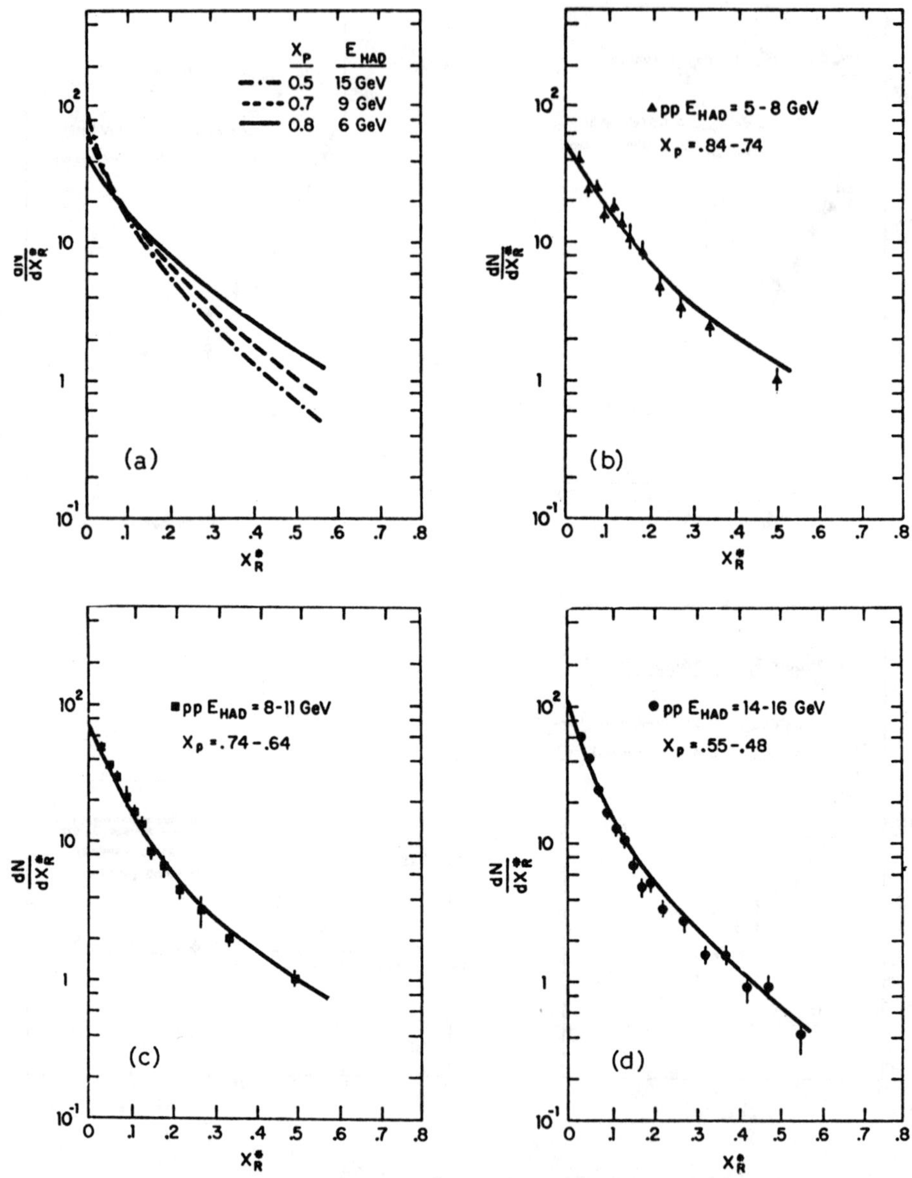

Fig.28a) Calculated inclusive hadron distributions for pp ≥ phX in the variable $x_R^*=x/(1-x_p)$ for leading proton momentum fraction $x_p=0.5$ (dot-dashed curve), $x_p=0.7$ (dashed curve), and $x_p=0.8$ (solid curve) for produced hadron energies 15 GeV, 9 GeV, and 6 GeV, respectively, where x is the fraction of the longitudinal momentum carried off by hadron h.

b-d) Model predictions for the $x_R^*=x/(1-x_p)$ distributions for charged hadrons produced in the same hemisphere as the leading proton compared with experimental data [Ba80a] for various leading proton momentum fractions (x_p) and hadron energies (E_{HAD}), as indicated [Ng80].

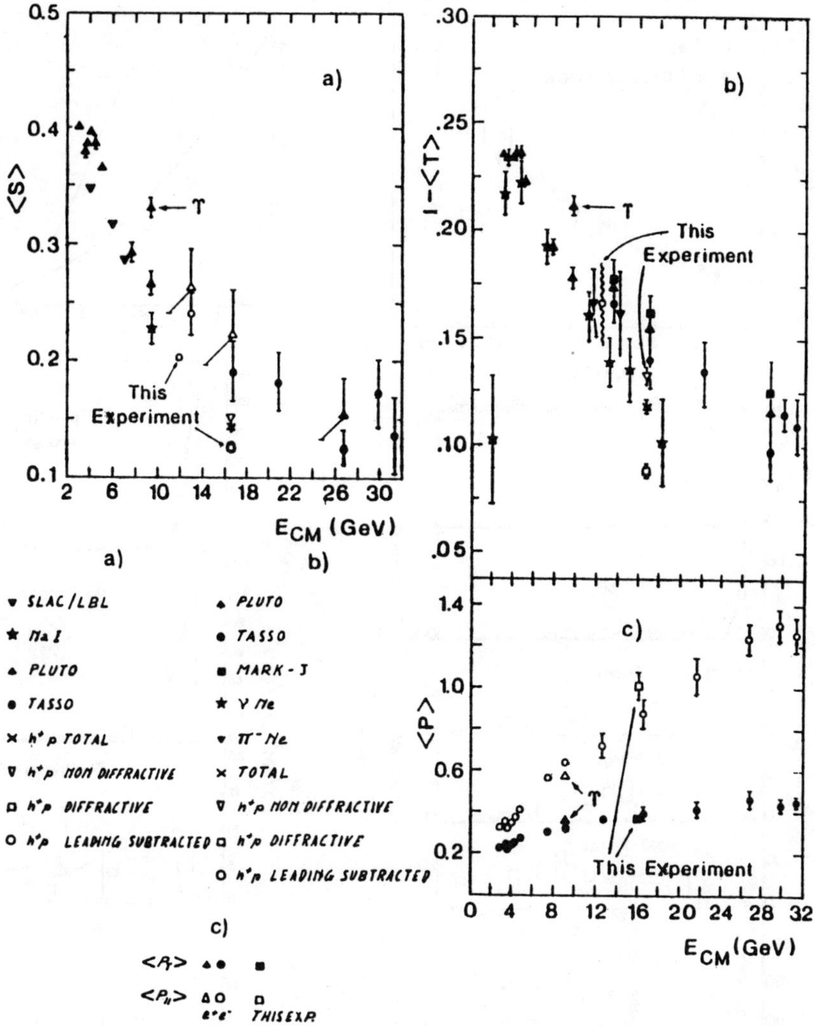

Fig.29 Average sphericity (a) and thrust (b) as a function of the cms energy of the hadronic system for e^+e^- annihilation and νNe data compared to the average sphericity and thrust for 147 GeV/c $\pi^+/K^+/pp$ events (total, diffractive and non-diffractive) at the full energy and at $\langle E_{had} \rangle$ after subtraction of the leading particles [Br81].

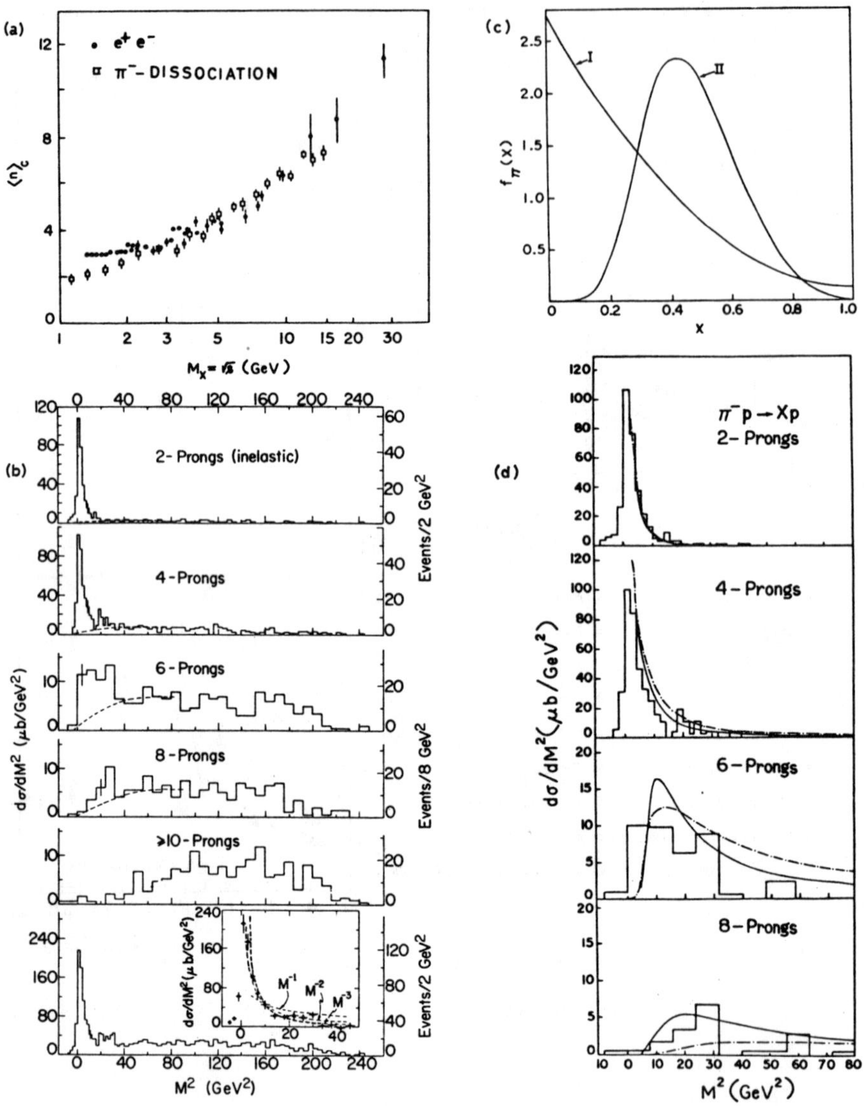

Fig.30a) Average charged multiplicities for e^+e^- annihilation and π^- dissociation against M_X or $\sqrt{s}$ [Mi80a,Wo79 ,Wi74].

b) Distributions in M^2 for $\pi^-+p \geq p+X$ at 205 GeV/c. Dashed curves are estimates of the nondiffractive background. The insert shows the inclusive M^2 distribution for pion diffraction only. The curves show M^{-1}, M^{-2}, and M^{-3} dependences normalized to the measured value of $d\sigma/dM^2$ at 10 GeV^2 [Wi74].

c) Two primordial fragmentation functions $f_\pi(x)$ used as phenomenological input [Mi80b].

d) $d\sigma/dM^2$ vs M^2 for 2-8 prongs. The continuous and the dot-dashed curves correspond to the primordial fragmentation functions I and II, respectively. The experimental data are from [Wi74] after background subtraction.

Chapter II: LEPTON-HADRON INTERACTIONS AND NUCLEON STRUCTURE

H.J. LUBATTI

CERN *

Geneva, Switzerland

Abstract:

Results from deep-inelastic, lepton scattering on protons and on nuclear targets are reviewed. Attention is given to the nucleon structure functions and quark fragmentation functions. The hadron final states produced in leptonic interactions are compared with those produced in hadronic interactions.

1. Introduction

The use of lepton beams as a probe of hadronic matter at very small distances,"deep inelastic scattering", has been extraordinarily successful in providing information about the structure of the nucleon and the strong interactions. Since the epochal SLAC-MIT experiment which showed that ep deep inelastic scattering (large Q^2) occurred as though there were single quasi-free constituents within the nucleon,we physicists have rapidly come to think of quarks as the quintessence of hadronic matter. The strong interactions are now discussed in terms of interactions between quarks. However, all experiments attempting to observe free quarks have given negative results; quarks are apparently confined within hadrons by a color field mediated by vector gluons. An attempt to "knock out" a quark results instead in the struck quark accreting additional quarks, "hadronization", to form the observable hadrons.

* On leave from the Univ. of Washington, Seattle, Washington, U.S.A.

The lectures will review what has been learned about the structure of the nucleon and the hadronization process from lepton-hadron interactions. Where possible comparison with hadron-hadron interactions will be made. Results are available for interactions of electrons, muons, and neutrinos on both proton and nuclear targets. Since there exist numerous, extensive reviews devoted to only one of the above beams, this paper will concentrate on recent results. In addition, in such a brief review it is not possible to discuss all of the very fine data which are available. In presenting the data the language of the quark-parton model is extensively used.

In section 2 there is a brief review of lepton nucleon cross sections and the relevant kinematic variables. The predictions of the simple quark parton model for the structure functions and comparison with data are given in section 3. A discussion of the hadron final states (jets) is given in section 4 and in section 5 specific nuclear effects are discussed.

2. Lepton-Nucleon Cross Sections

The general form of a lepton nucleon interaction is shown diagramatically in Fig. 1. The incoming lepton can be thought of as a source of virtual photons (γ) for muons and electrons and of virtual intermediate bosons ($W^{\pm}, Z^{o}$) for neutrinos and antineutrinos. The relevant kinematic variables are:

(i) the energy transferred to the hadronic system

$$\nu = p_N \cdot (p_\ell - p_\ell')/M_N = E_\ell - E_{\ell'} = E_h - M_N$$

(ii) the four momentum squared of the current

$$Q^2 = -q^2 = (p_\ell - p_\ell')^2 \simeq 4\, E_\ell\, E_\ell' \sin^2\frac{\theta}{2}$$

where E_ℓ and E_ℓ' are the laboratory energies of the incident and scattered lepton, respectively, and θ is the lepton scattering angle in the laboratory.

(iii) the scaling variables

$$x_{B_j} = Q^2/2M_N\nu$$

$$y = \nu/E_\ell$$

which range between 0 and 1.

(iv) the invariant mass of the hadronic system:

$$W^2 = p_h{}^2 = M_N{}^2 + Q^2\,(1/x_{B_j} - 1)$$

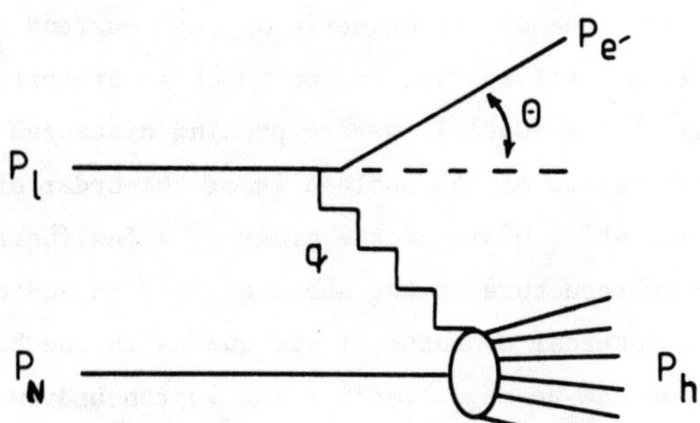

Fig. 1: Lepton-Nucleon Interaction.

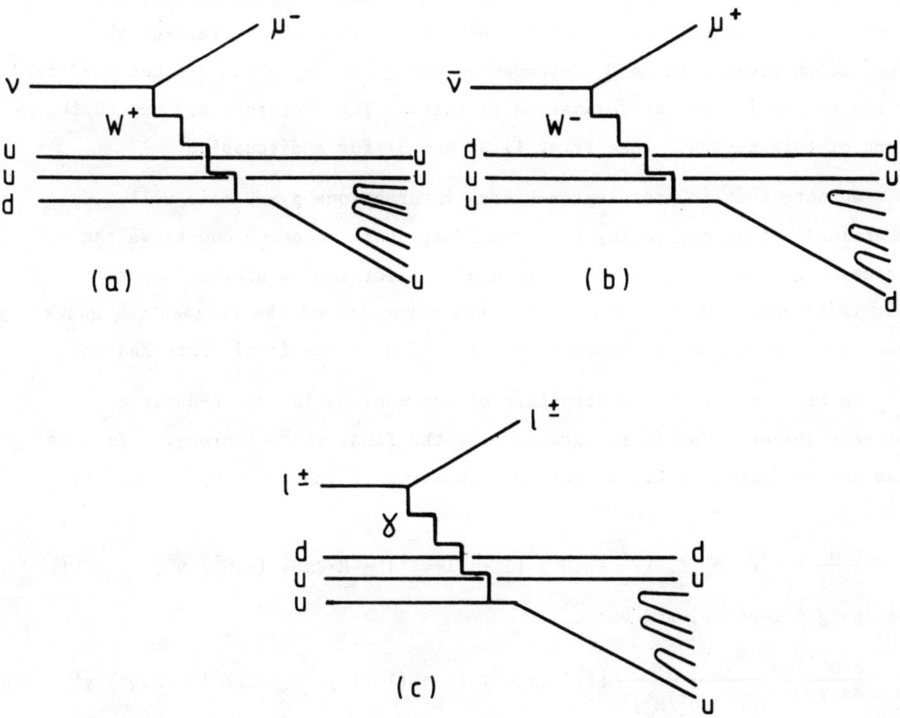

Fig. 2: Current-quark interaction showing most probable final states .

The interaction of the electromagnetic or weak current at sufficiently large Q^2 will sample a small spatial region which is proportional to $1/Q$. For example, taking $Q^2 > 4\ (\mathrm{GeV}/c)^2$ we are probing distances less than 0.1 fermi ; whereas the radius of the nucleon is of the order of one fermi. Thus interactions for which Q^2 is of the order of a few $(\mathrm{GeV}/c)^2$ will provide information on the substructure of the nucleon. If in addition the Q^2 is larger than the internal momentum of the quarks in the hadron,"primordial p_T", this quark motion can be neglected*. The lepton-hadron interaction can be treated as a current-quark interaction, Fig. 2, where the other quarks are spectators (impulse approximation). Such a picture requires modification because of the presence of the gluon field which results in quarks and antiquarks with which the incoming current can interact. In addition, the emission of a gluon (Bremsstrahlung) before or after the current-quark interaction gives rise to Q^2 dependences (non-scaling) which are not predicted by the sample impulse approximation picture. These effects are beyond the scope of this review. See refs. 1) 2) and 3) for a discussion.

We note that in most lepton-hadron interactions $p_\ell - p_\ell{}'$ is well determined; thus neglecting the primordial quark momentum one knows the direction of the struck quark. In contrast for e^+e^- annihilation there is no primordial momentum distribution but the direction of the fragmenting quark must be inferred from the momentum distribution of the final state hadrons.

In order to study the structure of the nucleon in lepton-nucleon interactions one usually integrates over the final state hadrons. In this case the differential cross sections, in terms of the above variables, are given by:

$$\frac{d^2\sigma^{\ell^\pm}}{dxdy} = \frac{4\pi\alpha^2}{Q^4} M_N E_\ell \left[F_2^\ell (x,Q^2) \left[1 + (1-y)^2\right] - R\cdot 2xF_1^\ell (x,Q^2)\, y^2\right] \tag{1}$$

for charged leptons, and by:

$$\frac{d^2\sigma^{\nu,\bar\nu}}{dxdy} = \frac{G^2 M_N E_\nu}{\pi[1 + Q^2/M_W^2]^2} \Big[F_2^{\nu,\bar\nu}(x,Q^2) \left[1 + (1-y)^2\right] - R\cdot 2xF^{\nu,\bar\nu} (x,Q^2)\, y^2 \pm \left[1 - (1-y)^2\right] xF^{\nu,\bar\nu} (x,Q^2)\Big] \tag{2}$$

for neutrinos and antineutrinos, where M_N and M_W are the masses of the target and W-boson, respectively, and the +(-) sign in eq. 2 is associated with the $\nu(\bar\nu)$ cross section. The parameter R is given by

$$R \equiv \sigma_L/\sigma_T = \left[F_2(1 + 4\, M_N x^2/Q^2) - 2xF_1\right] / 2xF_1 \tag{3}$$

* The primordial p_T is assumed to be a few hundred MeV/c.

where $\sigma_L (\sigma_T)$ is the cross section for a longitudinally (transversely) polarized photon or intermediate boson. For large Q^2 we have

$$R \simeq F_1/2xF_2 - 1 \tag{4}$$

Note that the contribution of $R(x,Q^2)$ in eqs. 1 and 2 is important only for large y; thus R need not play an important role in determining $F_2(x,Q^2)$. In addition R is small [1,2], of the order of 0.2 or less as can be seen in Fig. 3. For a good discussion of R see Ref. [4].

The inverse power of Q in eqs. 1 and 2 results from the propagator. For the ν beam presently available the Q^2 range is such that $Q^2/M_W^2 < 1$ and the cross section grows linearly with energy.

Information about the structure of the nucleon is contained in the structure functions $F_i\ (x,Q^2)$ defined in eqs. 1 and 2. In general the structure functions obtained from charged lepton, neutrino and antineutrino interactions are different. However, simple relations between these structure functions are predicted by the quark parton model.

3. Comparison with Quark Parton Model

In the simple quark parton model (QPM) an interaction with a nucleon is assumed to be between the current and the spin ½ valence or sea quark. The scaling variable x_{B_j} is the fraction of the nucleons momentum carried by the interacting quark. The quarks are assumed to be point-like. Hence the quarks and charged leptons have the same electromagnetic current and the strength of this coupling is given by the product of the charges of the quark and the incident charged lepton. For the weak interactions it is also assumed that the weak current of the quark is V-A (left-handed like the neutrino) and the antiquark's is V+A (right-handed like the antineutrino). A consequence of the different helicities of the couplings is found in the expected y-distributions; νq and $\bar{\nu}\bar{q}$ interactions have no y-dependence (isotropic in y) while $\bar{\nu}q$ and $\nu\bar{q}$ interactions have a $(1-y)^2$ distribution.

For neutrino interactions there is a further complication because the charged current interaction changes the flavour of the struck quark. In charged current interactions the $\nu(\bar{\nu})$ can be thought of as a "raising" ("lowering") operator for the doublets shown in Fig. 4. If the transition

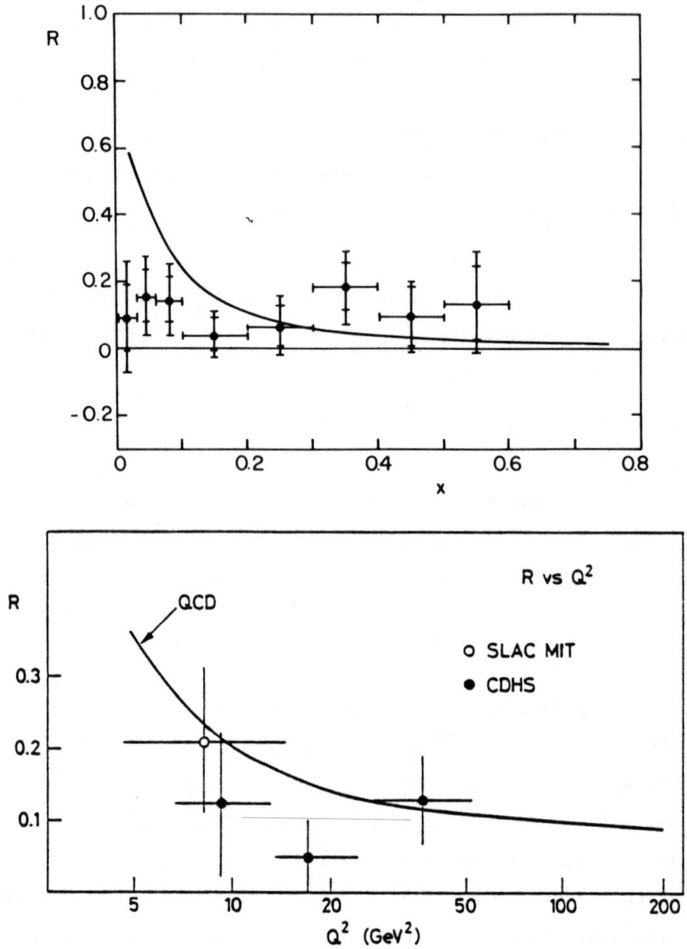

Fig. 3: Dependence of R on x and Q^2 determined from ν interactions (Ref. 1,5).

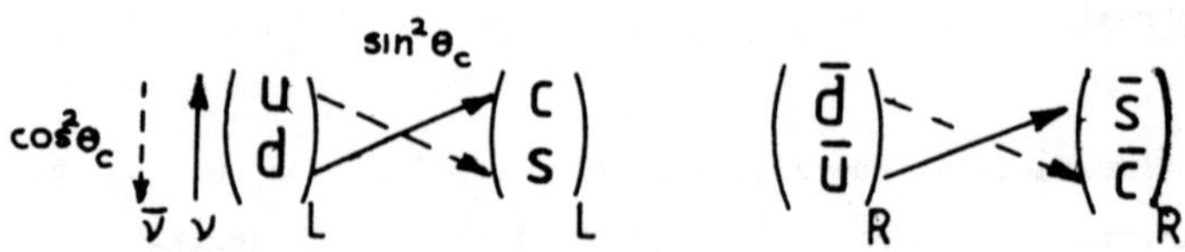

Fig. 4: Flavour changing nature of charged current interactions.

is within a doublet (Cabbibo favoured) there is a $\cos^2\theta_c \simeq 0.95$ in the cross section; whereas, if the transition is from one doublet to another (Cabbibo unfavoured) then there is a $\sin^2\theta_c \simeq 0.05$ in the cross section.

Within the framework described above the structure functions can be parameterized by the quark and antiquark distribution functions u(x), $\bar{u}(x)$, d(x), $\bar{d}(x)$, s(x), $\bar{s}(x)$, c(x), $\bar{c}(x)$.

$$F_2^{\ell p}(x) = 2xF^{\ell p}(x) = \frac{4}{9}x\,[u(x)+\bar{u}(x)+c(x)+\bar{c}(x)] + \frac{1}{9}x[d(x)+\bar{d}(x)+s(x)+\bar{s}(x)] \quad (5)$$

$$F_2^{\ell n}(x) = 2xF^{\ell n}(x) = \frac{4}{9}x[d(x)+\bar{d}(x)+c(x)+\bar{c}(x)] + \frac{1}{9}x[u(x)+\bar{u}(x)+s(x)+\bar{s}(x)] \quad (6)$$

$$F_2^{\nu p}(x) = 2xF_1^{\nu p}(x) = 2\,x[d(x)+s(x)+\bar{u}(x)+\bar{c}(x)] \quad (7)$$

$$F_2^{\nu n}(x) = 2xF_1^{\nu n}(x) = 2\,x[u(x)+s(x)+\bar{d}(x)+\bar{c}(x)] \quad (8)$$

$$F_2^{\bar{\nu} p}(x) = 2xF_1^{\bar{\nu} p}(x) = 2\,x[u(x)+c(x)+\bar{d}(x)+\bar{s}(x)] \quad (9)$$

$$F_2^{\bar{\nu} n}(x) = 2xF_1^{\bar{\nu} n}(x) = 2\,x[d(x)+c(x)+\bar{u}(x)+\bar{s}(x)] \quad (10)$$

$$F_3^{\nu p}(x) = d(x)+s(x)-\bar{u}(x)-\bar{c}(x) \quad (11)$$

$$F_3^{\nu n}(x) = u(x)+s(x)-\bar{d}(x)-\bar{c}(x) \quad (12)$$

$$F_3^{\bar{\nu} p}(x) = u(x)+c(x)-\bar{d}(x)-\bar{s}(x) \quad (13)$$

$$F_3^{\bar{\nu} n}(x) = d(x)+c(x)-\bar{u}(x)-\bar{s}(x) \quad (14)$$

The structure functions which are obtained from interactions on an isoscalar target (equal number of neutrons and protons) are particularly informative. These follow directly from averaging the appropriate neutron and proton structure functions, where the notation $q(x) = u(x)+d(x)+s(x)+c(x)$ and $\bar{q}(x) = \bar{u}(x)+\bar{d}(x)+\bar{s}(x)+\bar{c}(x)$ is used.

$$F_2^{\ell N}(x) = \frac{5}{18}\,x[q(x)+\bar{q}(x) - \frac{6}{5}\,s(x) + \frac{6}{5}\,c(x)] \quad (15)$$

$$F_2^{\nu N}(x) = F_2^{\bar{\nu} N}(x) = 2xF_1^{\nu N}(x) = 2xF_1^{\bar{\nu} N}(x) = x\,[\bar{q}(x) + q(x)] \quad (16)$$

$$F_3^{\nu N}(x) = x[q(x) - \bar{q}(x) + 2\,s(x) - 2\,c(x)] \quad (17)$$

$$F_3^{\bar{\nu} N}(x) = x[q(x) - \bar{q}(x) - 2\,s(x) + 2\,c(x)] \quad (18)$$

$$\frac{1}{2}\,[F_3^{\nu N}(x) + F_3^{\bar{\nu} N}(x)] = x\,[q(x) - \bar{q}(x)] \quad (19)$$

From eqs. 15, 16 and 17 it follows that

$$F_2^{\nu,\bar{\nu}N}(x) = \frac{18}{5} F_2^{\ell N}(x), \tag{20}$$

if antiquarks do not contribute (i.e. for $x_{B_j} \gtrsim 0.2$ as we show below). Thus from the structure functions extracted from interactions on nuclear (isoscalar) targets we can obtain the momentum distribution of quarks in the nucleon.

The results from ℓN and $\nu(\bar{\nu})N$ interactions give $F_2(x)$ distributions which are in good agreement. In Figs. 5, 6 and 7 the $F_2(x)$ obtained from ep, ed (SLAC), μp (FNAL and CERN) and μFe, νFe, $\bar{\nu}$Fe (CERN) interactions are compared. From the CDHS data, Fig. 7, one concludes that the fraction of the nucleon's momentum carried by:

all quarks $(q + \bar{q}) = 0.45 \pm 0.01$

valence quarks $(q - \bar{q}) = 0.32 \pm 0.01$

sea quarks $(2\,\bar{q}) = 0.13 \pm 0.01$.

From which we conclude that approximately 50% of the nucleon's momentum is carried by the gluons.

Within the QPM there are many sum rules which can be derived (see Ref. 1). A useful one is that obtained from the sum of the total νN and $\bar{\nu}$N cross sections

$$\frac{3\pi}{4G^2ME}\left[\sigma_{tot}^{\nu N} + \sigma_{tot}^{\bar{\nu}N}\right] = \int x\,[q(x) + \bar{q}(x)]\,dx \tag{21}$$

which gives the fraction of momentum carried by all quarks. The cross sections measured by the CDHS and Caltech groups are given in Fig. 8. A more recent result by the CHARM collaboration[9] gives

$$\sigma_{tot}^{\nu N}/E_\nu = 0.604 \pm 0.032 \times 10^{38}\ \mathrm{cm}^2/\mathrm{GeV}$$

$$\sigma_{tot}^{\bar{\nu}}/E_{\bar{\nu}} = 0.301 \pm 0.018 \times 10^{38}\ \mathrm{cm}^2/\mathrm{GeV}$$

With these last results the momentum sum rule eq. 21 gives 0.44 ± 0.02 for the momentum fraction carried by the quarks and antiquarks in the nucleon. However, a new measurement of the ν and $\bar{\nu}$ total cross sections by the CFRRN collaboration (FNAL) gives [10]

$$\sigma_{tot}/E_\nu = 0.733 \pm 0.005 \pm 0.035$$

$$\sigma_{tot}/E_{\bar{\nu}} = 0.371 \pm 0.004 \pm 0.037,$$

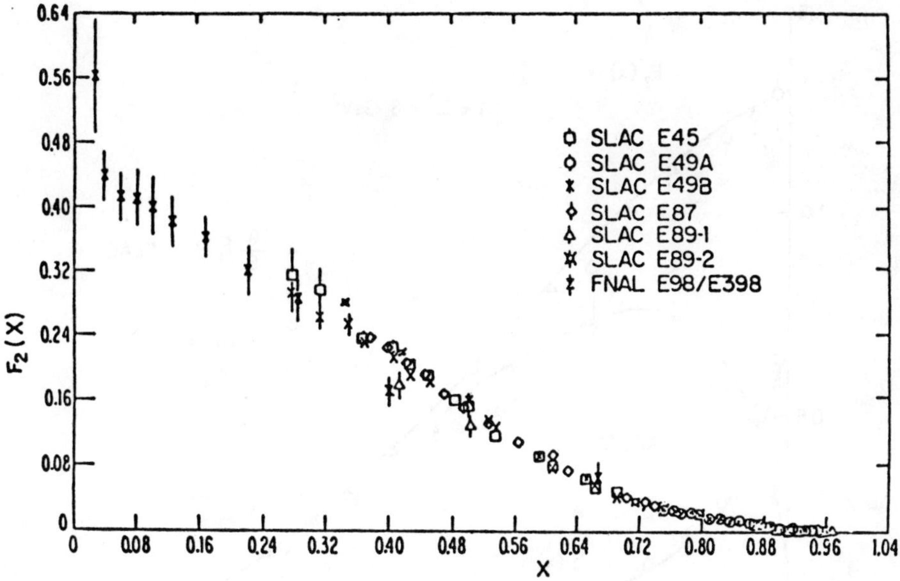

Fig. 5: $F_2(x)$ obtained in ep and μp interactions (Ref. 6,7).

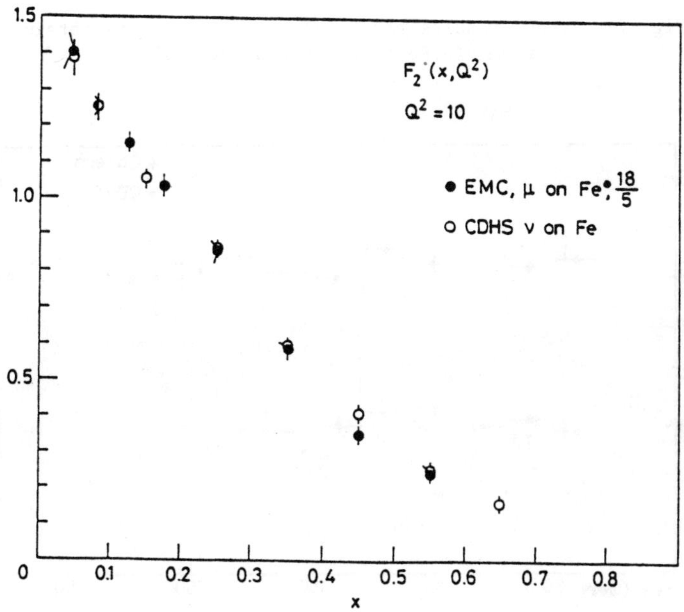

Fig. 6: $F_2(x)$ from νF_e and μF_e interactions with the latter scaled by 18/5 as predicted by the QPM (eq. 20).

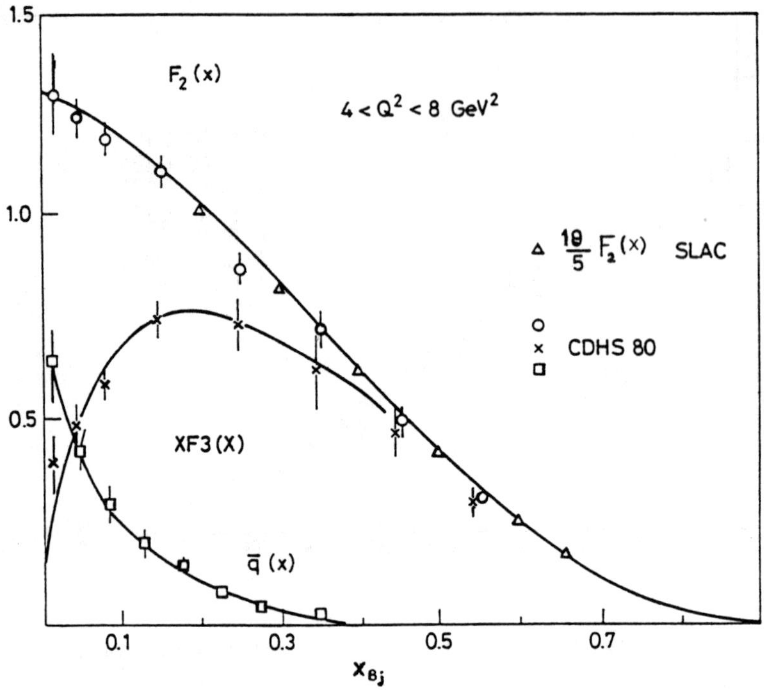

Fig. 7: Recent results for $F_2(x)$, $xF_3(x)$ and $\bar{q}(x)$ obtained from $\nu(\bar{\nu})$Fe interactions (Ref. 1).

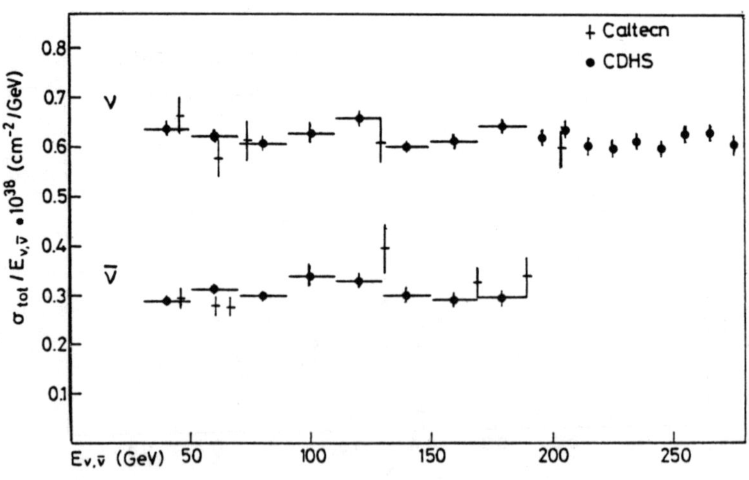

Fig. 8: Neutrino and antineutrino total cross sections scaled by the neutrino energy (Ref. 1, 8).

where the quoted errors are statistical and systematic, respectively. This result, which is preliminary, gives a 20% larger value for the momentum fraction carried by the quarks. Since the ratio $\sigma^{\nu}/\sigma^{\bar{\nu}}$ obtained by all of the above experiments agrees well, the differences come from the beam flux determinations.

The above discussion has neglected the Q^2 dependence (non-scaling) of the structure functions. The interested reader can find good discussions in Refs. [1,2,3,4,11].

Because of the flavour changing nature of the charged current coupling, an additional measure of the sea and valence quark distributions can be obtained from studying opposite sign di-muon production. The second muon is supposed to come from the decay of a charmed quark which is produced from interactions with either the strange sea or a d-valence quark for neutrino interactions

$$\nu_\mu + \{{}^{s}_{d}\} \rightarrow \mu^- + c \qquad (22)$$

and only from the sea in the case of $\bar{\nu}$ interactions

$$\bar{\nu}_\mu + \{{}^{\bar{s}}_{\bar{d}}\} \rightarrow \mu^+ + \bar{c} \qquad (23)$$

Thus from the $\bar{\nu}$ data one can obtain directly the sea quark x-distribution which, when subtracted from the x-distribution obtained for ν interactions, yields the d-quark x distributions. The result of such an analysis performed by the CDHS group is shown in Fig. 9 [12].

For the region $x \gtrsim 0.2$, where sea quark contributions are minimal, the ratio

$$F_2^{\nu p}(x)/F_2^{\ell p}(x) = \frac{18\ d/u}{4 + d/u} \qquad (24)$$

gives a measure of $d(x)/u(x)$. A recent determination by the ABCMO collaboration[13] using their νp data (BEBC) and a combined ep and μp data set gives the results shown in Fig. 10. As x increases the u-quark becomes increasingly dominant. Only at the smallest x-values does one observe the simple quark parton model prediction of 1/2. Various model predictions are also given in Fig. 10.

In a recent series of papers Brodsky et al. [14,15] have shown that the wave function of the proton can be expanded into a set of amplitudes which can include heavy quark pairs such as $c\bar{c}$.

$$|p\rangle = \alpha|uud\rangle + \beta|uudg\rangle + \gamma|uudq\bar{q}\rangle + \ldots$$

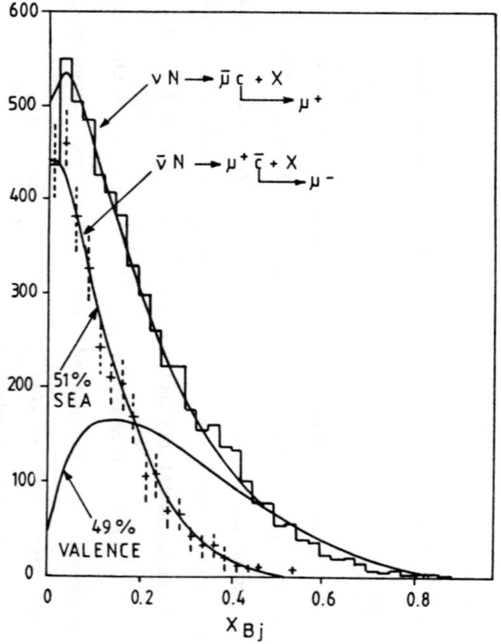

Fig. 9: Valence and sea quark x-distributions as observed in $\nu(\bar{\nu})$ production of dimuons (Ref. 12).

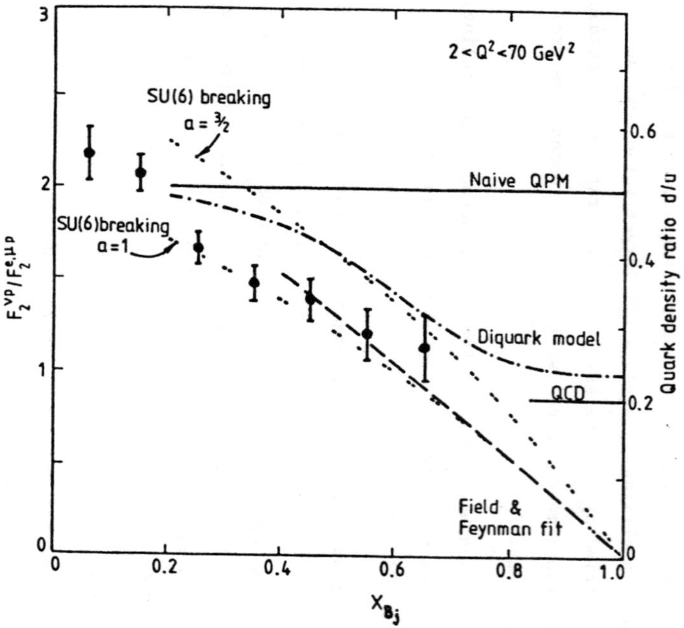

Fig. 10: The dependence of the ratio d(x)/u(x) on x_{B_j} obtained by the ABCMO collaboration (Ref. 13).

A consequence of the intrinsic heavy quark states is the diffractive production of charmed hadrons. They claim that γ of the order of 1% can account for the large forward production of charm observed at the ISR. In addition the large number of like-sign dimuon events observed in $\nu(\bar{\nu})$ interactions can be accounted for in this picture by the reactions

$$\bar{\nu}c \to \mu^+ b$$
$$\quad \lfloor\!\to c \to s\mu^+\nu$$

$$\nu\bar{c} \to \mu^- \bar{b}$$
$$\quad \lfloor\!\to \bar{c} \to \bar{s}\mu^-\bar{\nu}$$

The predictions of these authors is compared with the available data in Fig. 11. The agreement with data is rather good. Thus the wave function of the proton may also contain a non negligible fraction of intrinsic heavy quark states whose consequences are discernable.

4. Hadron Final States (Jets)

The hadronic system resulting from lepton nucleon interactions at large Q^2 is expected to have a two-jet structure along the direction of the current in the center of mass of the hadronic system. One, the current jet, results from the hadronization of the struck quark, the other, the target jet, results from the hadronization of the spectator di-quark system. Thus we expect to find similarities between the hadronic jets resulting from different classes of interactions (lepton-nucleon, e^+e^- annihilation, hadron-nucleon) provided that the flavor of the hadronizing quark (di-quark) systems which are compared is the same.

For leptonic interactions the flavour of the originator of the jet is rather well specified. Because the charged lepton coupling is proportional to the square of the charge u and ud fragmentation accounts for 8/9 of the proton cross section. While for neutrinos (antineutrinos) u and uu (d and ud) fragmentation accounts for ≈ 90% of the cross section. Interactions with sea quarks and Cabbibo unfavoured interactions account for the rest. Thus forward (backward) jets in charged lepton interactions are comparable to forward (backward) jets in neutrino (antineutrino) interactions. For e^+e^- annihilations above a few GeV all quark flavours can occur with the relative strength proportional to the square of the charge. The flavours of the fragmenting systems are summarized in Table 1.

Table 1: Flavour of fragmenting system

Interaction	Target	Current
νp	(uu)	u
$\bar{\nu} p$	(ud)	d
$\ell^{\pm} p$	(ud)	u
e^+e^-	-	$q(\bar{q})$

The longitudinal direction of the fragmenting quarks for leptonic interactions is determined by the current direction, which is well determined in e, μ interactions. However, for ν, $\bar{\nu}$ interactions, where the small cross sections and hadron detection dictate that the experiments be done with a wide band beam and bubble chamber, one usually uses the vector sum of the hadronic momentum to estimate the current direction. The bubble chamber is an imperfect calorimeter, 15 to 20 per cent of the hadronic energy is not detected. Various energy balance techniques are used in order to estimate the missing hadronic energy and on the average these work very well. Thus the center of mass of the hadronic system is not determined precisely for each event in neutrino interactions whereas for interactions of charged leptons the hadronic mass, W, is very well determined. In the case of e^+e^- annihilation the direction of the fragmenting quark must be determined by a statistical procedure such as the calculation of a thrust or sphericity axis.

The inclusive hadron spectra are often parametrized by a quark fragmentation function $D_i^h(z)$ which gives the average number of hadrons of type h produced in the interval dz by a quark of flavour i

$$D_i^h(z) \equiv \frac{1}{\sigma_{tot}} \frac{d\sigma}{dz} \tag{25}$$

where z is the fraction of the quarks energy taken by the hadron h; z is usually approximated by E^h/ν or $p^h/|\Sigma p^h|$. A similar function $D_{ij}(z)$ can be defined for the di-quark system. If the hadrons are pions, isospin conservation gives a number of relations between the quark fragmentation functions

$$D_u^{\pi^+} = D_d^{\pi^-}, \quad D_u^{\pi^-} = D_d^{\pi^+}, \quad D_i^{\pi^+} + D_i^{\pi^-} = 2\, D_i^{\pi^o}$$

$$D_{uu}^{\pi^+} = D_{dd}^{\pi^-}, \quad D_{uu}^{\pi^-} = D_{dd}^{\pi^+}, \quad D_{ud}^{\pi^+} = D_{ud}^{\pi^-}$$

The current quark fragmentation functions can be extracted from interactions on nuclei, since the rescattering effects, particularly in light nuclei, are not too important (see section 5). However, for the diquark system which is slow in the target frame, there can be considerable rescattering in the nucleus. Thus diquark fragmentation must be studied in ℓp or $\nu(\bar{\nu})p$ interactions. As a consequence there is currently very little information available for D_{qq}. The only data come from $\nu(\bar{\nu})p$ interactions.

The above view of hadron production assumes that W is sufficiently large to separate well the target and current fragmentation regions; i.e. that there is a large rapidity range available. At present energies this criterion is only marginally satisfied. In Fig. 12 the W distribution obtained in $\bar{\nu}p$ [16] interactions shows that most of the data occur for $W \lesssim 6$ GeV. The rapidity distributions obtained for these data for several W ranges are shown in Fig. 13. We note that there is no good evidence for a plateau.

In Fig. 14 the νp(BEBC) data [17] also illustrate this point. Only for $W \gtrsim 8$ GeV does one begin to observe an indication of a rapidity plateau and the charge difference between the current and target fragmentation that we expect from the quark parton model. With this caveat in mind we will continue.

In Fig. 15 the rapidity distribution obtained for ± particles in $\bar{\nu}p$ interactions for $4 < W < 12$ GeV and pp interactions at 24 GeV/c [18] are compared Identified protons have been removed in both samples. The available energy is the same in both reactions: $\langle W \rangle - m_p = 4.8$ GeV and $\sqrt{s} - 2\,m_p = 4.9$ GeV. For $y \lesssim -0.7$, the two distributions are similar. In the central region however the distributions are quite different. This may be due to the range of c.m. energies covered by the neutrino data, as the integral of the two distributions agree well. The small differences in the average charge multiplicities can be accounted for by the missing one prong events in the $\bar{\nu}p$ data.

The normalized invariant cross section obtained in $\bar{\nu}p$ interactions is shown in Fig. 16 for both positive and negative particles where identified protons have been removed and an $x_{B_j} > 0.2$ selection is made. Also shown on Fig. 16 are the results obtained from $pp \to \pi^{\pm}x$ at 200 GeV/c [19] and $\pi^{+}p \to \pi^{-}X$ at 16 GeV/c [20], where for the hadronic reactions the π^+ and π^- data are scaled by 0.75 and 1.5, respectively, as suggested by Andersson et al.[21]

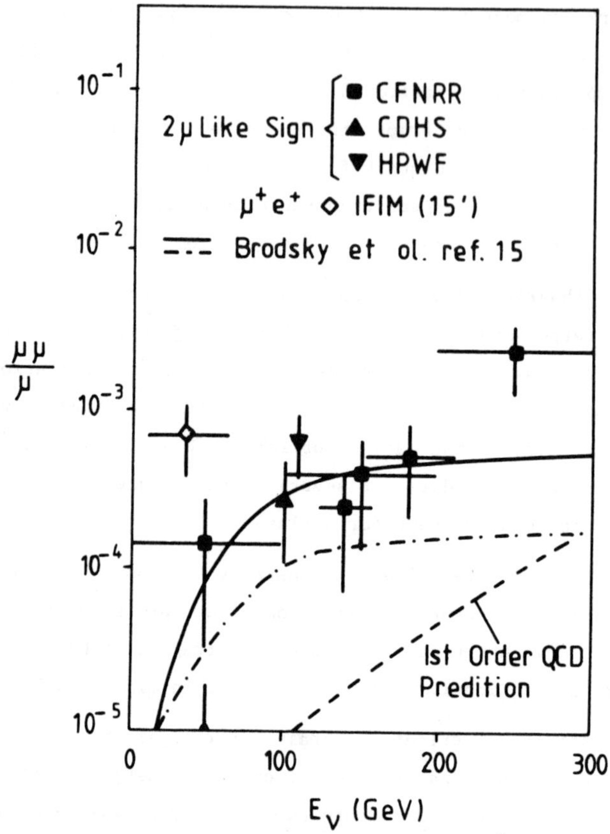

Fig. 11: Ratio of same sign dimuon events to the normal charged current events (Ref. 15).

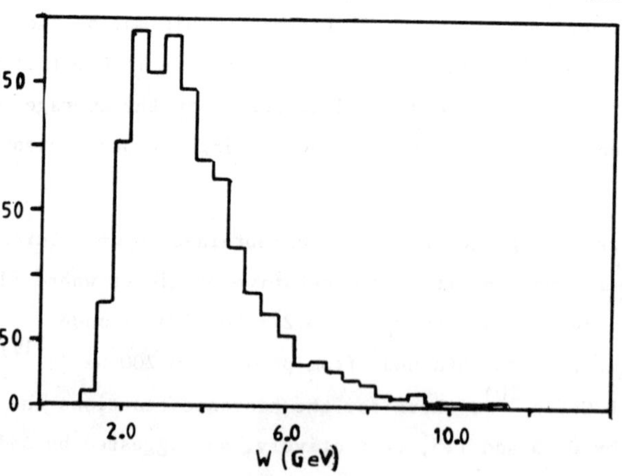

Fig. 12: W distribution observed in $\bar{\nu}p$ interactions (Ref. 16).

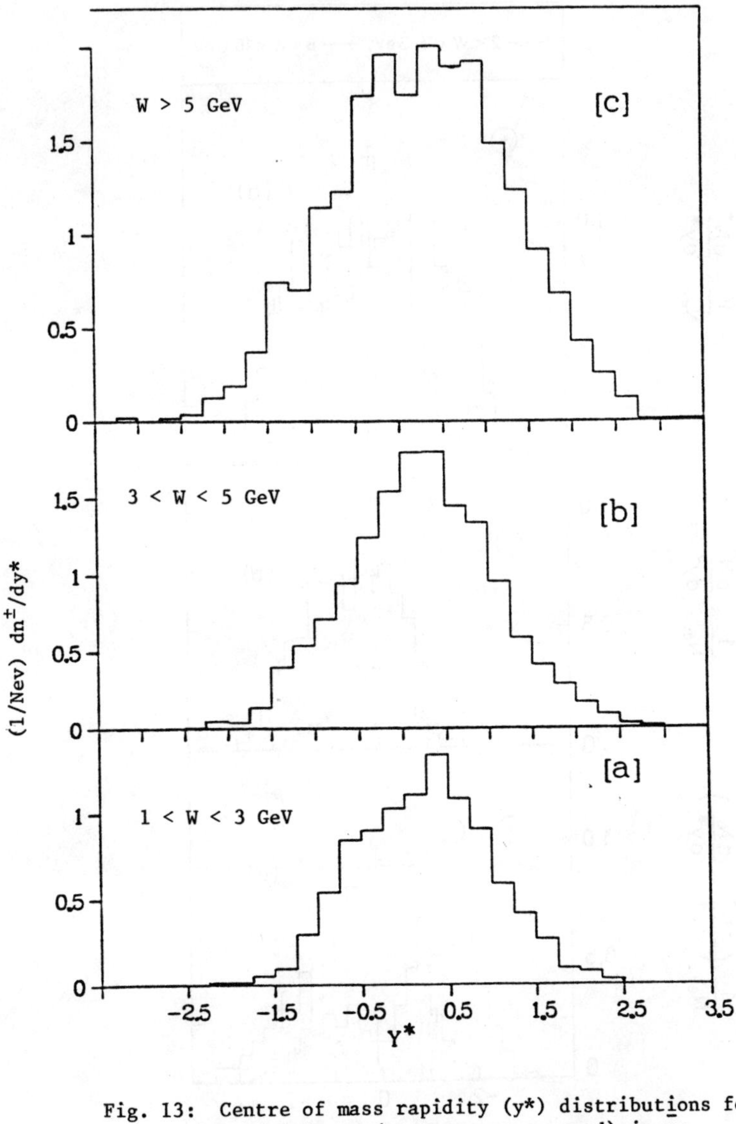

Fig. 13: Centre of mass rapidity (y*) distributions for charged hadrons (protons not removed) in νp interactions (Ref. 16) with $Q^2 < 45$ $(GeV/c)^2$.

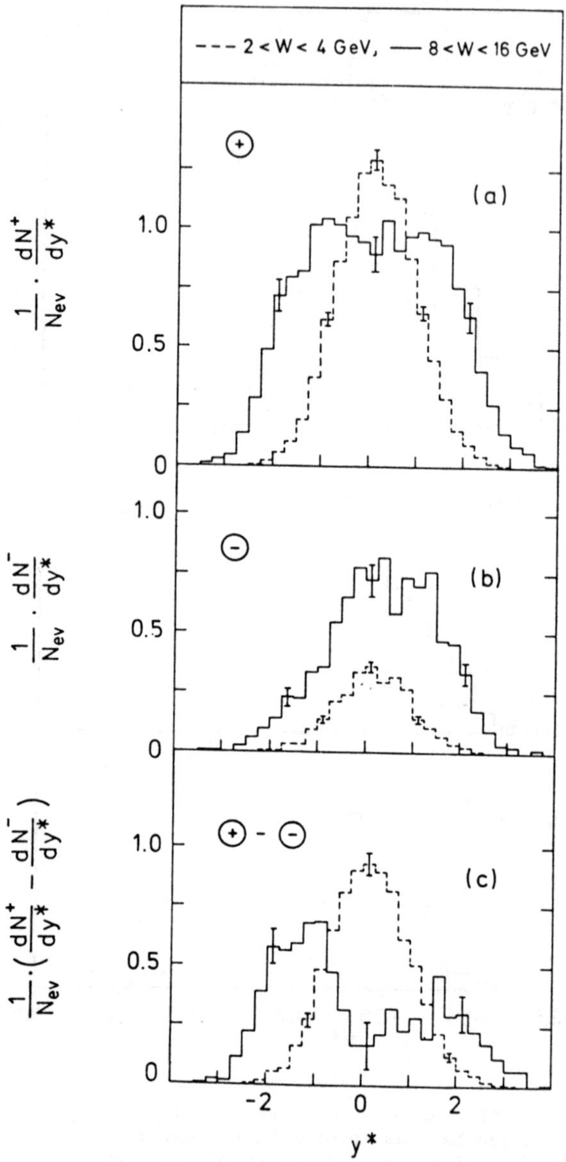

Fig. 14: The center of mass rapidity (y*) distributions for positive and negative hadrons and their difference observed in νp interactions (Ref. 17).

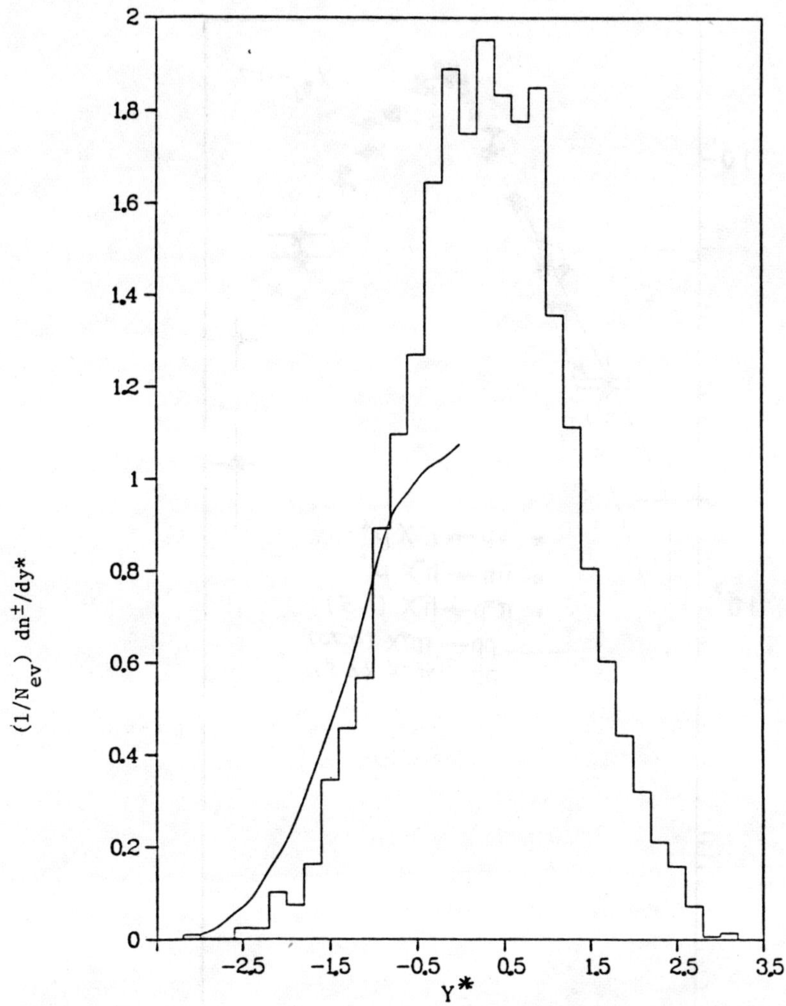

Fig. 15: The centre of mass rapidity (y*) distribution for charged hadrons observed in $\bar{\nu}p$ interactions (identified protons removed) (Ref. 16). The solid line is the equivalent distribution obtained from $pp \to h^{\pm}x$ at 24 GeV/c (Ref. 18).

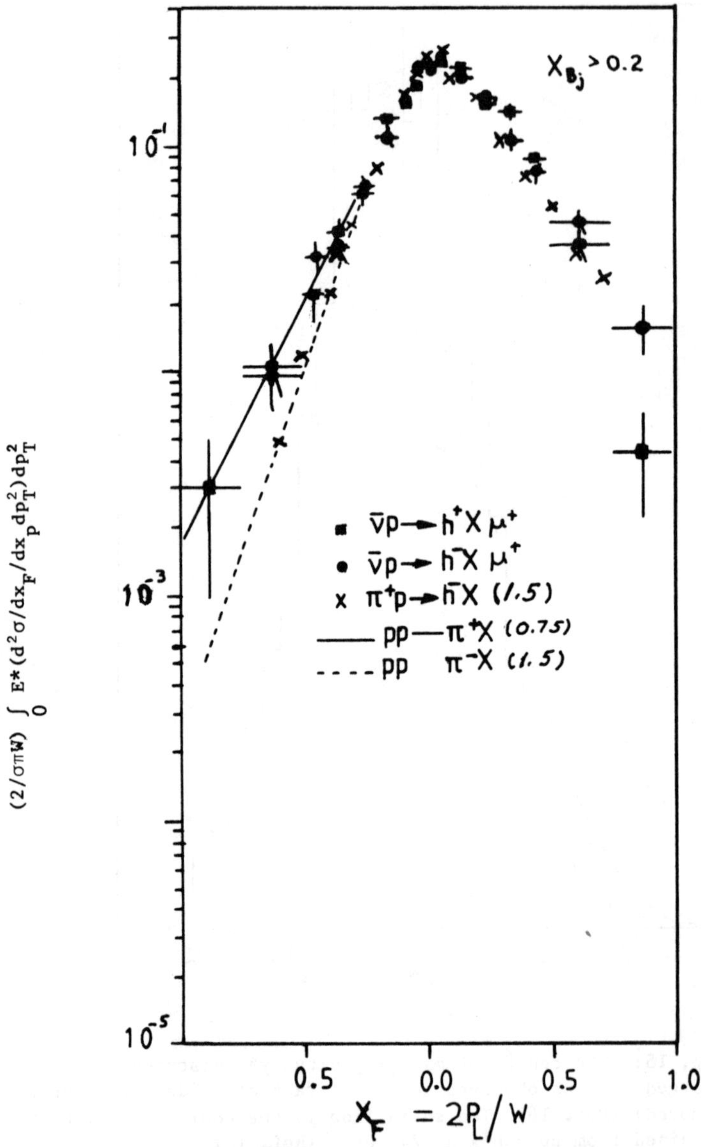

Fig. 16: The normalized invariant x_F distribution for positive (identified protons removed) and negative hadrons for the following itneractions: $\bar{\nu}p$ (Ref. 16), pp (Ref. 19), and π^+p (Ref. 20).

Their essential assumption for the target fragmentation region is that

$D^{\pi+}_{ud} = D^{\pi-}_{ud} \simeq D^{\pi+}_{un} > D^{\pi-}_{un}$.

A fit of the $\bar{\nu}p$ data to the form $A(1-|x_F|)^n$ yields the results summarized in Table 2*. While the errors are large the results suggest n = 3 behaviour for the target fragmentation region.

Table 2: Fit of the $\bar{\ }$p invariant x_F distribution to $A(1-|x_F|)^n$

x_F	charge	n
< − 0.05	+	3.0 ± 0.3
< − 0.05	−	3.3 ± 0.5
> 0.05	+	2.0 ± 0.2
> 0.05	−	1.6 ± 0.2

In Fig. 17 the fragmentation functions $D^{h^+}_{ud}$ and $D^{h^-}_{ud}$ are compared, and we note that they have a similar x_F dependence. The identified protons have been removed.

Some further interesting comparison between hadronic interactions and neutrino interactions can be made if we choose specific triggers. For example, by studying the reaction $K^-p \to \pi^{\pm}\bar{K}^o x$ and selecting a fast $\bar{K}^o$ B. Buschbeck [21] argues that one selects the production mechanism in which the s-quark in the incoming K^- is fast and therefore the slow ū-quark has a large probability of annihilating with a u-quark in the proton. Thus in the target fragmentation region one should observe the fragmentation of a ud diquark system. She finds that the π^+/π^- ratio is approximately 1 when a fast $\bar{K}^o$ is selected. In Fig. 18 the normalized invariant cross section is compared with that obtained in $\bar{\nu}p$ interactions. The agreement is good for events with a fast $\bar{K}^o$. A more general result has been obtained by the Paris-Saclay-Rutherford collaboration [23] studying K^-p interactions at 70 GeV/c. The target fragmentation can be expressed as:

$$\frac{1}{\sigma}\frac{d\sigma^{\pi^\pm}}{dz} \simeq \frac{1}{3} D^{\pi^\pm}_{uu}(z) + \frac{2}{3} D^{\pi^\pm}_{ud}(z) \qquad (26)$$

Thus by using $2\, D^{\pi+}_{ud} \simeq D^{\pi+}_{uu} + D^{\pi-}_{uu}$ proposed by Anderson et al.[20], one can extract $D^{\pi+}_{uu}$ and $D^{\pi-}_{uu}$ from the π^+ and π^- $(1/\sigma)\, d\sigma/dz$ distributions. The distributions thus obtained, Fig. 19, agree well with the νp and $\bar{\nu}p$ results.

---

where $x_F \equiv 2P_L/W$.

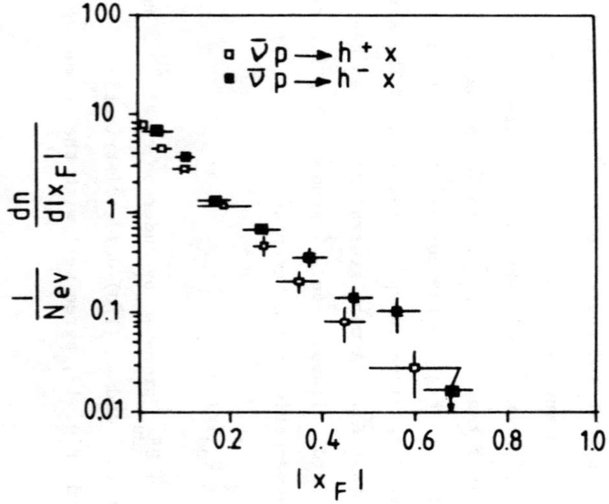

Fig. 17: The diquark (ud) fragmentation functions (Ref. 16)

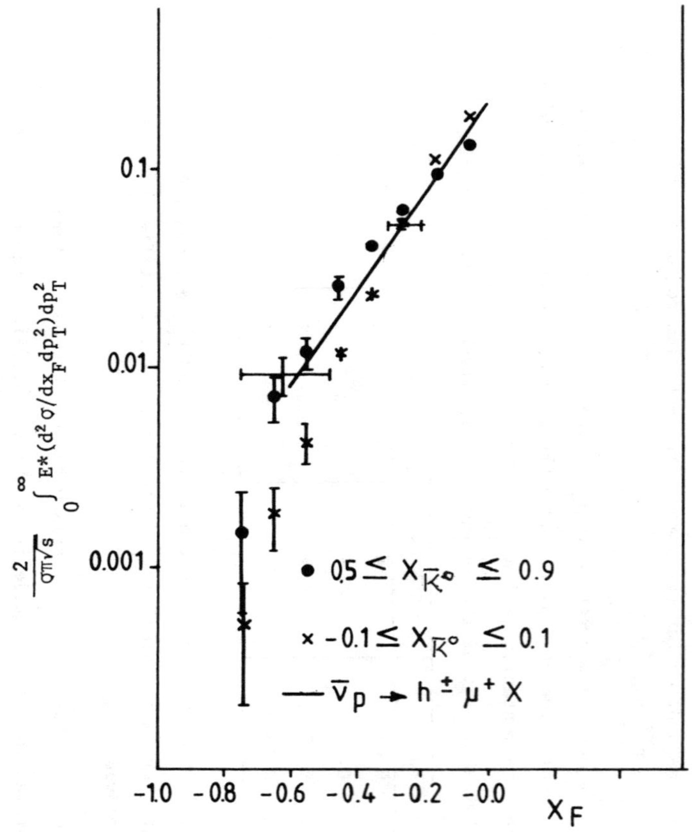

Fig. 18: Invariant cross section for $K^-p \rightarrow \pi^-\bar{K}^o x$ for fast and slow $\bar{K}^o$'s

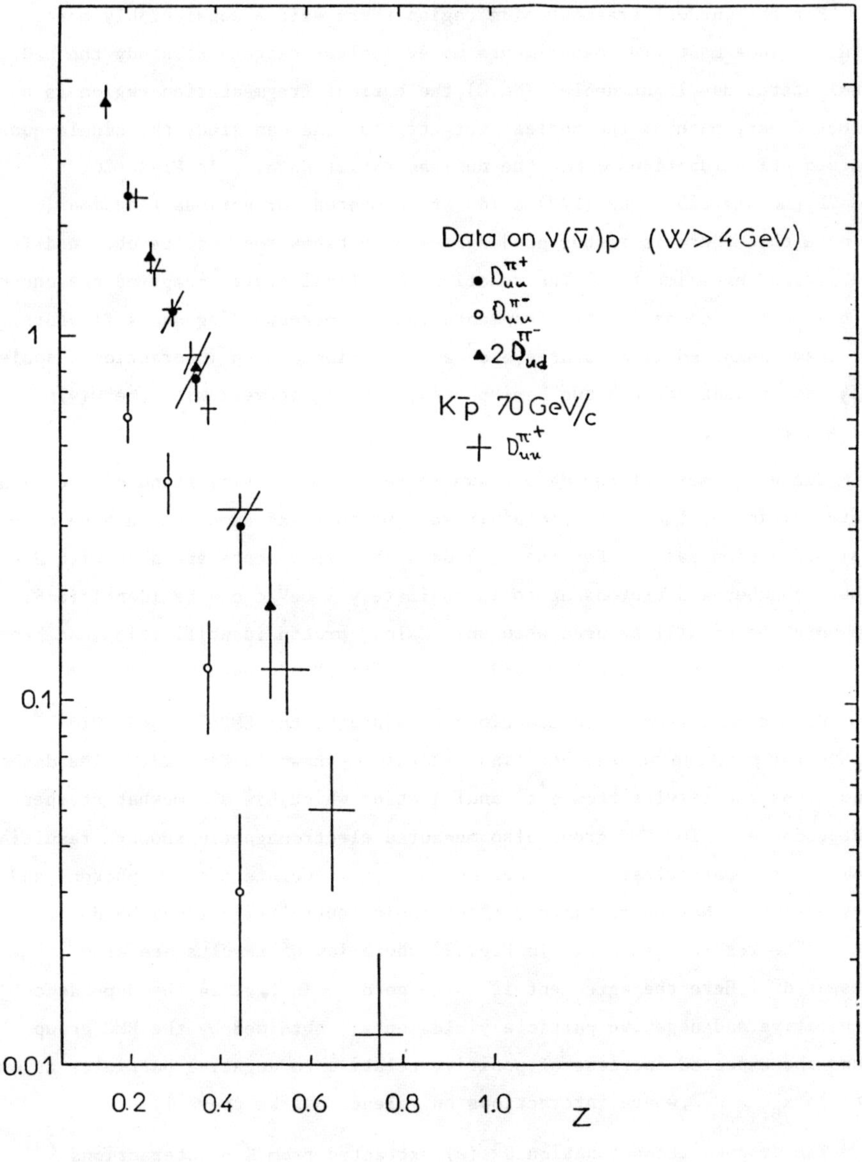

Fig. 19: Diquark fragmentation function obtained in K^-p interactions at 70 GeV/c compared with $\nu(\bar{\nu})p$ results (Ref. 23).

For the current fragmentation region there exists considerably more data. Since most $\nu(\bar{\nu})$ experiments using nuclear targets to study the hadron final states use light nuclei (Ne,C), the current fragmentation region is not affected very much by the nuclear targets; thus one can study the single-quark fragmentation functions with the nuclear target data. In Figs. 20, and 21 the distributions $(1/\sigma)\ d\sigma/dz$ are compared for various reactions. We note that there is rather good agreement between the results obtained from the various experiments. The reactions and final states compared are chosen so that one is comparing the fragmentation of corresponding quark flavours. For the ± combined final states, e^+e^- annihilation and pp interactions, scaled by ½ to account for the two hemispheres, also agree well with the $\nu(\bar{\nu})p$ and e^-p results.

Since for most of the data shown there is no identification of the final state particles, there are certainly some protons and kaons which have been assigned a pion mass. For the $\nu(\bar{\nu})$ data the experiments are done with a bubble chamber and protons up to approximately 1 GeV/c can be identified*. The notation $h^{\pm}$ will be used when only "slow" proton identification has been made. We will discuss below evidence for "fast" protons.

Recent data from μp interactions obtained by the EMC [25)] and CHIO [26)] groups for positive and negative final states are shown in Fig. 22. The dashed band gives the results from e^+e^- annihilation which has a somewhat steeper z-dependence. The EMC group also measured electromagnetic showers resulting from neutral particles. They are not able to separate the two photons, and they estimate that approximately 6% of their neutral signal may be due to η's. The results are shown in Fig. 23 where low Q^2 results are also presented. Here the agreement is quite good. In Fig. 24 the dependence of positive and negative particle yields on x_{B_j} obtained by the EMC group shows the expected increase of positive relative to negative particles as one goes to $x_{B_j} > 0.2$, where interactions on valence quarks prevail.

The fragmentation function $D_u^{\pi^-}(z)$ extracted from K^-p interactions [23)] at 70 GeV/c is shown in Fig. 25, where it is compared with $\nu(\bar{\nu})p$ results and the Field-Feynman parametrization. The agreement is rather good. The $D_d^{\pi^+}$ distribution has an unidentified proton contribution [29)], particularly at low z, which can account for it being somewhat higher. In extracting

* The exact value depends on the liquid and chamber conditions.

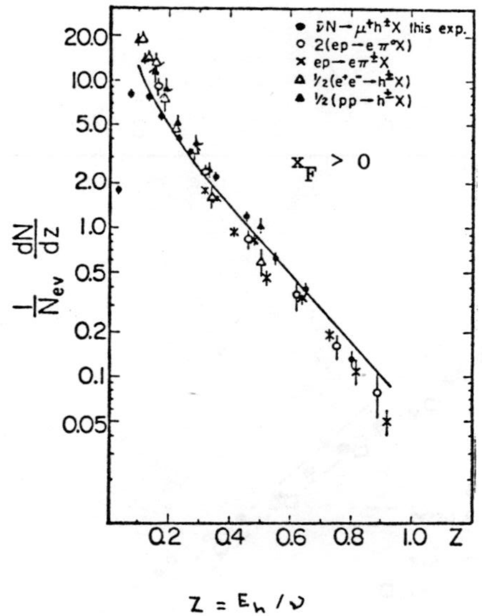

Fig. 20: Quark fragmentation functions from several reactions (Ref. 24)

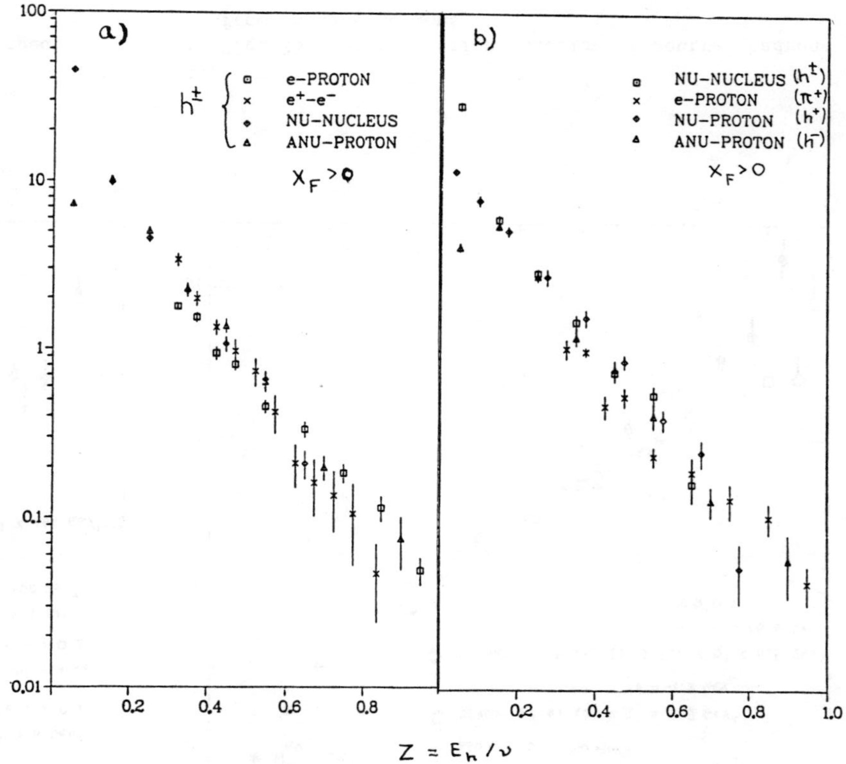

Fig. 21: Quark fragmentation functions for a) $\pm$ combined and b) comparable charge states (Ref. 16).

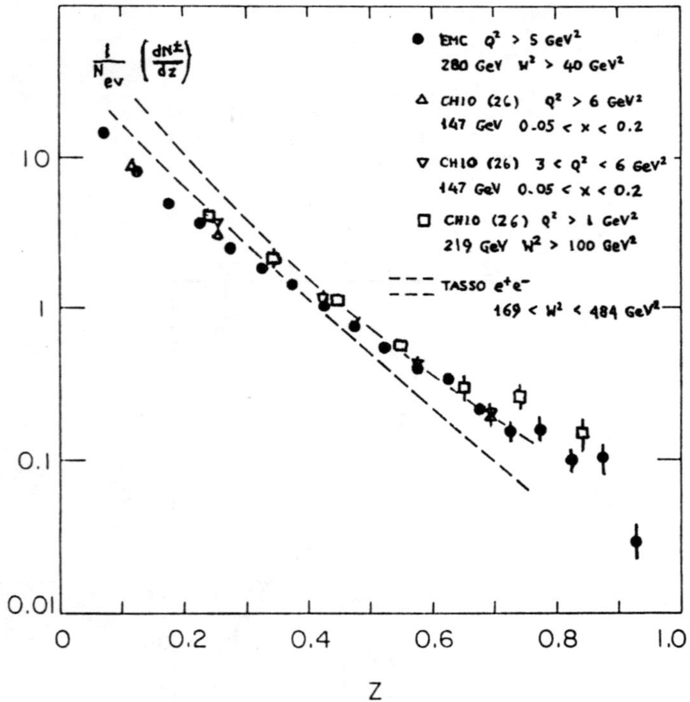

Fig. 22: Fragmentation function of charged hadrons from μp interactions.

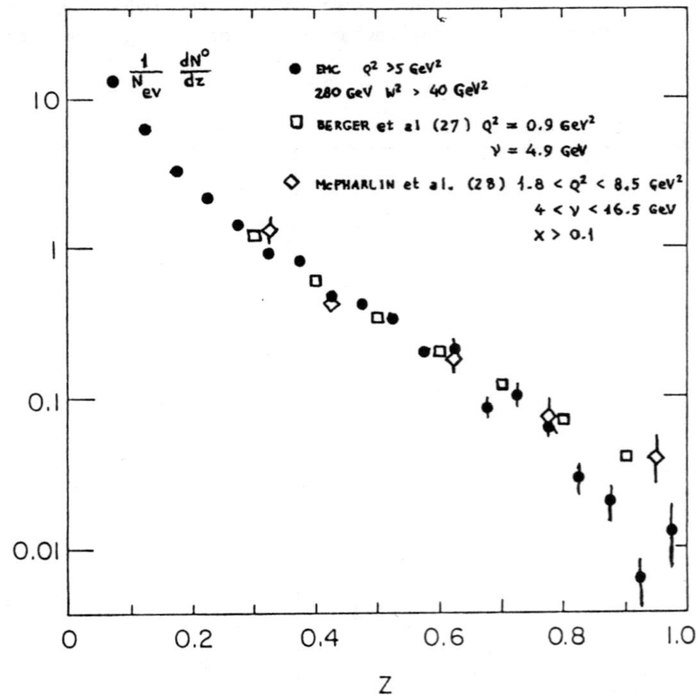

Fig. 23: Fragmentation function of neutral hadrons from μp interactions

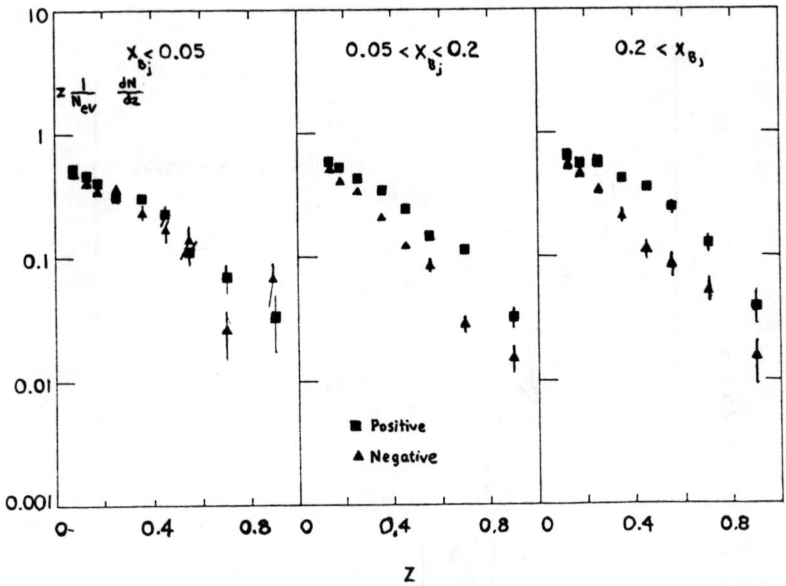

Fig. 24: Dependence of the fragmentation function on x_{B_j} for positive and negative particles

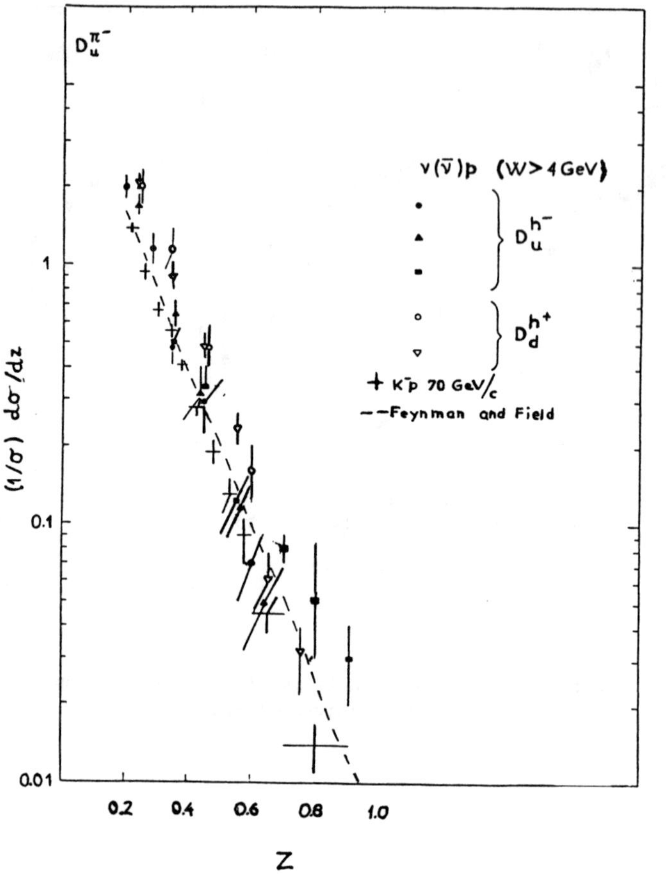

Fig. 25: $D_u^{\bar{h}^-}$ (z) obtained from K^-p interactions compared to $D_u^{h^-}$ from νp (●,▲,■) and $D_d^{h^+}$ from $\bar{\nu}p$ (○,▼) interactions and Field-Feynman parametrization (from Ref. 22).

the fragmentation function from the $K^- p$ data the authors assume $D_u^{\pi^-} \simeq D_s^{\pi}$ and obtain

$$D_u^{\pi^-}(x) \simeq \frac{1}{\sigma}\frac{d\sigma}{dx}^{K^- \to \pi^+} . \qquad (27)$$

In a nice series of papers Basile et al. [30] have recently compared hadron final states observed in pp collisions at the ISR ($\sqrt{s}$ = 62 GeV) with results observed in e^+e^- annihilation. In order to make this comparison they correct for the leading proton by removing the fastest particle in either hemisphere which is positive and has $0.4 < x_p < 0.8$ $(x) \equiv 2|P_L|/\sqrt{s}$ and P_L is along the direction of the beams in the centre of mass). Only ∿ 10% of the events have at least one shuch + particle. All other particles in the same hemisphere have energy $E_{had} = \frac{\sqrt{s}}{2}\ [1 - |x_p|]$. Then this system is compared with e^+e^- data at $E_{beam} = E_{had}$ using the variable $x_R^* = P/E_{had}$, where all particles are assumed to be pions.

In Fig. 26 the comparison is shown for two different energies. The TASSO (e^+e^-) data have been normalized relative to one half hemisphere. A further comparison between e^+e^- and pp final states which gives remarkable agreement is the fraction of total available energy taken by the charged particles $\alpha \equiv \langle E_{chg}\rangle/\langle E_{tot}\rangle$, Fig. 27, where $\langle E_{tot}\rangle_{pp}$ is the invariant mass of the hadronic system after removal of the leading proton.

Thus within statistics and relative normalization uncertainties the inclusive spectra of hadrons observed in leptonic, hadronic, and annihilation interactions show the remarkable similarities which are expected from the quark-parton model. This qualitative and often quantitative agreement occurs despite the fact that for $\nu(\bar{\nu})$ interactions we are often working in a region of W where the separation between current and target fragmentation is not too large.

The inclusive production of ρ-mesons in leptonic and hadronic interactions also show similarities. A recent result from the BEBC νp experiment [31] compared to $\pi^+ p$ interactions at 16 GeV/c [32] is shown in Fig. 28. The same νp data are compared with results from μp interactions [33] in Fig. 29.

There are also similarities, and differences, in the charged particle multiplicity distributions observed in νp and hadronic interactions. These results have been reviewed by Schmitz [34] and in a recent publication by the BEBC collaboration [17] and will not be repeated here.

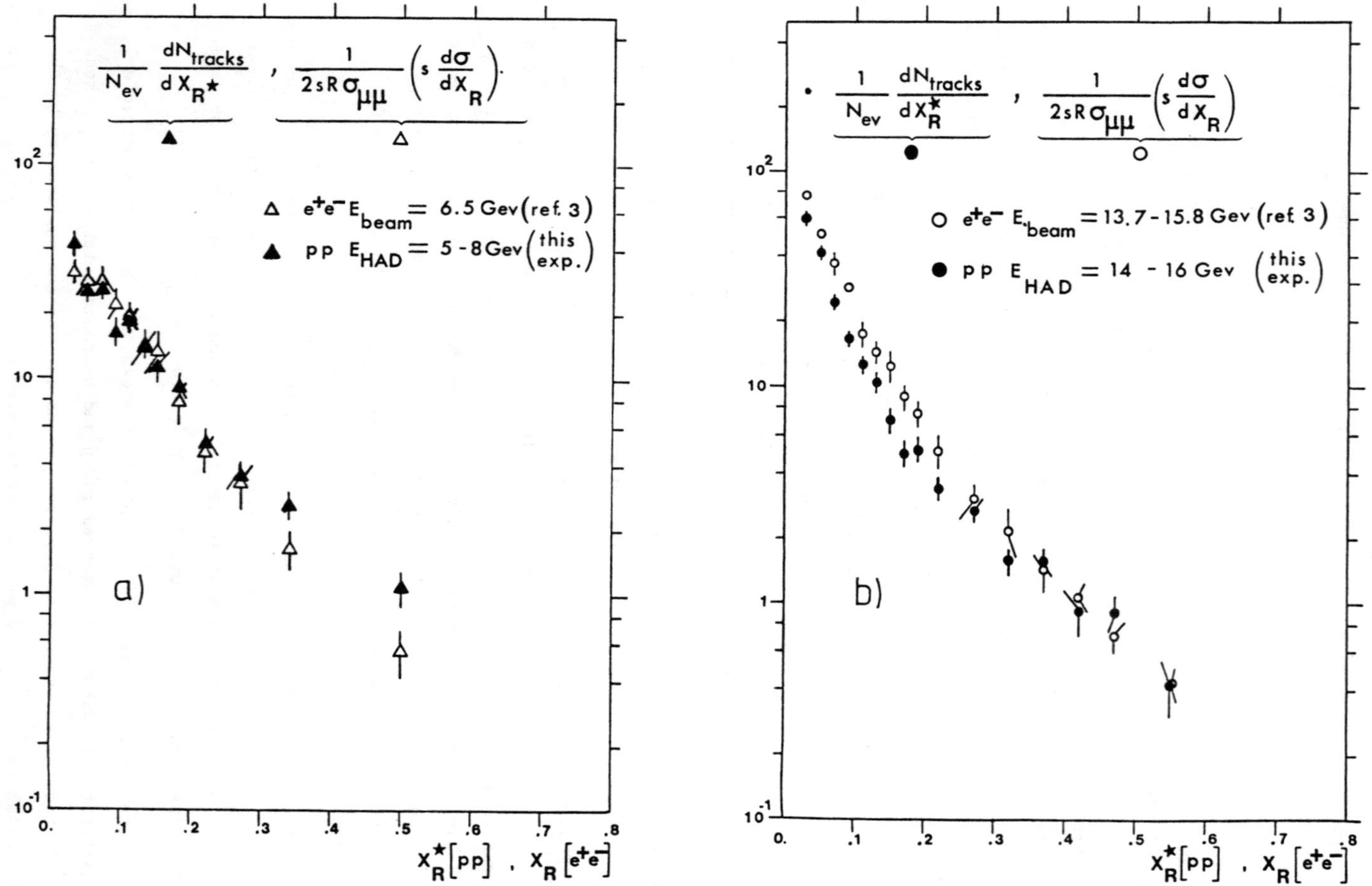

Fig. 26: Comparison of $(1/Nev)(dN/dx_R^*)$ obtained in pp interactions (Ref. 30) with e^+e^- annihilation results.

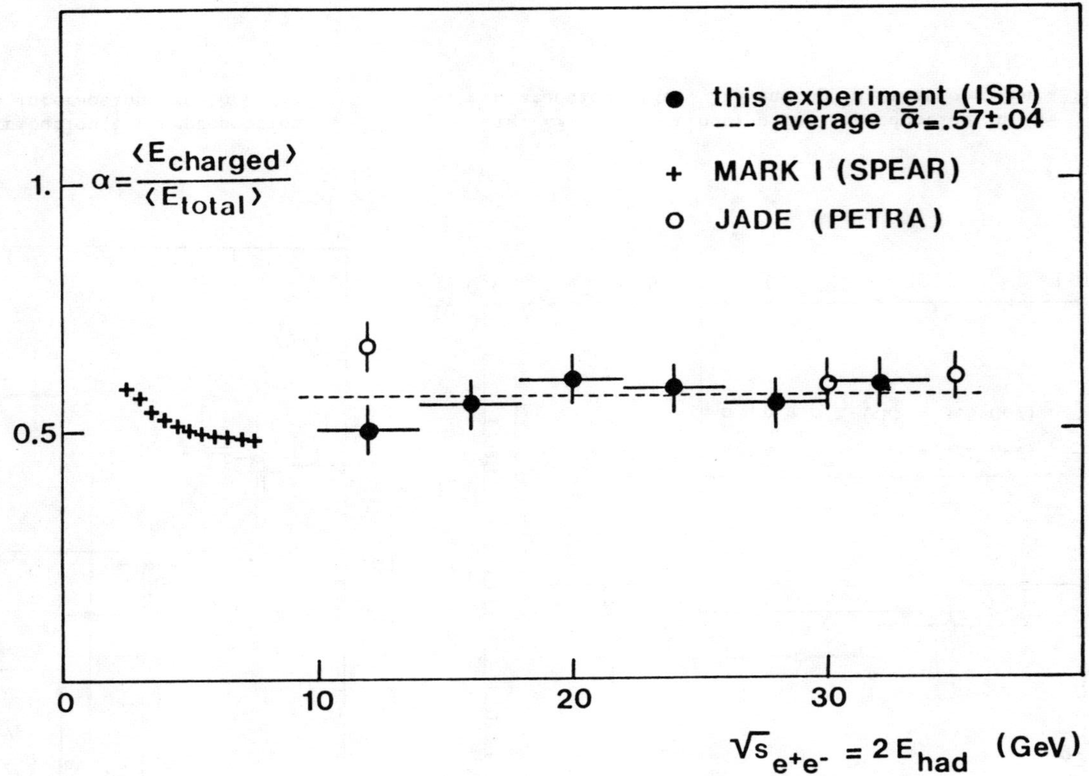

Fig. 27: Ratio of charged to neutral energy in pp interactions (Ref. 30) and e^+e^- annihilation.

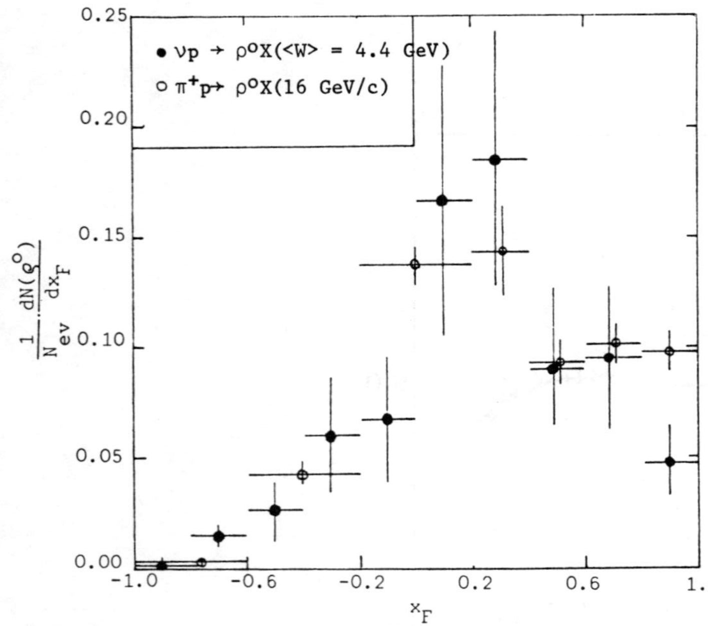

Fig. 28: Feynman x distribution for ρ-production in νp (Ref. 31) and π^+p interactions at 16 GeV/c (Ref. 32)

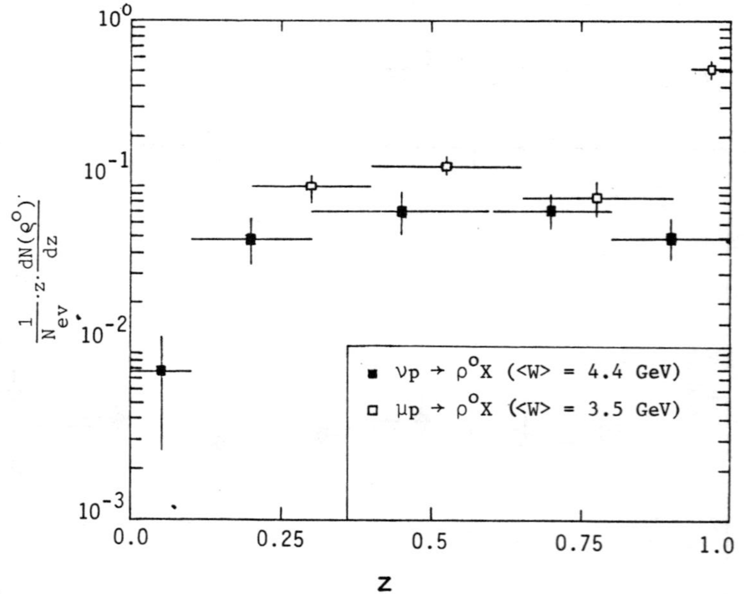

Fig. 29: z-distribution for ρ-production in νp interactions (Ref. 31) and μp interactions (Ref. 33)

Recent results from e^+e^- annihilation indicate significant production of protons [35]. A similar, preliminary result, has been reported by the EMC collaboration [36]. Andersson et al. have incorporated baryon-antibaryon production [37] in the Lund model for jet fragmentation. Their predictions and comparison with data are shown in Fig. 30. Further evidence for baryon production is presented in section 5.

5. Lepton Nucleus Interactions

Nuclear targets allow the possibility that the evolving hadronic system will interact with the nucleons in the target nucleus [38]. Because leptonic interactions give a quark of known momentum, it is conjectured that with nuclear targets one can study the interaction of a quark with the nucleons in the target nucleus. Thus comparing with leptonic production on nucleons may give us information about quark-quark interactions.

Experiments on nuclear targets are in general difficult. The necessity of using bubble chambers for neutrino interactions in order to measure the momentum of each final state hadron limits one to relatively light nuclei such as Carbon, Neon, or Deuterium. Thus there are only a few experiments which have studied lepton-nucleus interactions. In this section we present some of the results. We will not discuss the slow protons which always attend interactions on nuclear targets [39].

In Fig. 31 the mean of the νNe charged particle multiplicity distribution [40] versus W is compared with that observed in $\pi^{\pm}$Ne and π^-C interactions, where $\sqrt{s}$ is used for the hadronic interactions. The general trend is quite similar. From Fig. 32 we see that the $\nu(\bar{\nu})$Ne data also approximately satisfy KNO scaling and fall on the same universal curve observed for $\pi^{\pm}$Ne and π^-C interactions.

The rapidity distribution of the final state hadrons observed in $\nu(\bar{\nu})$Ne interactions and $\pi^{\pm}$Ne interactions are shown in Fig. 33. The agreement here is quite good. For the neutrino data $3 \leqslant W \leqslant 6$ GeV/c^2 and for the $\pi^{\pm}$Ne data $\sqrt{s} = 4.4$ GeV/c^2. The rapidity for neutrino interactions is defined with respect to the direction of the visible hadron momentum.

In Fig. 34, the pseudo rapidity distribution observed in μ-emulsion interactions [41] for the range $9 \leqslant W \leqslant 14$ GeV/c^2 ($\langle W \rangle \simeq 10$ GeV/c^2) is compared with π^--emulsion and p-emulsion interactions at 60 and 67 GeV/c, respectively. While the target and fragmentation regions agree well, there

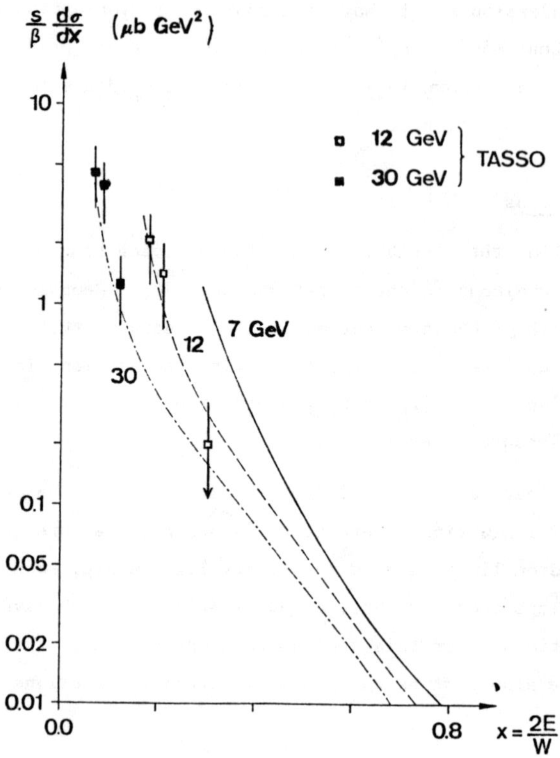

Fig. 30: Scaling cross-section for proton production in e^+e^- annihilation and predictions of the Lund model (Ref. 37).

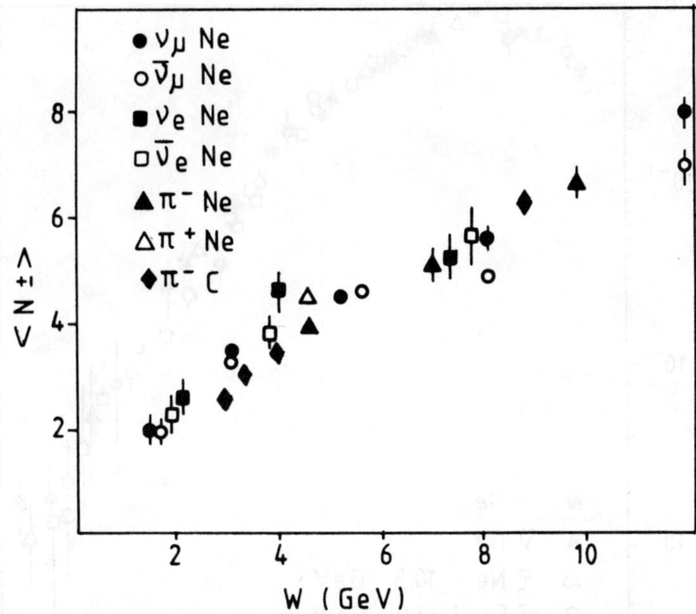

Fig. 31: Mean of the charged particle multiplicity distribution versus W. for $\nu(\bar{\nu})$Ne, $\pi^{\pm}$Ne, and π^{-}C interactions (Ref. 40).

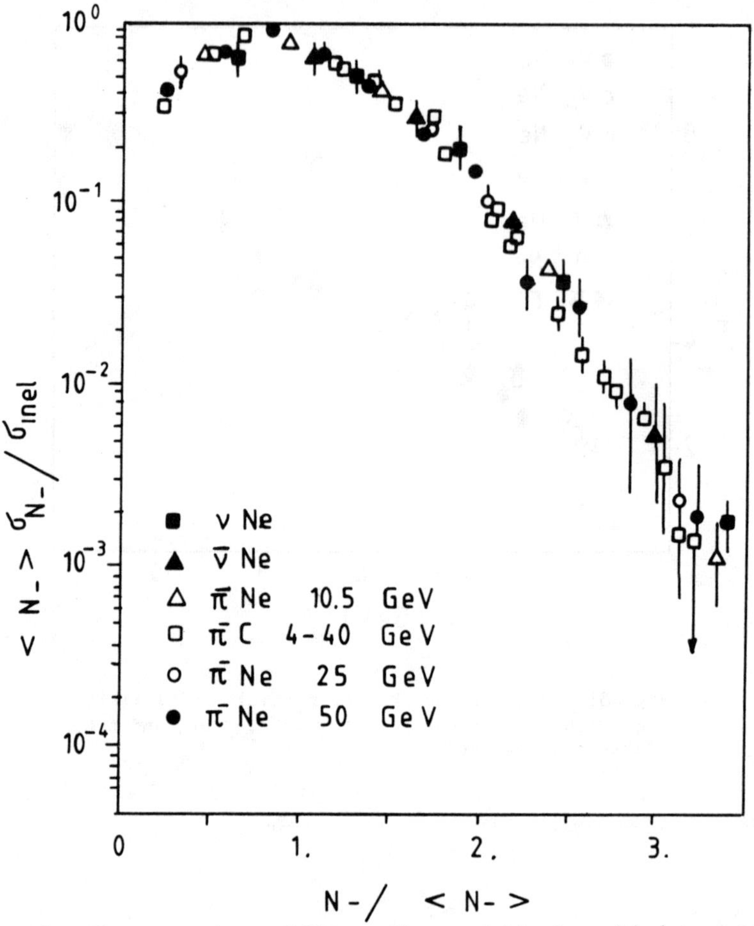

Fig. 32: Comparison of KNO scaling variable for $\nu(\bar{\nu})$ interactions and hadron-nucleus interactions.

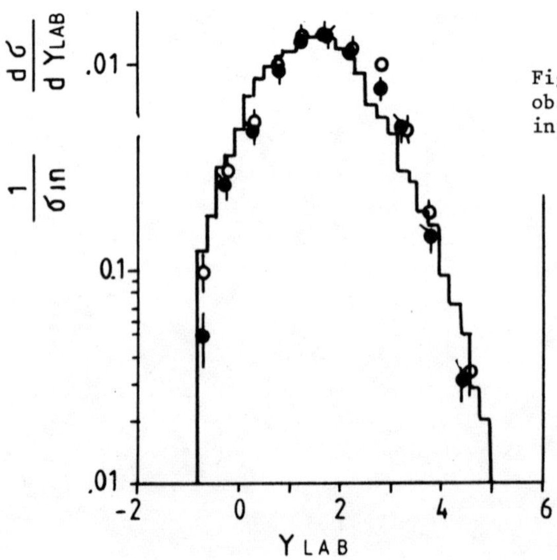

Fig. 33: Rapidity distributions observed in $\nu(\bar{\nu})$Ne and $\pi^{\pm}$Ne interactions.

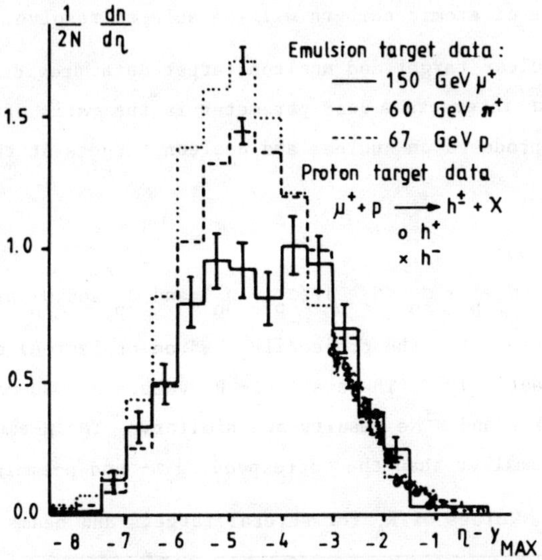

Fig. 34: Distribution of pseudorapidity (rapidity) relative to the maximum possible rapidity (Ref. 41).

are significantly fewer particles produced by leptonic interactions in the central region. The emulsion target is mainly Ag (A = 108) whereas in Fig. 33 the target was Ne (A = 20). Perhaps there is a dependence on atomic number? Better data over a range of atomic numbers will be able to resolve this question.

A comparison of nuclear target and nucleon target data provides information on how a hadronizing quarks interacts. A good parameter is the ratio of average charged multiplicities produced on nuclear and nucleon targets at the same c.m. energy

$$R_A = \langle N_\pm\rangle_A / \langle N_\pm\rangle_{\text{Nucleon}}, \tag{28}$$

where $\langle N_\pm\rangle_{\text{Nucleon}} \equiv (\sigma_p \langle N_\pm\rangle_p + \sigma_n \langle N_\pm\rangle_n)/(\sigma_p + \sigma_n)$ and σ_p and σ_n are the inelastic cross sections for the projectile (hadron or lepton) on a "free" proton or neutron, respectively. The available R_A values are summarized in Table III. The $\nu(\bar{\nu})$Ne and $\pi^\pm$Ne results are similar. The μ-emulsion value is significantly smaller than the corresponding π- and p-emulsion values.

Table 3: Values of R_A for several targets and beams

Reaction	W(GeV)	R_A
ν Ne	3-6	1.24 ± 0.04
$\bar{\nu}$ Ne	3-6	1.22 ± 0.05
π^+ Ne	4.4	1.20 ± 0.05
π^- Ne	4.4	1.11 ± 0.04
μ-emulsion	9-14	1.55 ± 0.13
π-emulsion	10.7	2.10 ± 0.10
p-emulsion	11.3	2.53 ± 0.07

In Fig. 35 the dependence of the ratio

$$(1/\sigma_{inel})(d\sigma/dy)_{Ne}/(1/\sigma_{inel})(d\sigma/dy)_{\text{Nucleon}} \tag{29}$$

shows that most of the $R_A > 1$ contribution for the Ne data comes from the target fragmentation region. In Fig. 35 favoured (unfavoured) refers to particles with the same (opposite) charge as the beam or exchanged W-meson.

An ep experiment at SLAC [42)] has studied electro-production from several nuclear targets. The kinematic ranges covered in this experiment are $0.35 \leqslant Q^2 \leqslant 5.0\ (\text{GeV/c})^2$, $7 < W^2 < 31\ (\text{GeV})^2$, and $0 < z_{cm} < 1$. They have also obtained data with a deuterium target which they use to compare with heavier nuclear targets. The ratio $(1/\sigma)\ (d\sigma/dz_{cm})_A\ /\ (1/\sigma)\ (d\sigma/dz_{cm})_{\text{deuterium}}$

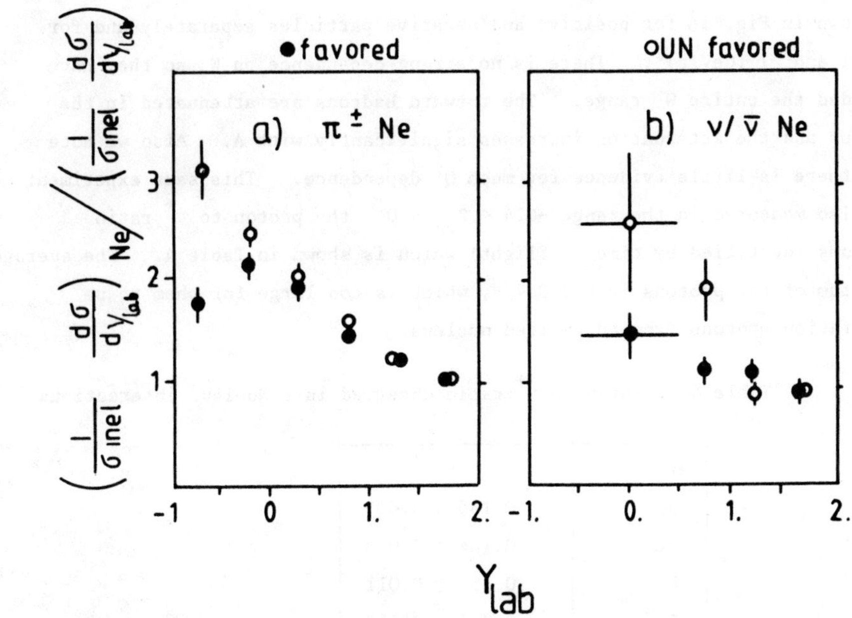

Fig. 35: Comparison of the ratios of the laboratory rapidity distributions for a)$\pi^{\pm}$Ne ($\sqrt{s}$ = 4.4 GeV) and and b) $\nu(\bar{\nu})$Ne interactions (3 < W < 6 GeV) for favoured and unfavoured particles.

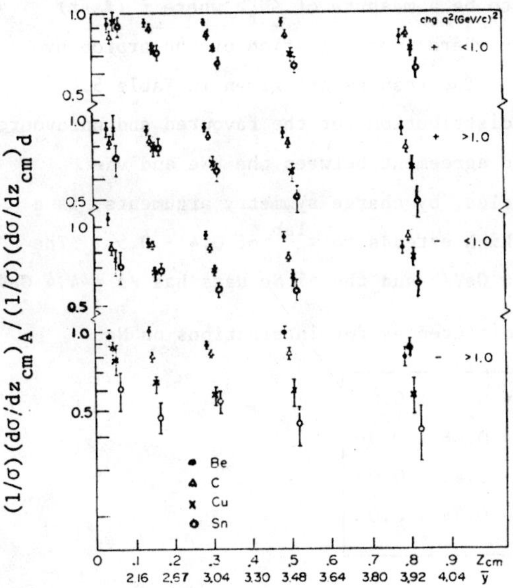

Fig. 36: Dependence of the ratio $(1/\sigma)(d\sigma/dz_{cm})_A/(1/\sigma)(d\sigma/dz_{cm})_d$ for interactions of electrons on Be, C, Cu, and Sn. The average rapidity, $\bar{y}$, based on $\langle p_T\rangle$ = 0.3 GeV/c is also indicated.

is shown in Fig. 36 for positive and negative particles separately and for $Q^2 < 1$ and >1 $(GeV/c)^2$. There is no strong dependence on W, so they have included the entire W range. The forward hadrons are attenuated in the nucleus and the attenuation increases significantly with A. Also we note that there is little evidence for much Q^2 dependence. This same experiment has also measured in the range $-0.4 < Z_{cm} < 0$ the proton to π^+ ratio (protons identified by time of flight) which is shown in Table 4. The average momentum of the protons is 1.3 GeV/c, which is too large for them to be evaporation protons from an excited nucleus.

Table 4: Proton to π^+ ratio observed in e-Nucleus interactions

Target	P/π^+
D_2	0.135 ± 0.003
Be	0.144 ± 0.008
C	0.159 ± 0.011
Cu	0.207 ± 0.015
Sn	0.232 ± 0.015

Evidence for fast protons has also been observed in $\nu(\bar{\nu})$Ne and $\pi^{\pm}$Ne interactions [29,43]. In these experiments, done with bubble chambers, charge symmetry is used to obtain the average number of fast positive particles which are not pions; i.e., the differences $\langle N_+\rangle_{\pi^{\pm}Ne} - \langle N_-\rangle_{\pi^{\mp}Ne}$ and $\langle N_+\rangle_{\nu(\bar{\nu})Ne} - \langle N_-\rangle_{\bar{\nu}(\nu)Ne}$ are assumed to be a measure of $\langle N_f^p\rangle$ where f (fast) is greater than approximately 1 GeV/c where identification of the proton by ionization becomes very inefficient. The results are given in Table 5. In Fig. 37 the laboratory Feynman x distribution for the favoured and unfavoured minimum ionizing particles shows good agreement between the πNe and νNe data. The excess of positive particles, by charge symmetry arguments, is a measure of the proton distribution which extends to x_F^{lab} of 0.4 - 0.5. The $\nu(\bar{\nu})$ data are for the range $3 \leqslant W \leqslant 6$ GeV/c and the $\pi^{\pm}$ Ne data has $\sqrt{s} = 4.4$ GeV/c^2.

Table 5: Charge symmetric differences for interactions on Ne

$\langle N_+\rangle_{\pi^+} - \langle N_-\rangle_{\pi^-}$	= 0.78 ± 0.04
$\langle N_+\rangle_{\nu} - \langle N_-\rangle_{\bar{\nu}}$	= 0.68 ± 0.06
$\langle N_+\rangle_{\pi^-} - \langle N_-\rangle_{\pi^+}$	= 0.46 ± 0.03
$\langle N_+\rangle_{\bar{\nu}} - \langle N_-\rangle_{\nu}$	= 0.34 ± 0.06

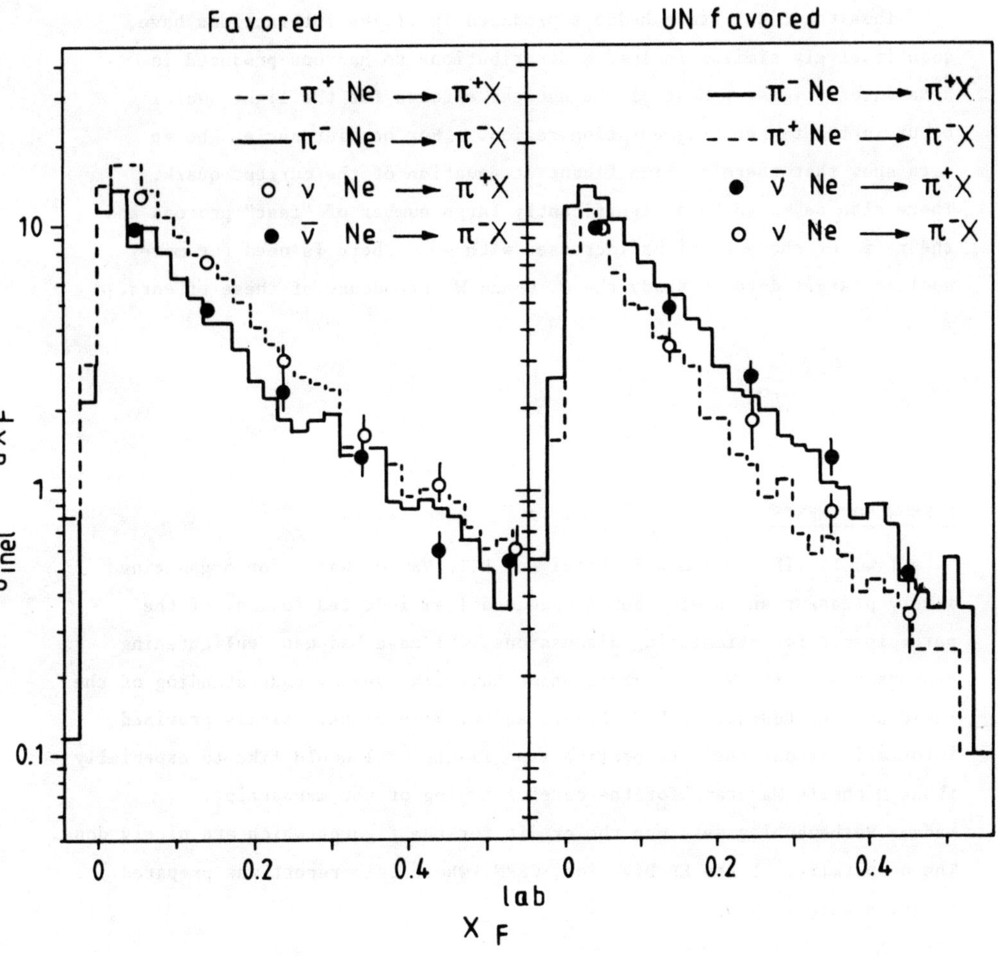

Fig. 37: Comparison of the x_F^{lab}-distributions from $\pi^{\pm}Ne \rightarrow h^{\pm}Ne$ ($\sqrt{s}$ = 4.4 GeV) with the z-distributions from $\nu(\bar{\nu})NE \rightarrow hNe$ (3 < W < 6 GeV)

Thus the final state hadrons produced in $\nu(\bar{\nu})$Ne interactions have quantitatively similar inclusive distributions to hadrons produced in $\pi^{\pm}$Ne interactions. Most of the nuclear effects for the light nuclei occur in the target fragmentation region. For heavier nuclei the ep data show that there is significant attenuation of the current quark. There also seems to be a significantly large number of "fast" protons and the ratio of these to pions increases with A. There is need for more nuclear target data to study the Q^2, ν and W dependence of these effects.

ACKNOWLEDGEMENTS

I would like to thank S. Ratti and R.T. Van de Walle for organizing a very pleasant and useful conference, and I am indebted to many of the participants for stimulating discussions. I have had many enlightening conversations with N.A. McCubbin which have improved my understanding of the subject. B. Buschbeck, B.E. Eckart and G. Valenti have kindly provided information about their respective experiments. I would like to especially thank Michelle Mazerand for the careful typing of the manuscript. Madame Rothenbuhler deserves the credit for the figures which are nicely done. The hospitality of the EP Division, CERN, where this report was prepared is gratefully acknowledged.

REFERENCES

1. J. Steinberger "Lepton-Hadron Inclusive Scattering and QCD", CERN-EP 80-222 (1980). To appear in the Proceedings of 18th School of Subnuclear Physics, Erice, Sicily (1980).

2. H.E. Montgomery "Results from CERN Muon Scattering Experiments" To appear in Proceedings of the SLAC Topical Conference July 28-August 8, 1980.

3. F. Sciulli, Proceedings of the XXth Intern. Conf. on High Energy Physics, Madison, Wisc, (1980), p. 1278 (Ed. L. Durand and L.G. Pondrom).

4. C.Y. Prescott "Deep Inelastic Phenomena" SLAC, PUB-2625, October (1980). To appear in the proceedings of the 18th School of Subnuclear Physics, Erice, Sicily (1980).

5. J.G.H. de Groot (CDHS collaboration), Proceedings of XXth Intern. Conf. on High Energy Physics, Madison, Wisc. (1980) p. 735 (Ed. L. Durand and L.G. Pondrom).

6. J.S. Poucher et al., Phys. Rev. Lett. 32, 118 (1974);
W.B. Atwood et al., Phys. Lett. 64B, 479 (1976).

7. B.A. Gordon et al., Phys. Rev. D20, 2645 (1979).

8. J.G.H. de Groot et al., Z. Phys. C1, 143 (1979).

9. M. Jonker et al., Phys. Lett. 99B, 265 (1981).

10. M. Shaevitz et al., "Recent Results from the CFNRR Neutrino Experiments" Presented at the Int. Conf. of High Energy Physics, Madison, Wisc. July (1980).

11. E. Reya "Deep Inelastic Structure Functions", Univ. of Dortmund DO-TH 80/27. To appear in the Proceedings of the Intern. School of Elementary Particle Physics, Kupari-Dubrovnik, Sept. 1980.

12. J. Knobloch et al., "The Structure and Amount of the Sea and Charm Jet Broadening". Presented at the XXth Intern. Conf. on High Energy Physics, Madison, Wisc. July 1980.

13. P. Allen et al., "Determination of the Quark Density Ratio d(x)/u(x) in the Proton", OUNP 81-22, Submitted to Phys. Lett.

14. S.J. Brodsky, P. Hoyer, C. Peterson and N. Sakai, Phys. Lett. 93B, 451 (1980).

15. S.J. Brodsky, C. Peterson and N. Sakai "Intrinsic Heavy Quark States" SLAC-PUB-2660, Submitted to Phys. Rev. D, and references therein.

16. M. Derrick et al., "Hadron Production Mechanisms in Antineutrino-Proton Charmed-Current Interactions" ANL-HEP-PR-80-54.

17. P. Allen et al., Nucl. Phys. B181, 385 (1981).

18. V. Blobel et al., Nucl. Phys. B69, 454 (1974).

19. T. Kafka et al., Phys. Rev. D16, 1261 (1977).

20. P. Bosetti et al., Nucl. Phys. B54, 141 (1973).

21. B. Andersson et al., Phys. Lett. 69B, 221 (1977) and 71B, 337 (1977).

22. B. Buschbeck, private communication.

23. L. de Billy et al., (Paris-Saclay-Rutherford Collaboration) Submitted to Nucl. Phys. B.
A.M. Touchard, Private Communication.

24. V.V. Ammosov et al., "Quark Jets from Antineutrino Interactions II Inclusive Multiplicities in the Quark Jets" Fermilab-PUB-80/96-Exp. Submitted to Nucl. Phys.

25. J.J. Aubert et al., CERN-EP/80-130.

26. W.A. Loomis et al., Phys. Rev. D19, 2543 (1979);
C.S, Tao, Ph. D. Thesis, Harvard University, Cambridge, Mass (1979).

27. C.H. Berger et al., Phys. Lett. 89B, 471 (1977).

28. T.P. McPharlin et al., Phys. Lett 90B, 479 (1979).

29. H.J. Lubatti, Proceedings of the Neutrino'79 Conference p. 543 (Ed. A. Haatuft and C. Jarlskog).

30. M. Basile et al., Phys. Lett. 92B, 367 (1980), and 95B, 311 (1981), and references therein.

31. B.E. Eckart, Diplomarbeit University of Bonn, Bonn-IR-81-4 (1981).

32. M. Deutschmann et al., Nucl. Phys. B103, 426 (1976).

33. C. del Papa et al., Phys. Rev. Lett. 40, 90 (1978).

34. N. Schmitz Acta Phys. Pol. B11, 913 (1980).

35. R. Brandelik et al., Phys. Lett. 94B, 444 (1980).

36. J. Favier, Presented at the Rencontre de Moriond, (1981).

37. B. Andersson, G. Gustafson, and T. Sjöstrand, "A Model for Baryon Production in Quark and Gluon Jets" University of Lund LU TP 81-3 (1981).

38. A. Dar, These Proceedings.

39. J.P. Berge et al., Phys. Rev. D18, 1367 (1978);
V.I. Efremenko et al., Phys. Rev. D22, 2681 (1980).

40. T.H. Burnett et al., Phys. Lett. 77B, 443 (1978).

41. L. Hand et al., Z. Phys. C1, 139 (1979).

42. L.S. Osborne et al., Phys. Rev. Lett. 40, 1624 (1978).

43. W.M. Yeager et al., Phys. Rev. D17, 1294 (1977).

Chapter III: MODELS FOR COLOR JETS AND FRAGMENTATION FUNCTIONS.

UDAY P. SUKHATME
Physics Dept., Univ. of Illinois at Chicago Circle
Chicago, Illinois, USA.

Abstract: This article contains a review of various ideas and models which have been suggested for the fragmentation of colored systems into jets of hadrons due to the color confinement mechanism. Particular emphasis is given to plausible models for gluon and diquark jets. At present energies, gluon jets are found to have essentially the same multiplicity (but a softer momentum distribution) than quark jets. The composite nature of a diquark seems to play an important role during its fragmentation, and the two quarks in a diquark do not always become part of a single baryon. The important problem of devising an algorithm for establishing the identity of the jet-initiating parton is discussed in detail.

1. INTRODUCTION.

It is now generally accepted that whenever a colored system tries to escape freely by itself, it gets converted via the confinement mechanism into a jet of hadrons. In this article, we shall discuss various ideas and models for the formation and properties of jets initiated by a variety of colored systems (quarks, gluons, and diquarks).

In order to set the stage and establish the terminology, we will first give a brief discussion of well-known qualitative ideas (Sec. II) and quickly review the recursive (or cascade) model for quark jets (Sec. III). By incorporating color confinement in a phenomenological manner, the recursive model provides useful, quantitative answers for specific measurable jet properties.

Then, we will focus on several more recent topics in greater detail. Sec. IV contains a long discussion of an important question which an experimentalist always asks (and a theorist rarely answers)--- given a jet of hadrons, what is the identity of the jet-initiating parton ? We will discuss some of the existing suggestions for establishing the identity of a jet, and then describe the best way of making the identification. Sec. V is devoted to plausible models (motivated by our understanding of quantum chromodynamics) for gluon fragmentation functions and a discussion of their properties. This is followed by a new approach for obtaining parton fragmentation functions when the initial parton is far off its mass shell (Sec. VI). This technique is based on initial branching as given by the jet calculus and subsequent recombination of partons into hadrons. The resulting Q^2 dependence of fragmentation functions is different from that predicted by standard QCD evolution equations.

The last section (Sec. VII) is devoted to the new topic of diquark fragmentation. This topic has recently begun to attract considerable experimental and theoretical interest due to the unique properties of a diquark --- it is a color source which is manifestly composite and spatially extended. One can therefore ask questions about the dynamics of partons during fragmentation; for example, into which hadrons do the initial quarks go during fragmentation? Such questions can be answered by specific models for diquark jets. Present indications are that the composite nature of a diquark is important, and the diquark cannot naively be treated as a single unit which becomes part of a baryon.

2. QUALITATIVE IDEAS.

(i) Explicit example of jet formation: Consider a quark-antiquark system making up a meson. The system is held together by the color force (represented by a string in Fig. 1). If the meson is knocked hard (for example in a high energy collision), one obtains a situation in which the q and $\bar{q}$ are flying apart, say with momentum P. (This is also the situation in e^+e^- collisions). If there had been no color confinement, one would have observed colored systems (q and $\bar{q}$), which have not been directly seen. Qualitatively, it is thought that the confinement mechanism acts as follows. The separating qq system leads to the production of a "chromoelectric flux tube" or stretched string, which eventually breaks with the creation of a $q\bar{q}$ pair. At this stage, one has two $q\bar{q}$ systems flying apart with a reduced momentum (less than P). This process of repeated breakups stops when the quark momentum falls below

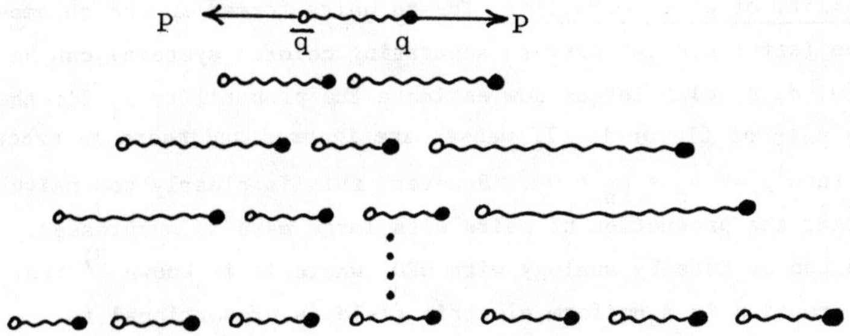

Fig. 1 : Pictorial representation of a quark-antiquark pair producing a jet of mesons.

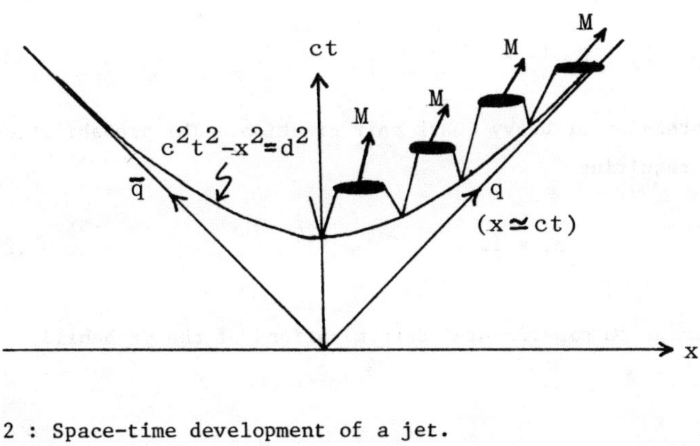

Fig. 2 : Space-time development of a jet.

a critical value (presumably of order m_π), the final result being a jet of mesons (see Fig. 1).

The space-time development of a jet[1,2] in the overall CM frame is shown in Fig. 2. The initial quark and antiquark are roughly on the trajectories $x \simeq ct$. String breaking and $q\bar{q}$ pair formation in the color flux tube occurs on a hyperbola where the invariant quantity $c^2t^2 - x^2 \simeq d^2$, d being a typical hadronic dimension. Horizontal time slices of Fig. 2 show that the slower mesons in a jet are formed first (in the overall CM frame). This is often referred to as inside-outside jet formation.

(ii) Probability of pair formation: The $q\bar{q}$ pairs formed in the chromo-electric flux tube (stretched out between separating colored systems) can be of many flavors (u, d, s, c). Let us now estimate the probability p_i for the formation of a $q\bar{q}$ pair of flavor i. If masses are ignored and there is exact flavor symmetry, then $p_u = p_d = p_s = p_c$. However, this is clearly too naive since one knows that the production of pairs with large mass is suppressed. A simple estimate can be made by analogy with QED, where it is known[3] that the rate of pair creation in a uniform electric field is proportional to $\exp(-Cm^2)$. Therefore, for QCD, a reasonable guess for the rate of production[4] of a $q\bar{q}$ pair is $\exp(-\pi^2\alpha' m^2)$, where m = quark mass, α' = Regge trajectory slope $\simeq 1\ \text{GeV}^{-2}$. For reasonable values of quark masses, this yields

$$p_u : p_d : p_s : p_c :: 1 : 1 : 0.33 : 10^{-15}, \qquad (1)$$

showing a strong suppression of heavy quark pair creation. The probabilities p_i are normalized by requiring

$$\sum_i p_i = 1. \qquad (2)$$

Later on, we will comment on experimental determinations of the probabilities p_i.[5]

(iii) The two stages of jet evolution: Since jet formation is a consequence of confinement, and this is a region in which QCD calculations cannot yet be made, one needs phenomenological models (like the recursive model discussed in the next section) suggested by QCD ideas. However, if the parton which starts jet formation has a large momentum P and is also far off its mass shell (Q^2 large), then perturbative QCD is initially applicable and the parton branches repeatedly via the QCD vertices quark → quark + gluon, gluon → quark + antiquark and gluon → gluon + gluon. This initial branching produces many partons of lower momentum which are less far off the mass shell; that is the momentum and offshellness decrease with each successive branching. Eventually, a point $Q^2 = Q_0^2$ is reached where the QCD running coupling constant is no longer small and one needs a non-perturbative model for hadronization. These two stages of jet evolution are shown in Fig. 3.

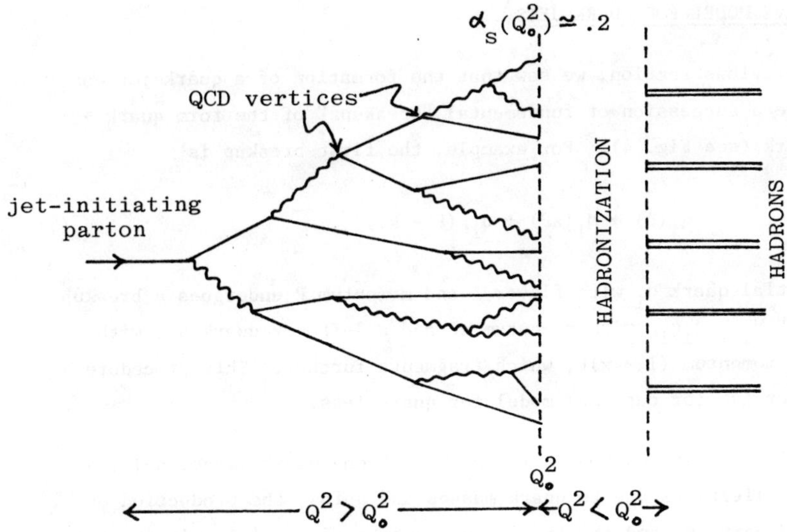

Fig. 3: The perturbative ($Q^2 > Q_o^2$) and non-perturbative ($Q^2 < Q_o^2$) stages of jet evolution.

The perturbative QCD stage corresponds to planar diagrams in the leading logarithm approximation. It can be handled by the formalism of the jet calculus [(6)] and leads to large multiplicities $n \sim \exp(C\sqrt{\ln s})$ and a transverse momentum which increases like $p_T^2 \sim Q^2$. In contrast, the final non-perturbative stage, which can be described by the cascade model[(7-9)] (for example) gives logarithmic multiplicity growth $n \sim C \ln s$ and limited transverse momentum.

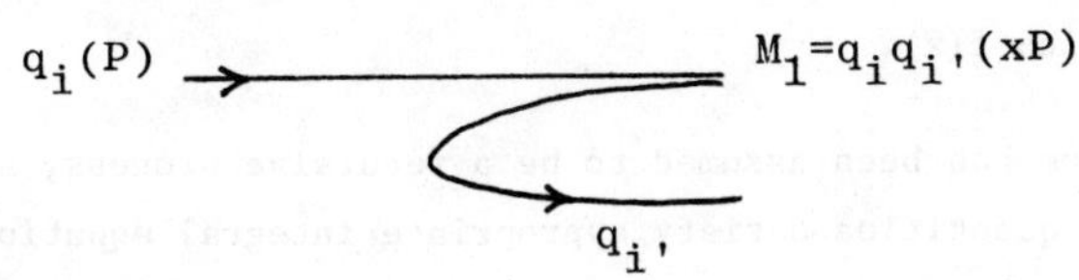

Fig. 4: The fundamental breakup quark → meson + quark which specifies the recursive model for quark jets.

3. RECURSIVE MODEL FOR QUARK JETS.

In the previous section, we saw that the formation of a quark jet can be taken to be a succession of fundamental "breakups" of the form quark → "meson" + quark (see Fig. 4). For example, the first breakup is

$$q_i(P) \rightarrow M_1(xP) + q_{i'}(P - xP). \tag{3}$$

Here, the initial quark q_i with flavor i and momentum P undergoes a breakup into a "meson" $M_1 \equiv q_i\bar{q}_{i'}$ with momentum xP and a leftover quark $q_{i'}$ with flavor i' and momentum (1 - x)P, which fragments further. This procedure gives the recursive (or cascade) model for quark jets.

For simplicity, let us consider the case of one quark flavor only, and ignore the complications due to quark masses and spins, the production of resonances and baryons, and the transverse momenta of particles in the jet. These complications can all be satisfactorily handled.[7,8] For our simplified, one-dimensional jet, the breakup process is fully described by a single function f(x), for which we have assumed scaling. f(x)dx is the probability that a breakup yields a "meson" M with momentum fraction in the interval (x,x + dx). Since f(x) describes how the incident quark momentum is shared between the meson and leftover quark, we shall henceforth call it the "momentum sharing function". Clearly,

$$\int_0^1 f(x)dx = 1. \tag{4}$$

Given the breakup function f(x), the jet is fully specified. In fact, in the recursive model, our ignorance of the confinement mechanism (and inability to do non-perturbative QCD) is all buried in the phenomenological choice of the momentum sharing function f(x).

Since jet formation has been assumed to be a recursive process, all physically interesting quantities satisfy appropriate integral equations. The meson multiplicity n(P) in a jet initiated by a quark with momentum $P \geq \mu$ satisfies the equation

$$n(P) = 1 + \int_{\mu/P}^{1} dx' f(1-x')n(x'P). \tag{5}$$

The term unity on the right hand side corresponds to the meson produced in the first breakup, and the integral corresponds to the multiplicity coming from the leftover quark, which has momentum x'P.

Similarly, the fragmentation function D(x), which is the inclusive single-meson momentum distribution dn/dx, is given by

$$D(x) = f(x) + \int_x^1 \frac{dx'}{x'} f(1-x')D(x/x'). \tag{6}$$

Again, the two terms in eq.(6) have a simple physical interpretation. The term f(x) corresponds to the possibility that a meson with momentum fraction x is formed in the first breakup. The integral corresponds to the possibility that the first breakup produces a quark with momentum fraction $x' \geq x$, and this quark yields a meson at x in a subsequent breakup. Note that integrating eq. (6) over x yields the multiplicity integral eq. (5).

The same type of arguments can be used to write down integral equations for fragmentation functions into more than one meson.

We now present the salient features of the solutions to integral eqs. (5) and (6). The general solution to the multiplicity integral equation (5) is

$$n(P) = \alpha \ln(\frac{P}{\mu}) + O(\frac{\mu^2}{P^2}) \; ; \; \alpha = \frac{-1}{\int_0^1 dx' \; f(1-x') \ln x'} \tag{7}$$

Note the very general result that <u>one always obtains logarithmic multiplicities</u> at large energies for any choice of the momentum sharing function f(x). The slope α depends on f(x), and one can see that if f(x) is such that the meson tends to take a larger share of momentum in a breakup, then a smaller slope α results. Since $\ln x'$ is large near $x' = 0$, the slope α is sensitive to the behavior of f(x) near x = 1.

Next, we solve integral eq. (6) for the fragmentation function D(x). It is a linear Volterra equation of the second kind and a convergent solution is obtained by successive iteration.

$$D(x) = \sum_{j=1}^{\infty} D^{(j)}(x) \tag{8}$$

where,

$$D^{(1)}(x) = f(x),$$

$$D^{(j+1)}(x) = \int_x^1 \frac{dy}{y} f(1-x/y) D^{(j)}(y); (j=1,2,...) . \tag{9}$$

The index j in eq. (8) has a physical meaning, since $D^{(j)}(x)$ is the contribution to the fragmentation function from the jth breakup.

The jet-initiating quark is of course a part of the "meson" M_1 produced in the first breakup. Let us note that if the contribution of the first breakup $D^{(1)}(x)$ behaves like $(1-x)^\beta$ as $x \to 1$, then it is easy to see that the contribution of the second breakup $D^{(2)}(x) \xrightarrow{x\to 1} (1-x)^{\beta+1}$, and in general $D^{(j+1)}(x) \xrightarrow{x\to 1} (1-x)^{\beta+j}$. Thus, one gets an extra factor of (1-x) at each breakup.

In order to get a feel for the behavior of the fragmentation function, let us consider the following simple but realistic choice for the momentum sharing function:

$$f(x) = (1+\beta)(1-x)^\beta \qquad (\beta \geq 0). \tag{10}$$

For this example, the multiplicity and momentum distributions are:

$$n(P) = (1+\beta)\ell nP + \text{constant} \tag{11}$$

$$dn/dx \equiv D(x) = (1+\beta)(1-x)^\beta/x \tag{12}$$

For concreteness, take the case $\beta = 1$ (which is roughly in agreement with e^+e^- inclusive meson distributions). Fig. 5 shows how the full meson distributions $D(x) = 2(1-x)/x$ is built up form the contributions of successive breakups: $D^{(1)}(x) = f(x) = 2(1-x)$, $D^{(2)}(x) = 4(1-x + x\ell nx)$, etc. Note that for $x \gtrsim .5$, truncation of the series solution (eq. (8)) after 2 breakups is adequate.

In the discussion above, we have solved integral eq. (6) by successive iteration, since each iteration corresponds physically to an additional breakup. An alternative mathematical procedure is to take Mellin transforms and treat the resulting algebraic equations.[9]

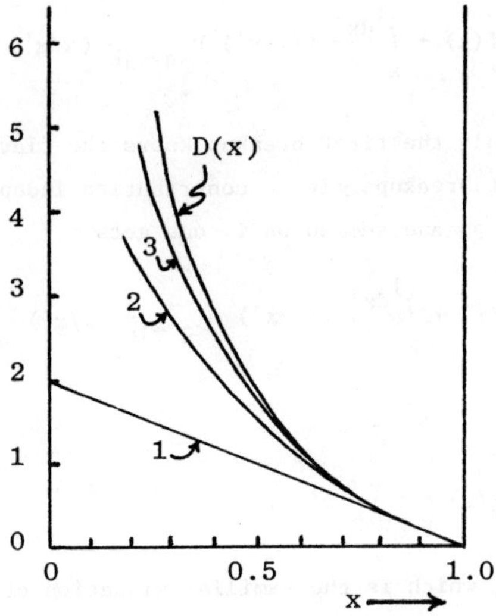

Fig. 5: A plot of the quark fragmentation function into mesons for the choice f(x) = 2(1-x) of the momentum sharing function. The figure illustrates how successive breakups build up the full fragmentation function D(x) = 2(1-x)/x. The first breakup gives curve 1, the first two breakups give curve 2, etc.

Let us now include quark flavors in the recursive model. $D_{i\to j\bar{k}}(x)$, the fragmentation function for a quark of flavor i into meson $q_j\bar{q}_k$ clearly satisfies the integral equation

$$D_{i\to j\bar{k}}(x) = \delta_{ij}p_k\ f(x) + \int_x^1 \frac{dx'}{x'}\ f(1-x')\sum_{i'} p_{i'}D_{i'\to j\bar{k}}\ (x/x') \tag{13}$$

The term $\delta_{ij}p_k\ f(x)$ is the contribution from the first breakup. This leaves behind a mean (or thermal) quark <q>, which is quark q_i, with probability $p_{i'}$. Define the fragmentation function of a mean quark by

$$D_{<q>\to j\bar{k}}(x) \equiv \sum_{i'} p_{i'}D_{i'\to j\bar{k}}(x) \tag{14}$$

Then, eq. (13) can be re-written as

$$D_{i\to j\bar{k}}(x) = \delta_{ij}\, p_k\, f(x) + \int_x^1 \frac{dx'}{x'}\, f(1-x')\, D_{\langle q\rangle\to j\bar{k}}\,(x/x') \tag{15}$$

This explicitly shows that only the first breakup knows the flavor i of the initial quark. Subsequent breakups give a contribution independent of i. If eq. (15) is multiplied by p_i and summed on i, one gets

$$D_{\langle q\rangle\to j\bar{k}}(x) = p_j p_k\, f(x) + \int_x^1 \frac{dx'}{x'}\, f(1-x')\, D_{\langle q\rangle\to j\bar{k}}\,(x/x') \tag{16}$$

Define the function D(x) by

$$D_{\langle q\rangle\to j\bar{k}}(x) = p_j p_k\, D(x). \tag{17}$$

Then, D(x) satisfies eq. (6), which is the familiar situation of a fragmentation function when flavor is ignored! Thus, previous results ignoring flavor are not wasted, and the final solution is[8] (combining eqs. (15), (16), (17), and (6))

$$D_{i\to j\bar{k}}(x) = \delta_{ij}\, p_k\, f(x) + p_j p_k \left[D(x) - f(x)\right]. \tag{18}$$

Similar analytic solutions can be written down for quark fragmentation functions into two or more mesons.

Resonances can be readily included in the recursive model[8], but one needs to fix a parameter R = resonance fraction. Current indications from experiment are $R \cong .7$[5]. Also, one has the probability parameters p_u, p_d and p_s for pair production. Since $p_s = 1 - p_u - p_d$ and $p_u = p_d$ from isospin, only one unknown parameter (say p_s/p_u) remains. This can be measured experimentally, for example in the d quark fragmentation region in $\bar{\nu}p \to \mu^+ + X$. At large x, only the first breakup dominates and the K^o/π^- ratio is essentially p_s/p_u.

$$\frac{D_{d\to K^o}(d\bar{s})}{D_{d\to\pi^-}(d\bar{u})} \simeq \frac{p_s}{p_u} \quad . \tag{19}$$

These measurements[10] yield $p_s/p_u = 0.27 \pm .04$. Thus a nice simple choice of probabilities (consistent with experiment) is

$$p_u = p_d = \frac{4}{9} , \ p_s = \frac{1}{9} . \tag{20}$$

In general, the recursive model provides a nice description of available experimental data on quark fragmentation. However, it is important to keep in mind that the recursive model deals throughout with probabilities, rather than quantum mechanical amplitudes. Also, it is a phenomenological model since the momentum sharing function is not derived from a fundamental approach.

4. IDENTIFYING THE JET-INITIATING PARTON.

If the identity of the jet-initiating parton is known, one can study its fragmentation and construct plausible models for jet production. However, one is often confronted with the inverse problem - given a jet event, what is the identity of the jet-initiating parton? For this, one needs an algorithm which is necessarily based on a model for fragmentation. The efficiency of the algorithm can be tested in reactions where the flavor of the fragmenting quark is known (for example νp and $\bar{\nu} p$ interactions). This in turn should yield better fragmentation functions which should improve the parton identification algorithm further. Hopefully, the procedure is convergent!

We will now discuss various possible ways of identifying the initial parton, and show that there is an optimum procedure based on fragmentation functions.

(i) Fastest particle in a jet: It has often been suggested that the leading particle in a jet provides a way of identifying the initial parton. Let us make this statement more quantitative. Consider a sample of jets which contain a particle with $x \geq 0.5$, and let us ask for the probability that the leading particle came from the first breakup and hence contains the jet- initiating quark. This probability r is given by (see Fig. 5)

$$r = \int_{\frac{1}{2}}^{1} f(x)dx \Big/ \int_{\frac{1}{2}}^{1} D(x)dx. \tag{21}$$

For the realistic choice $f(x) = 2(1-x)$ of the previous section, $r = .6$. Therefore, in 4 events out of 10, the $x \geq \frac{1}{2}$ particle came from the second or subsequent breakup, and does not contain the quark which initiated the jet. By a similar calculation, it is easy to check that there is a high probability (≃85%) that the jet-initiating quark is contained in the fastest particle only if it has $x \geq .75$. This is a useful, practical criterion which experimentalists can use for identifying the quark origin of a jet by looking at the fastest meson. Note, however, that only a small sample of jet events contain a particle with $x > .75$. (The fraction can be estimated from $\int_{.75}^{1} dx\, f(x)$ and is about 6%). Secondly, the identification of the jet-initiating parton has been made at the expense of getting a very abnormally biased jet! Nevertheless, this can be put to good use in e^+e^- annihilation where one has back to back $q\bar{q}$ jets. For example, if a biased jet on one side is found to be a u-jet, then one can study the fragmentation of the unbiased $\bar{u}$-jet on the opposite side.

(ii) Charge measure using ranking in rapidity: This method[11] uses the charges on the fastest N particles in a jet in order to try and identify the jet-initiating parton. In some sense, it is a generalization of the preceding method based only on the fastest particle. Let us label the particles 1,...,N by their rapidity ranking.

$$y_1 > y_2 > \dots > y_N \tag{22}$$

If Q_j denotes the charge of the fastest j particles, then one can construct the quantity $Q(q)$ called the charge measure.

$$Q(q) \equiv \frac{1}{N}\sum_{j=1}^{N} Q_j \tag{23}$$

The numerical value of the charge measure in an event is an indicator of the initial parton identity. This can be seen by considering a cascade model for the fragmentation of a quark of charge q (Fig. 6). If the successive breakups are assumed to be ordered in rapidity, then the charge measure is

$$Q(q) = q + \frac{1}{N}\sum_j \bar{q}_j = q + \langle\bar{q}\rangle \tag{24}$$

where $\langle\bar{q}\rangle$ = average charge on an antiquark. For the probabilities given in eq. (20),

$$\langle q\rangle = -\langle\bar{q}\rangle = \sum_i p_i q_i = \frac{1}{9} \ . \tag{25}$$

Thus, one expects $Q(u) = 5/9$ and $Q(d) = Q(s) = -4/9$. If there are rapidity inversions in the jet, and this will usually be the case, the charge measure

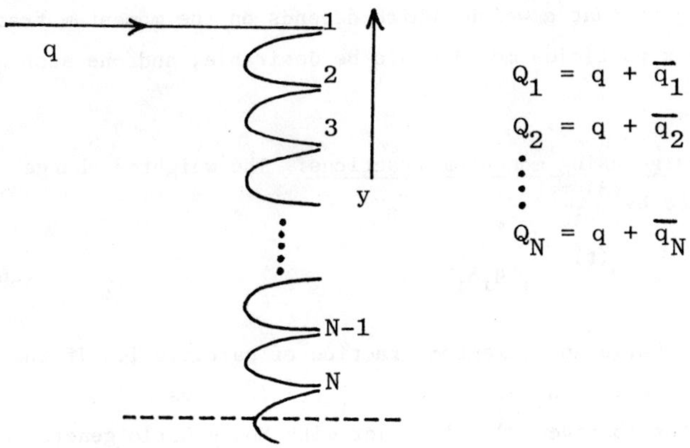

Fig. 6: A jet event in which successive breakups are ordered in rapidity. The charge measure for such a jet is given by eq. (24).

acquires corrections of $0(\frac{1}{N})$.[11] Thus the accuracy of the charge measure improves as N increases.

The above method is easy to use and weights the charge of the fastest particle more than the slower ones. In fact, from eq. (23) we see that the charge of particle 1 is counted N times, that of particle 2 is counted N - 1 times, etc. Effectively the charge measure method takes a given event with N particles and artificially treats it as N events with 1, 2, ..., N particles respectively. The charges in these events are then averaged to give the charge measure.

The method has two disadvantages. It does not give the probability of correct identification of the initial quark. Also, it does not make use of the momentum fractions of various constituent particles:

Jet 1:	$\pi^+(0.8)$,	$\pi^-(0.1)$,	$\pi^0(0.08)$,	$\pi^+(0.02)$
Jet 2:	$\pi^+(0.4)$,	$\pi^-(0.25)$,	$\pi^0(0.2)$,	$\pi^+(0.15)$.

In jet 1, clearly the leading π^+ has a very large probability of containing the jet-initiating quark whereas this is not the case for jet 2. Nevertheless, both jets are treated in exactly the same way when the charge measure is computed. This suggests that a weight which depends on the momentum fraction x (weighting large x particles more) would be desirable, and one such method is given below.

(iii) Weighted charge using momentum fractions: The weighted charge of a jet event is defined by[8]

$$Q^{(r)} \equiv \sum_i q_i x_i^r \tag{26}$$

where q_i and x_i are the charge and momentum fraction of particle i. If the parameter r is chosen greater than zero, large x particles have a larger weight. Comparison of the observed $Q^{(r)}$ in a jet with Monte Carlo generated curves for the probability of having this value from different initial partons gives an indication of the flavor of the jet-initiating parton. Such a method has been tested in $\bar{\nu}p$ interactions to verify d quark fragmentation.[12] In general, the method of weighted charge does not give reliable identification on an event by event basis. It is also not easy to use nor is the weight x^r or choice of the parameter r motivated from physics considerations.

Several other criteria for identifying the initial parton have been discussed in the literature. A survey[8] shows that most of them have poor reliability.

(iv) The optimum method of identifying the jet-initiating parton: Suppose that we have a jet event with N mesons having momentum fractions $x_1, x_2, \ldots, x_N$. From this information alone, we cannot identify the jet-initiating parton with complete certainty, since several partons could give rise to the observed event. Thus, the best one can do is to calculate the probability P_i that the given jet event was initiated by parton i. Clearly, P_i is proportional to the N-hadron fragmentation function.

$$P_i = \frac{D_{i\to h_1 \ldots h_N}(x_1,x_2,\ldots,x_N)}{\sum_i D_{i\to h_1 \ldots h_N}(x_1,x_2,\ldots,x_N)} \tag{27}$$

Thus, our algorithm for identifying the initial parton is to compute P_i for various partons i. The parton which maximized P_i is likely to have started the jet in question. This method, by definition, has the maximum possible reliability for jet identification.[13]

In order to use eq. (27) in practice, one needs analytic expressions for N-hadron fragmentation functions. As mentioned in Sec. III, these expressions are available in the recursive model and agree well with Monte Carlo calculations using a somewhat more sophisticated model.[8]

What is a good choice for N? It should be large enough so that we make sure that the initial parton is included in the analysis. However there is no point in taking N to be too large, since we have seen that after the first breakup, it is the "mean quark" which fagments further. Thus an unnecessarily large value of N only complicates the calculations in eq. (27) without including additional information. Since the mean breakup producing the fastest particle is 2.3[8], N = 3 or 4 is presumably a good choice.

The above algorithm has been implemented for N = 3 into a computer program. The inputs are the three fastest particles in a jet and their momentum fractions. The output gives the probabilities P_i that the given event was initiated by parton i.[13]

5. GLUON JETS.

Since the recursive cascade model of Sec. III has proven so successful in describing quark jets, it is natural to construct a similar model for gluon jets too. A plausible set[14] of breakups (vertices) can be motivated from QCD. These three breakups are quark → meson + quark, gluon → meson + gluon and gluon → quark + antiquark, and the momentum sharing functions are denoted by f(x), g(x), and h(x) (see Fig. 7). Note that f(x) is the same vertex which was used in Sec. III for quark jets. Clearly, since the momentum sharing functions represent probabilities, they are normalized to one.

$$\int_0^1 dx\, f(x) = \int_0^1 dx\, g(x) = \int_0^1 dx\, h(x) = 1. \tag{28}$$

A quark fragments by emitting a meson (Fig. 4 or 7a). However, a gluon has a choice of emitting a meson (Fig. 7b) or of converting into a $q\bar{q}$ pair (Fig. 7c). If the probabilities for these two choices are taken to be 'a' and '1 - a' respectively, it is straightforward to write down the recursive model equations for gluon jets. The multiplicity $n_G(P)$ of a gluon jet of momentum P satisfies the equation

$$n_G(P) = a\left\{1 + \int_{\mu/P}^1 dx\, g(1-x)n_G(xP)\right\} + 2(1-a)\int_{\mu/P}^1 dx\, h(x)n_Q(xP), \tag{29}$$

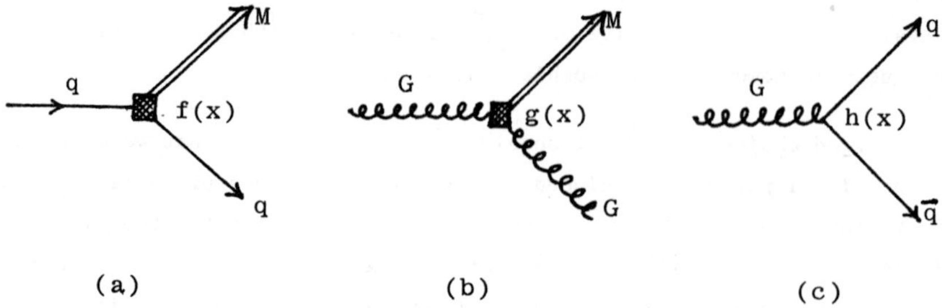

Fig. 7: The three fundamental breakups which define a plausible recursive model for quark and gluon jets.

where n_Q is the multiplicity of a quark jet, which has been computed before (eqs. (5) and (7)). The last term in eq. (29) is the contribution coming from $G \to q\bar{q}$ (Fig. 7c).

Similarly, the integral equation for the gluon fragmentation fuction into mesons can be readily written down.

$$D_G(x) = a\left\{g(x) + \int_x^1 \frac{dx'}{x'}\, g(1-x')\, D_G(x/x')\right\}$$
$$+2(1-a) \int_x^1 \frac{dx'}{x'}\, h(x')\, D_Q(x/x'). \qquad (30)$$

The solution of equation (29) is

$$n_G(P) = 2\alpha \ln(P/\mu c) + O(\mu/P),$$

$$\ln c = \frac{a(2\alpha-\gamma)}{2\alpha\gamma(1-a)} - \int_0^1 dx\, h(x)\ln x, \quad \frac{1}{\gamma} = -\int_0^1 dx\, g(1-x)\ln x. \qquad (31)$$

Note that the asymptotic multiplicity is always logarithmic, independent of the specific form of the momentum sharing functions. For $a \neq 1$, $n_G(P)/n_Q(P) \to 2$ asymptotically, but the ratio is much lower at smaller energies. The energy scale for attaining asymptopia is set by $c\mu$, and this increases as $a \to 1$. Fig. 8 shows a comparison of the multiplicities in a quark and gluon jet for reasonable choices[14] of the momentum sharing fuctions.

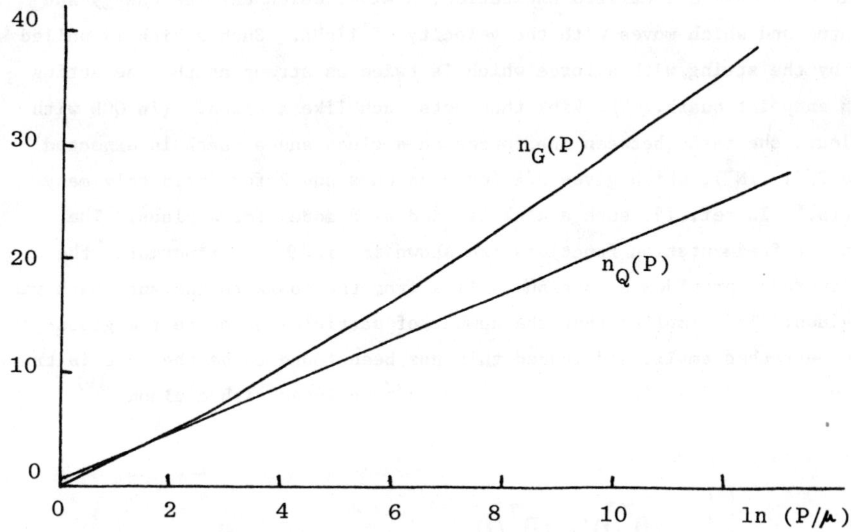

Fig. 8: The energy dependence of the total multiplicity of produced particles in quark and gluon jets. At present energies the multiplicities are comparable, whereas asymptotically, $n_G/n_Q = 2$.

Note that even when the total multiplicity in one quark jet is 10 (and this easily takes care of the highest PETRA, PEP energies), the multiplicity in a gluon jet of the same energy is approximately 11. This also explains why there is only at most a 20% jump seen in the e^+e^- multiplicity as one goes through the Υ resonance. Just off the Υ, $n_{e^+e^-} = 2n_Q(P) \simeq 9$. At the Υ, assuming that the 3 gluon decay mode dominates, one gets $n_Υ \simeq 3n_G(2P/3) \simeq 11$, which is indeed about 20% above the off-resonance multiplicity.

Eq. (30) can be solved to give the gluon fragmentation function into mesons. At some stage during the recursive cascade the gluon branches into a quark-antiquark pair via Fig. 7c, and this makes gluon jets softer than quark jets.[14]

In many respects similar to the above-described recursive model for fragmentation is the massless relativistic string model approach with a quark and antiquark at the end points (see Sec. II). On such a string it is possible to have a localized excitation, a kink, which carries energy and momentum and which moves with the velocity of light. Such a kink is pulled back by the string with a force which is twice as strong as the one acting on an endpoint quark. The kink thus acts much like a gluon. (In QCD with N colours the ratio between the forces on a gluon and a quark is expected to be $2/(1-1/N^2)$, which gives 9/4 for 3 colours and 2 for infinitely many colours.) In ref. 15, such a kink is used as a model for a gluon. The gluon jet fragmentation functions are shown in Fig. 9 . Furthermore, the kink approach provides a continuous link from the quark to the antiquark via the gluon. This implies that the number of particles opposite the gluon jet would be rather small, and indeed this has been found to be the case in triple jet events at PETRA, if the softest jet is associated with a gluon.[16]

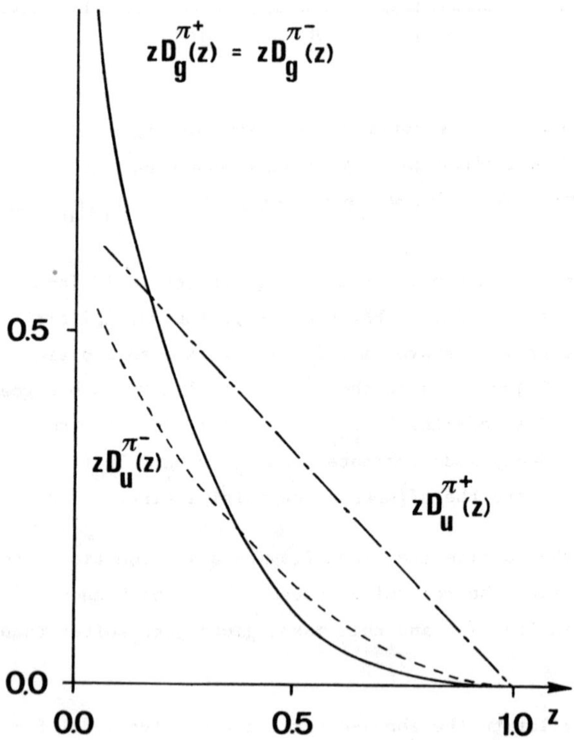

Fig. 9: A comparison of gluon and quark fragmentation into pions.

6. FRAGMENTATION FUNCTIONS AT LARGE Q^2

As discussed in Sec. II, the recursive cascade model describes the fragmentation of partons having a large momentum P but low Q^2. In this section, we consider the case when the jet-initiating parton not only has a large momentum, but is also far off its mass shell (large Q^2). Until the offshellness is degraded to a value Q_o^2 (see Sec. II (iii) and Fig. 3) the branching can be described by perturbative QCD. Beyond Q_o^2, there are at least two ways of hadronization, which we shall refer to as fragmentation and recombination. These terms refer to the processes shown in Fig. 10 - fragmentation of a parton with momentum P refers to a union of the parton with a self-generated partner to give a meson M with momentum $p < P$, whereas, the term recombination refers to the union of a parton with other available partners at Q_o^2, giving a meson M with momentum $p > P$.

(i) Fragmentation of partons at Q_o^2: If only fragmentation occurs at Q_o^2, then the fragmentation fuction of parton 'i' into hadron type h is given by

$$D_{i\to h}^{FRAG}(x;Q^2) = \sum_a \int_x^1 \frac{dx'}{x'} D_{i\to a}(x';Q^2,Q_o^2)\, D_{a\to h}(x/x'), \tag{32}$$

where $D_{i\to a}(x';Q^2,Q_o^2)$ is the jet calculus[6] perturbative QCD result for obtaining parton 'a' at offshellness Q_o^2 from an initial parton 'i' at Q^2.

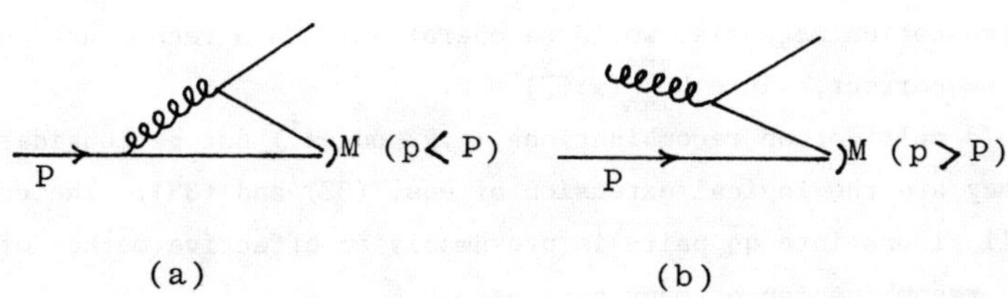

Fig. 10: (a) Fragmentation and (b) recombination of a parton of momentum P to give a meson M of momentum p.

$D_{a\to h}$ is the non-perturbative fragmentation function for parton 'a', which can be found from a recursive model or can be taken from experiment. The Q^2 evolution of D^{FRAG} is the standard QCD result obtained from the evolution equations.[17]

(ii) Recombination of partons at Q_o^2: If all the partons crossing Q_o^2 only combine in groups of two (or three) to form mesons (or baryons), then the fragmentation function of parton 'i' is given by [18]

$$xD^{REC}_{i\to h}(x;Q^2) = \sum_{a_1a_2}\int dx_1dx_2\, D_{i\to a_1a_2}(x_1,x_2;Q^2,Q_o^2)R_{a_1a_2\to h}(x_1,x_2,x) \tag{33}$$

where, $D_{i\to a_1a_2}(x_1,x_2\,;Q^2,Q_o^2)$ is the two parton inclusive cross section for $i\to a_1a_2$, which is known from the jet calculus.[6] The recombination function $R(x_1,x_2x)$ is known from successful phenomenology[19] in soft hadronic collisions.

$$R(x_1,x_2,x) = \frac{x_1x_2}{x^2}\delta\left[\frac{x_1}{x}+\frac{x_2}{x}-1\right] . \tag{34}$$

In making the calculation,all gluons at Q_o^2 are converted into $q\bar{q}$ pairs by artificially inserting an additional perturbative QCD vertex. The calculations of eq. (33) are found to be in moderately good agreement with currently available e^+e^- data.[18]

Several comments on the recombination prescription can be made:

(a) The initial perturbative QCD branching of parton 'i' created partons at Q_o^2 for recombination. Clearly, if parton i was at Q_o^2 to start with, then only the fragmentation mechanism would be operative. Thus recombination alone cannot be correct, since $D^{REC}_{i\to h}(x;Q_o^2) = 0$.

(b) Why should multiparton recombinations (e.g. $u\bar{d}g\to\pi^+$) not be considered? After all, they are the logical extension of eqs. (32) and (33). The conversion of all gluons into $q\bar{q}$ pairs is presumably an effective method of incorporating recombination of many partons.

(c) During the initial QCD evolution stage, $p_T^2 \sim Q^2$. Should one recombine partons widely separated in p_T? Maybe only the pre-confined color singlet clusters[19] should be recombined.

(d) Note that the Q^2 evolution of the recombination approach is governed by the Q^2 evolution of the two parton inclusive distribution $D_{i\to a_1a_2}$, whereas in the more usual fragmentation approach the Q^2 dependence is that of the one parton distribution $D_{i\to a}$. Thus there is a possibility (at least in principle) of distinguishing between the two approaches by looking at the Q^2 evolution of fragmentation functions.

(e) The boundary between the perturbative and non-perturbative regimes of jet evolution is clearly fuzzy and cannot be sharply represented by Q_o^2 alone. This makes the distinction between fragmentation and recombination less sharp, and physically one expects both mechanisms to be operative.

(f) Finally, let us note that at the present time, there is no clear cut experimental data showing the Q^2 dependence of fragmentation functions.

7. DIQUARK FRAGMENTATION

In this section, the work diquark refers to a nucleon from which a current quark has been "knocked out". Clearly, a diquark is a unique object, since it is both manifestly composite (essentially having a spatially extended structure similar to a hadron) as well as colored. A careful study of diquark fragmentation can offer new insight into how composite colored objects evolve into hadron jets. For example, some questions of obvious interest are:

(a) How is the baryon formed in a diquark jet and what is its momentum distribution?

(b) Do the two quarks in a diquark eventually become part of the same baryon?

(c) Is the composite character of a diquark important, or can the diquark be Simply treated as a color $\bar{3}$ point source?

(d) What are the diquark fragmentation functions into mesons, and what is the flavor dependence?

We shall first describe several experiments which can be (or have been) used to study diquark fragmentation, and then discuss various theoretical models.

(i) Experiments on Diquark Fragmentation: Some experiments in which diquark fragmentation can be studied are listed in Table I. The flavor content of the diquark in each experiment is given. We now discuss the relative merits of these different approaches.

(a) Charged current interactions with neutrino and antineutrino beams have an advantage, because the flavor content of the fragmenting diquark is known with reasonable certainty (neglecting $q\bar{q}$ sea effects). Data from bubble chamber studies[20,21] of (uu) and (ud) fragmentation (using ν and $\bar{\nu}$

Type of Reaction	Examples	Diquark System
ν reactions	$\nu p \to \mu^- x$ $\nu n \to \mu^- x$	uu ud
$\bar{\nu}$ reactions	$\bar{\nu} p \to \mu^+ x$ $\bar{\nu} n \to \mu^+ x$	ud dd
Electroproduction	$ep \to ex$ $\mu p \to \mu x$ $en \to ex$ $\mu n \to \mu x$	$\frac{8}{9}$ ud + $\frac{1}{9}$ uu $\frac{2}{3}$ dd + $\frac{1}{3}$ ud
Hadronic collisions with $\mu^+\mu^-$ trigger	$\pi^+ p \to (\mu^+\mu^-) x$ $\pi^- p \to (\mu^+\mu^-) x$ $\bar{p} p \to (\mu^+\mu^-) x$	uu ud $\frac{2}{3}$ ud + $\frac{1}{3}$ uu
Hadronic collisions with large-p_T hadronic trigger	$pp \to \pi^+ x$ $pp \to \pi^- x$	≃ud ≃uu
Hadronic collisions with low-p_T multiparticle production	p fragmentation $\bar{p}$ fragmentation n fragmentation	$\frac{2}{3}$ ud + $\frac{1}{3}$ uu $\frac{2}{3}\,\bar{u}\bar{d} + \frac{1}{3}\,\bar{u}\bar{u}$ $\frac{2}{3}$ ud + $\frac{1}{3}$ dd

Table I: Diquark Systems occurring in various reactions. The assumptions required to establish diquark flavor content are described in the text.

beams) into charged hadrons are plotted in Figs. 11 and 12. Note that the ratio r of (uu) fragmentation functions into π^+ and π^- (see Fig.11) is considerably bigger than unity particularly at large x, where x is the fraction of the diquark momentum carried by the detected pion.

(b) Electroproduction in the deep inelastic region in ep and μp reactions results in a fragmenting diquark which is essentially (ud), whereas for en and μn reactions one gets a mixture of (ud) and (dd). Some data from μ^+p scattering is available,[22] and it is plotted in Fig. 12.

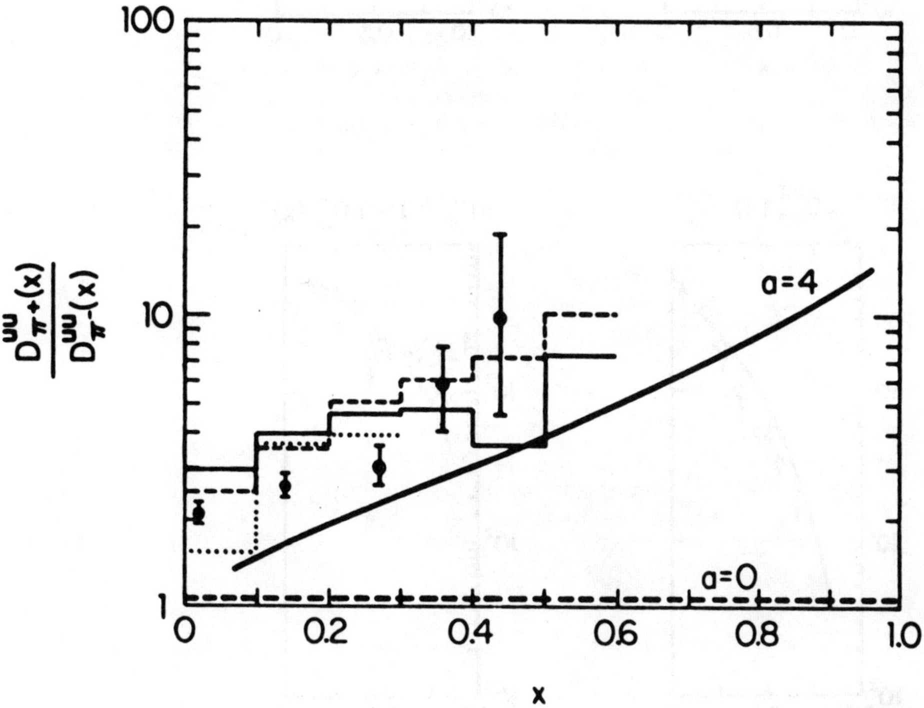

Fig. 11: The ratio r of (uu) diquark fragmentation functions into positive and negative pions. The recursive model prediction (a = 4) is shown too.

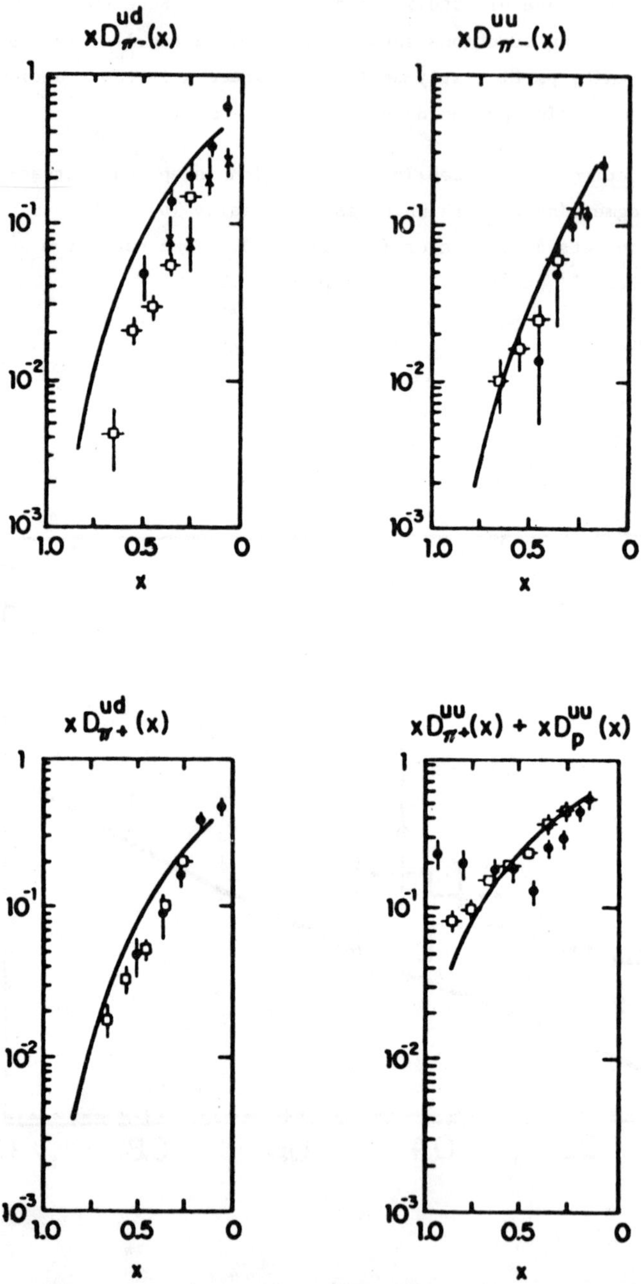

Fig. 12: A comparison of currently available data on diquark fragmentation with the predictions (solid lines) of the recursive model described in the text.

(c) Hadronic collisions producing $\mu^+\mu^-$ pairs are frequently due to the Drell-Yan mechanism. If this is the case, then $\pi^{\pm}p$ collisions with a $\mu^+\mu^-$ trigger offer a good way of studying uu, ud fragmentation. Likewise, $\bar{p}p$ interactions triggered on $\mu^+\mu^-$ pairs will provide diquark fragmentation information.

(d) Hadronic collisions (involving baryons) with a large-p_T trigger offer yet another way of studying the fragmentation of diquarks, *if* hard quark-quark scattering is the dominant mechanism for large-p_T particle production.[(23)] This scheme has three major drawbacks. First (as we have discussed in Sec. IV), the large-p_T trigger particle does not identify the jet-initiating parton with high probability;[(7)] consequently, the flavor content of the forward-moving diquark cannot be well-determined. Secondly, lowest-order QCD calculations show that there are a substantial number of large-p_T gluon jets coming from hard quark-gluon or gluon-gluon scattering.[(23)] If this happens, the forward jet is not necessarily coming from diquark fragmentation. Thirdly, there are indications that lowest order QCD calculations may have non-negligible higher order corrections[(24)] at present energies. Therefore, to extract diquark fragmentation functions from data on hadronic collisions with a large-p_T trigger requires a substantial number of model-dependent assumptions. Nevertheless, this has been done[(25)] using data on pp collisions at the CERN-ISR (see Fig. 12).

(e) Low-p_T (soft) multiparticle production in hadronic collisions like pp, $\bar{p}p$, πp, etc., almost certainly contain detailed information on diquark fragmentation. However, this is a model-dependent statement, since no hard partonic interaction is involved. For example, the two chain model based on dual topological unitarization has been shown to be remarkable successful in explaining available soft multiparticle production data.[(26)] In this scheme, the proton fragmentation region comes from a diquark of flavor 2/3 (ud) + 1/3 (uu).

Figs. 11-12 essentially contain all currently available data on diquark fragmentation functions. (There also exist some data for fragmentation into K°, Λ).[(21)] It should be pointed out that the data is not very precise and corresponds to relatively small values for the energy $\sqrt{W^2}$ of the fragmenting diquark. Nevertheless, there is reasonable agreement between the data from various sources, supporting the idea of universality of diquark fragmentation functions.[(27)] Clearly the quality of the data can be substantially improved and this can be done most profitably at larger values of $\sqrt{W^2}$ which will soon become available at the Fermilab Tevatron.

(ii) Models for Diquark Fragmentation: Although phenomenological fits[28] to leptoproduction data using dimensional counting rules [29] have been made in the past, it is only recently that theoretical models for diquark fragmentation have been investigated.

(a) Color tube model: This is essentially a model [30] for diquark fragmentation although it is developed for describing low-p_T baryon-hadron collisions. After the collision, the forward going system is taken to be a diquark in a color tube, which subsequently fragments into a jet of mesons. The baryon in the diquark jet is assumed to form at the location x_J of one of the quarks (called the junction or J-quark), whose distribution is assumed to be

$$\frac{dP}{dx_J} = 6x_J(1-x_J) \,. \tag{35}$$

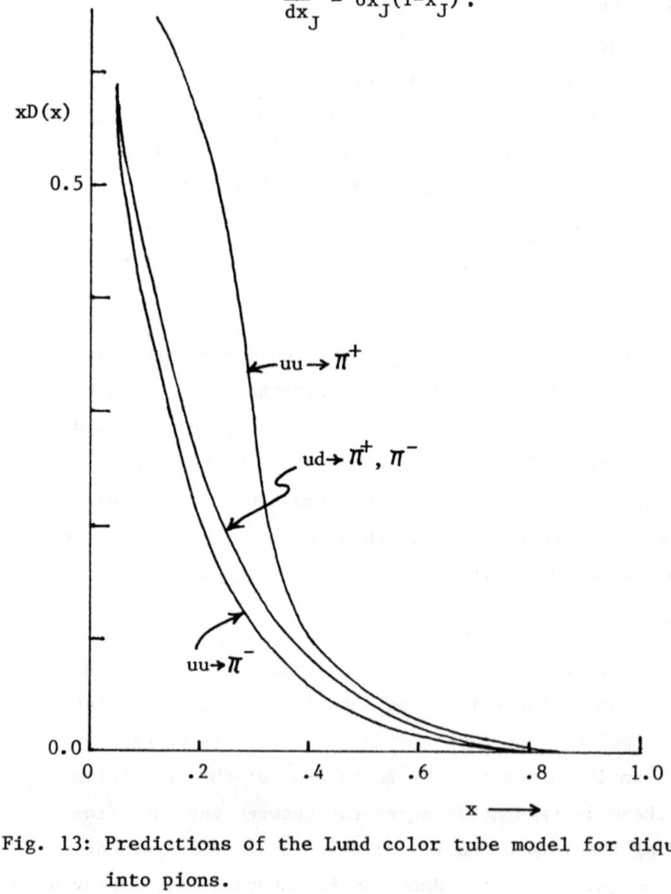

Fig. 13: Predictions of the Lund color tube model for diquark fragmentation into pions.

The diquark fragmentation functions into pions are shown in Fig. 13.

(b) Naive fragmentation model: If the extended structure of a diquark is ignored and it is simply treated as a $\bar{3}$ of color point source, then the most naive fragmentation model for a diquark is (see Fig. 14):

diquark jet = baryon + leftover antiquark jet .

Here the entire initial diquark becomes part of a single baryon, and such a scheme predicts the ratio r of Fig. 11 to be 1, in conflict with data. Thus, the data strongly suggests that the two quarks in a diquark do not always act like one unit and become part of the same baryon.

(c) Recursive model: Currently, this is the most complete model for diquark fragmentation.[32] If the two constituent quarks in a diquark do not always act like one unit, then it follows that sometimes they act like two quarks. This immediately suggests a recursive model for diquark jets. The two basic vertices (breakups) of the model[32] are diquark → baryon + leftover quark (Fig. 15a) and diquark → meson + leftover diquark (Fig. 15b). The momentum sharing functions for these vertices are fixed via the counting rules.[29]

$$f_B^{qq}(x) = 2(1-x) \ , \ f_M^{qq}(x) = 4(1-x)^3 \ . \tag{36}$$

The state of the diquark is specified by (P,δ) where P = diquark momentum and δ = scaled momentum difference.

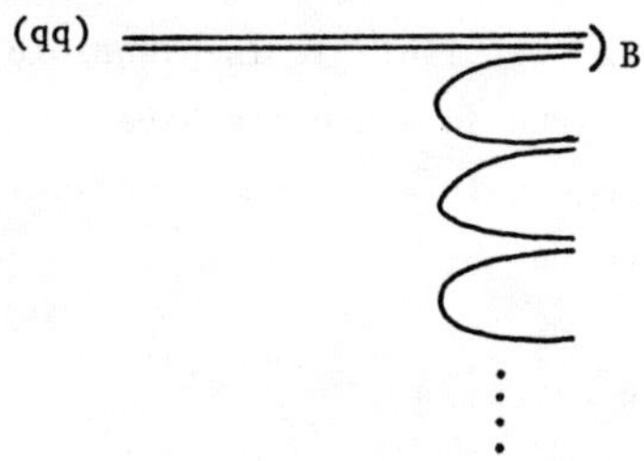

Fig. 14: Naive fragmentation model for a diquark assumes that the two quarks in the diquark both directly become part of the same baryon.

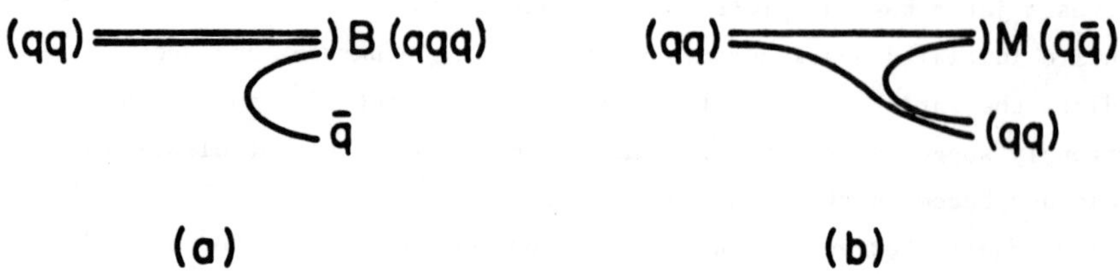

Fig. 15: The two basic vetices of a recursive model for diquark fragmentation. (a) $(qq) \longrightarrow B + \bar{q}$; (b) $(qq) \longrightarrow M + (qq)$.

$$\delta \equiv \frac{|p_1 - p_2|}{P} \quad , \; P = p_1 + p_2, \tag{37}$$

where, p_1, p_2 are the constituent quark momenta.

What is the relative probability of a diquark (P,δ) forming a baryon via Fig. 15a or forming a meson via Fig. 15b? It is plausible to expect that when δ is small, the constituent quarks in a diquark have roughly equal momenta and the diquark acts like one unit and forms a baryon. The probability P_B for baryon formation is taken to be

$$P_B(\delta) = e^{-a\delta^2} \quad . \tag{38}$$

Then, it follows that the probability P_M for the diquark to act like two separate quarks and form a meson is

$$P_M(\delta) = 1-P_B(\delta) = 1-e^{-a\delta^2} \quad . \tag{39}$$

The parameter 'a' can be roughly fixed numerically by noting (from past experience with hadronic collisions) that two particles "know" about each other (short range correlations) if their rapidity separation $|y_1-y_2|$ is less than one unit of rapidity. Since

$$\frac{x_1}{x_2} = \exp\,(y_1-y_2), \tag{40}$$

the criterion $|y_1-y_2| \simeq 1$ corresponds to $\delta_0 = (x_1-x_2)/(x_1+x_2) \simeq 1/2$. We shall therefore use the value $a = 1/\delta_0^2 = 4$ for the parameter 'a' in all subsequent numerical work.

It should be noted that if the parameter 'a' is taken to be zero, then $P_B(\delta) = 1$ which corresponds to the diquark always acting like a single unit and always forming a baryon. This is in fact the naive model of the previous subsection (b). For purposes of comparison, we shall examine $a = 0$ and $a = 4$ results and compare both of them with available diquark fragmentation data.

At this stage, one can write down the recursive cascade model equations for the fragmentation functions of a diquark specified by (P,δ). Assuming these functions scale, we obtain

$$D_B^{qq}(x;\delta) = P_B(\delta)f_B^{qq}(x) + P_M(\delta)\int_x^1 \frac{dx'}{x'}\, f_M^{qq}(1-x')D_B^{qq}(\tfrac{x}{x'};\delta'), \tag{41}$$

$$D_M^{qq}(x;\delta) = P_M(\delta)\left\{f_M^{qq}(x) + \int_x^1 \frac{dx'}{x'}\, f_M^{qq}(1-x')D_M^{qq}(\tfrac{x}{x'};\delta')\right\} + \\ + P_B(\delta)\int_x^1 \frac{dx'}{x'}\, f_B^{qq}(1-x')D_M^{\bar{q}}(\tfrac{x}{x'}) \ . \tag{42}$$

Each term on the right hand side of Eqs. (41) and (42) has a clear meaning. For example, the first term in Eq. (41) corresponds to the diquark forming B(x) (a baryon with momentum fraction x) immediately in the first breakup. The second term corresponds to the situation where the initial diquark first forms M(1-x') (a meson with momentum fraction (1-x')) and a leftover diquark with momentum fraction x', and, subsequently, the diquark fragments to baryon B(x).

Equations (41) and (42) are linear Volterra equations of the second kind and can be solved iteratively. As described in Sec. IV for quark jets, one must include flavor effects for diquark jets too. Furthermore, a physical diquark of momentum P has many possible allowed values of δ. The normalized distribution function $dP/d\delta$ can be obtained[32] from the structure function of a quark in a diquark, which can be guessed from counting rules.

$$\frac{dP}{d\delta} = 0.3(1+\delta)(1-\delta)^{-\frac{1}{2}} \tag{43}$$

which is peaked near $\delta = 1$. The average value of δ is

$$\bar{\delta} = \int_0^1 \delta \frac{dP}{d\delta} d\delta = .72. \tag{44}$$

Since $\bar{\delta} > \delta_o = .5$, clearly the diquark often acts like two separate quarks.

The fragmentation functions for physical diquarks are obtained by convoluting the solutions to eqs. (41) and (42) with $dP/d\delta$.

$$D_h^{qq}(x) = \int_0^1 D_h^{qq}(x;\delta)(dP/d\delta)d\delta. \tag{45}$$

Also, flavor can be included in the formalism.[32] A comparison of the recursive model with the data of Figs. 11 and 12 shows good agreement if the parameter $a = 4$. In contrast, the naive model ($a = 0$) discussed above fails to account for the π^+/π^- ratio r (Fig. 11). A sample of (ud) and (uu) fragmentation functions into mesons is shown in Fig. 16.

The above comparison with data suggests that the diquark cannot simply be treated as color $\bar{3}$ point source and in particular the diquark does not always become part of a baryon. Let us now ask the question: Given a diquark, how often does it directly go into a baryon? This probability is given by $\int_0^1 e^{-a\delta^2} P(\delta)d\delta \simeq .25$ when $a = 4$, whereas it is of course equal to one in the naive model when $a = 0$. Clearly, the precise value depends on the choice of the parameter 'a', and currently available data is not accurate enough to make a good determination. However it should be obvious that the field of diquark fragmentation has the potential to yield new insights into the behavior of partons during the jet formation process.

Acknowledgement: It gives me great pleasure to thank Professors R. Van de Walle and S. Ratti for organizing a lively and profitable conference at Erice, and allowing me the opportunity to participate. I would also like to thank other conference participants and my colleagues at the University of Illinois at Chicago Circle, Fermilab and Orsay for many useful discussions. The efficient typing of Ms. L. Perlini is gratefully acknowledged.

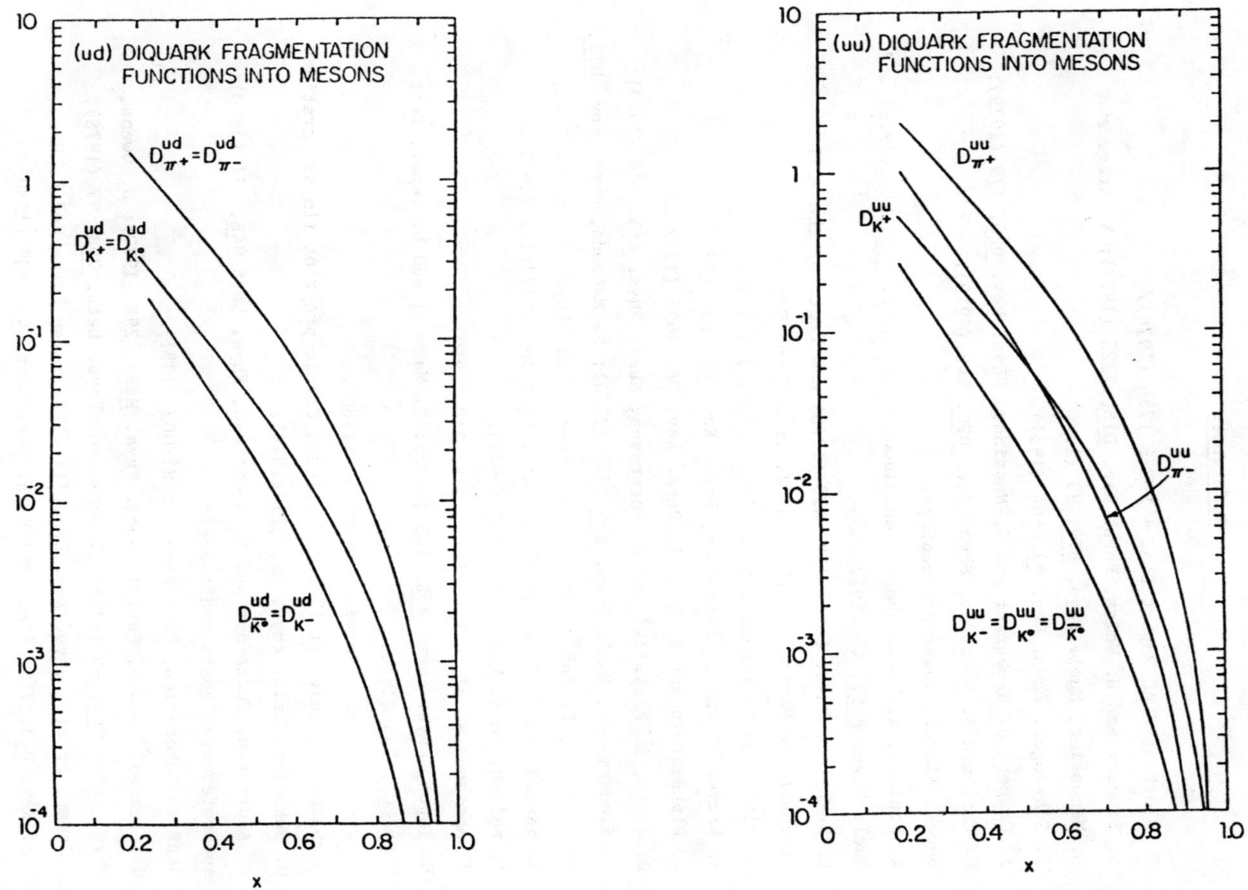

Fig. 16: Recursive model prediction for diquark fragmentation functions into mesons.

References

1. J. Bjorken, SLAC Report SLAC – PUB – 1756 (1976).
2. S. Brodsky and N. Weiss, Phys. Rev. D16, 2325 (1977); X. Artru and G. Menessier, Nucl. Phys. B70, 93 (1974).
3. J. Schwinger, Phys. Rev. 82, 664 (1951).
4. A. Casher, H. Neuberger and S. Nussinov, Phys. Rev. D20, 179 (1979); C. Chiu and S. Nussinov, Phys. Rev. D20, 945 (1979).
5. See W. Kittel, these proceedings.
6. K. Konishi, A. Ukawa and G. Veneziano, Phys. Lett. 78B, 243 (1978) and Nucl. Phys. B157, 45 (1979).
7. U. Sukhatme, Phys. Lett. 73B, 478 (1978) and Proc. of the XIII Rencontre de Moriond (1978), ed. J. Tran Thanh Van.
8. R. Field and R. Feynman, Nucl. Phys. B136, 1 (1978).
9. A. Krzywicki and B. Petersson, Phys. Rev. D6, 924 (1972); J. Finkelstein and R. Peccei, Phys. Rev. D6, 2606 (1972); R. Jengo, A. Krzywicki and B. Petersson, Nucl. Phys. B65, 319 (1973); F. Niedermayer, Nucl. Phys. B79, 355 (1974); S. Matsuda, Phys. Rev. D12, 1940 (1975); H. Fukuda and C. Iso, Prog. Theo. Phys. 57, 483 (1977); O. Sawada and G. Takeda, Tohoku Univ. Preprint TU/77/166 (1977); A. Seiden, Phys. Lett. 68B, 157 (1977).
10. V. Ammosov, et al., Phys. Lett. 93B, 210 (1980).
11. M. Teper, Phys. Lett. 90B, 443 (1980); C. Maxwell and M. Teper, Zeit. Phys. C7, 295 (1981).
12. R. Orava, et al., Fermilab preprint (1980).
13. U. Sukhatme, Univ. of Illinois Chicago Circle preprint (in preparation).
14. U. Sukhatme, Zeit. Phys. C2, 321 (1979).
15. B. Andersson, Gustafson and T. Sjostrand, Phys. Lett 94B, 211 (1980) and references contained therein.
16. JADE collaboration, DESY preprint 81-009 (1981).
17. G. Altarelli and G. Parisi, Nucl. Phys. B126, 298 (1977); J. Owens, Phys. Lett. 76B, 85 (1978); T. Uematsu, Phys. Lett. 79B, 97 (1978).
18. L. Jones, et al., Phys. Rev. D23, 717 (1981); Van Chang and R. Hwa, Phys. Rev. D23, 728 (1981) and Phys. Rev. Lett. 44, 439 (1980).
19. D. Amati and G. Veneziano, CERN report TH – 2620 (1979).

20. J. Bell, et al., Phys. Rev. D19, 1 (1979); D.R.O. Morrison, Proceedings of the XIX International Conference on High Energy Physics, Editors S. Hamma, M. Kawaguchi and H. Miyazawa (Phys. Soc. Japan, Tokyo, 1979) p. 354; M. Derrick, et al., Phys. Rev. D17, 1 (1978).

21. C.C. Chang, et al., Bull. Am. Phys. Soc. 25, 40 (1980). V.V. Ammosov, et al., preprint Fermilab-Pub-80/44-Exp., submitted for publication (1980).

22. C. del Papa, et al., Phys. Rev. D15, 2425 (1977).

23. R. Feynman, R. Field and G. Fox, Phys. Rev. D18, 3320 (1978); J. Owens, Phys. Rev. D19, 3279 (1979); A. Krzywicki, J. Engels, B. Petersson and U. Sukhatme, Phys. Lett. 85B, 407 (1979).

24. R.K. Ellis, M. Furman, H. Haber and I. Hinchliffe, Nucl. Phys. B173, 397 (1980).

25. D. Hanna, et al., to be published in Phys. Rev. Lett. (1981); H. Lubatti, these proceedings.

26. A. Cappella, U. Sukhatme and J. Tran Thanh Van, Zeit. Phys. C3, 329 (1980) and references contained therein; A. Capella these proceedings.

27. This observation was first made by D. Hanna, et al., CERN preprint (1981).

28. M. Fontannaz, B. Pire and D. Schiff, Phys. Lett. 77B, 315 (1978).

29. S. Brodsky and R. Blankenbecler, Phys. Rev. D10, 2973 (1974); S. Brodsky and J. Gunion, Phys. Rev. D17, 848 (1978).

30. B. Andersson, et al., Lund preprint (1980), to appear in Nucl. Phys. B.

31. A more sophisticated version of this model is discussed by D. Beavis and B. Desai, Rutherford preprint RL-80-057 (1980), to appear in Phys. Rev. D.

32. U. Sukhatme, K. Lassila and R. Orava, Fermilab preprint 81/20-THY (1981).

Chapter IV: HADRON STRUCTURE AND SOFT PROCESSES

RUDOLPH C. HWA
Institute of Theoretical Science and Department of Physics
University of Oregon, Eugene, Oregon 97403 USA

Abstract

The physics of soft hadronic processes is described in terms of the hadronic constituents. Careful distinction is made of the constituent and current quarks. The valon model is suggested as a way of describing hadron structure. The recombination model follows naturally as a description of the hadronization process. These ideas are applied to the hadronization problem in quark and gluon fragmentations in hard processes yielding good agreement with data. The description of inclusive reactions at low p_T in the fragmentation region is then formulated in the framework of valon and recombination models. A tight connection is thereby established between hadron structure and low-p_T inclusive processes. From the form factors it is also possible to extract information about the valon wave functions. In the case of the pion that information is used to calculate the pion decay constant with great success. The interaction of a fast quark with nuclear matter is examined, and a quantitative estimate of its radiation length is made. Finally, the recombination model is compared to the fragmentation model.

1. INTRODUCTION

The original title for these lectures was Recombination Model, which was chosen by the organizers of this School as the most succinct way to identify the approach to soft hadronic processes that I was to discuss. However, the scope of the material covered is far more than just recombination in low-p_T reactions. Indeed, the hadronization problem cannot be fully described without a proper description of the hadron-structure problem, which in turn must be related to structure functions and form factors of hadrons. In these lectures we examine all these problems, so the title of the written version is modified accordingly.

1.1 About Soft Hadronic Reactions

As a means of introducing the subject matter of these lectures, let me start with a rather unconventional beginning by arguing why low-p_T reactions are better than high-p_T reactions in learning about hadron structure. It should be stated at the onset, whether a reaction is soft (low momentum transfer) or hard (high momentum transfer), that it is the structure of hadrons in terms of their constituents that we seek to determine, and that knowing the structure facilitates to a large extent the prediction of various properties of hadronic reactions. For the moment it is sufficient to think of quarks and gluons as hadron constituents in the usual sense, although more careful definitions will be necessary later on.

In a high-p_T reaction there are many unknowns that stand in the way of interpreting the data. The three stages usually associated with such processes are: quark (or gluon) in hadron, hard-scattering subprocess, and fragmentation into detected hadron.[1] Considerable uncertainties exist in all three. In the first, the primordial k_T distribution of partons is theoretically unknown, yet it directly affects the p_T distribution of secondary hadrons detected at large angles. In the hard-scattering subprocess the higher-order terms in QCD turn out to be greater than the Born terms, so the hope of testing QCD by such means has for all practical purposes evaporated. In the fragmentation process, there are ambiguities associated with trigger bias, quark versus gluon jet, etc. With all these unknowns in the problem, one would be hard put to identify what has really been learned about hadron structure in high-p_T reactions after all these years of very costly experimentation.

By comparison, low p_T reactions in the fragmentation region are far simpler. The parton k_T distribution is irrelevant if we study inclusive cross sections integrated over hadron p_T. Instead of an uncalculable hard-scattering subprocess, fast partons in a projectile do not interact with the target hadron (to an accuracy of about 98% which will be discussed in Sec. 7). Hadronization of quarks can be described by a recombination process, the details of which turn out not to affect the inclusive cross sections in any sensitive way. Furthermore, the recombination function is by now no longer an unknown quantity. The only unknown in the whole process is the longitudinal-momentum distribution of the partons in the incident hadron. For proton it turns out to be very close to what is (incompletely) determined by electroproduction at low Q^2. Experiments on low-p_T reactions are ideally suited to determine those distributions, especially for such hadrons as pions and kaons that are unaccessible to probing by deep inelastic scattering. In that way we learn about the hadron structure and ultimately the confinement problem. There are many reactions that can be explored at an efficiency far greater than that of high-p_T experiments.

Having extolled the virtues of soft hadronic reactions, it is necessary to follow immediately with an explanation that such virtues are only virtual if one chooses to analyze the problems in terms of the Regge picture, or derivatives thereof. It is the understanding of the hadron structure that stands out as a worthy aim, and for that the description of soft reactions must be based in terms of quarks and gluons. Regge behavior should be an output of a fundamental theory, not an input that plays an essential role. We are in no position to talk about fundamental theory for soft processes at this point. But the approach which we shall present here is based on the parton model,[2] the justification of which would be very difficult without QCD.

The key parts of the problem in dealing with soft hadronic reactions are the hadron structure (from hadron to quarks) and the hadronization process (from quarks to hadron). In the fragmentation region these are the only important parts, whereas in the central region there are other complications. We shall concentrate mainly in the fragmentation region where we can learn more about hadron structure. The central region will be considered in Sec. 8. The mechanism for hadronization is recombination, which is, of course, more generally applicable than what is needed for low-p_T reactions. The recombination mechanism has survived phenomenological tests when applied to such processes as quark fragmentation. We shall cover in these lectures various topics that are all related: soft reactions, hard processes, form factors at low Q^2, and nuclear effects.

1.2 Simple Recombination Model

To further introduce the subject in more specific terms, let us focus on the simple recombination model suggested by Das and me a few years ago.[3] The inclusive distribution for pion production in the fragmentation region is

$$\frac{x}{\sigma}\frac{d\sigma}{dx} = \int F(x_1,x_2)R(x_1,x_2,x)\frac{dx_1}{x_1}\frac{dx_2}{x_2} \tag{1.1}$$

where $F(x_1,x_2)$ is the probability of finding in the incident hadron a quark with momentum fraction x_1 and an antiquark at x_2. $R(x_1,x_2,x)$ is the recombination function for forming a pion at x. Every quantity in (1.1) is defined with respect to invariant phase space. The physical basis for (1.1) is short-range correlation,[4] which justifies the neglect of the copious production of particles in the central region and of the independence on the other incident hadron.

In earlier times concerns have been expressed about the predictive power of (1.1) on account of the uncertainties related to F and R.[5] The answer to such concerns can be expressed in three different ways: historical, theoretical and experimental. Of course, a more important question is whether (1.1) contains the right physics, the answer to which is the subject of the first half of these lectures. Historically, (1.1) was suggested at a time when large-p_T physics was fashionable, and a large sector of the theoretical physics community was polarized to think of hadronization only in terms of quark fragmentation.[6] It was shown in Ref. 3 that a straight-forward application of quark fragmentation to low-p_T reaction would disagree badly with data, and that the recombination model based on rather naive (but first-try) assumptions about F and R .

$$F(x_1, x_2) = F_q(x_1)\, F_{\bar{q}}(x_2)\, \rho(x_1, x_2) \tag{1.2}$$

$$R(x_1, x_2, x) = \alpha \frac{x_1 x_2}{x^2}\, \delta\left(\frac{x_1}{x} + \frac{x_2}{x} - 1\right) \tag{1.3}$$

turned out to be remarkably successful. Admittedly, there are a number of unknowns in (1.2) and (1.3), like the factorizability, $\rho(x_1,x_2)$ and α. But by making reasonable estimates, we achieved our modest goal that the recombination idea is quantitatively sensible. With various directions of emphases, subsequent investigations[7-20] along the line of recombination all contribute to the establishment of a common agreement that (1.1) represents the essence of hadronization at low-p_T in the fragmentation region.

Theoretically, (1.1) is, of course, meaningless unless F and R are more firmly specified. The assumed forms in (1.2) and (1.3) need to be either justified on some acceptable basis or improved upon. In place of (1.2) the Kuti-Weisskopf model[21] has been used[9-11] to determine F. It is another model within the recombination model, designed to guarantee desirable kinematical properties, but with little dynamical content. In the next section we shall discuss both F and R in the framework of the valon model,[22-24] which focusses on the hadron structure. Using phenomenological results derived from other avenues of probing the hadron structure (notably hard processes), it is possible to calculate inclusive distributions without adjustable parameters. Thus the recombination model has now been formulated in a way that provides a direct relationship between hadron structure and soft processes, which is the kernel of our argument for doing low-p_T experiments.

Regarding experiments, they are not merely to check the predictions of the recombination model in the present context. If we remove the phenomenological input from hard processes, there are certain unknown functions about the parton distributions that must be determined, and a complete set of measurable low-p_T reactions would put severe constraints on those distributions, which would effectively test the validity of the model.

2. VALON MODEL

In order to formulate the problem of soft hadronic processes at the constituent level, it is necessary to define clearly what constituents we shall be discussing and to determine the hadron wave functions in terms of those constituents. That will be the subject of this section. It will appear to be a rather long digression from the point of view of recombination, but it is necessary not only for the determination of the recombination function but also for the establishment of a relationship between hadron structure and low-p_T reactions.

2.1 Constituent Versus Current Quarks

In the bound-state problem we regard the proton as consisting of three quarks and the pion a quark-antiquark pair. These are the constituent quarks. On the other hand, in deep inelastic scattering a hadron is regarded as having valence quarks, sea quarks, and gluons (collectively called partons hereafter). In fact, even at Q^2 as low as between 1 and 3 GeV^2 the gluons carry nearly half the nucleon momentum. How can this picture of the hadron reconcile with that of the bound-state problem? It must first be recognized that the quarks probed in deep inelastic scattering are current quarks, since they are coupled to electromagnetic or weak currents. Evidently, constituent quarks are different from current quarks. The failure to recognize this difference has led to much confusion and many mistakes in the literature. Here are a few examples.

(a) Since the total cross section for πp collision is about 2/3 that of pp, in the additive quark model one regards the constituent quarks as having independent cross sections so that $\sigma_{qp} \sim 1/3\ \sigma_{pp}$. Such quarks should therefore <u>not</u> be given a momentum distribution $q(x) \propto x^{-\frac{1}{2}}(1-x)^3$, which describes the (valence) current quarks.

(b) In the bag model[25] one is dealing with constituent quarks. But in esti-

mating the bag size it would be wrong to draw on the primordial quark transverse momentum (0.6–0.8 GeV) inferred from large-p_T reactions[1] as a phenomenological input. The estimated bag size (∿ 0.3 fm) comes out too small because in a hard scattering process it is the current quarks that are involved.
(c) In dual diagrams it is not clear what the lines correspond to exactly, except to keep track of quantum numbers. They have been regarded by some as constituent quarks since the three lines of a proton carry all the momentum, but then one cannot attach a fragmentation function (as determined in e^+e^- annihilation) to the end of one such line to describe the hadronization process since the $q\bar{q}$ created in e^+e^- annihilation are current quarks.
(d) What is a diquark? In a scattering problem it sounds like what remains in a proton after a quark is knocked out. If so, then those are constituent quarks since the totality is three. But how does one knock out a constituent quark? In a hard process it is the current quark that is struck, while in a soft process we know of no way a constituent quark can be totally removed (except in models where the distinction between constituent and current quarks is fuzzy).

The confusion is due partly to the usage of the same word "quark" for both entities and partly to the naive extension of the concept of constituent quarks, well defined in a bound-state problem, to the scattering situation. The former can be removed by simply using different names. To that end let us hereafter reserve the term quark (unmodified) only in reference to the current quark whose field appears in the basic Lagrangian in QCD, while for the other we call it valon. The other cause of the confusion can only be removed by a better understanding of how a valon differs from a quark, the elucidation of which is the goal of this whole section.

2.2 Physical Picture of Valons

A valon is a valence quark plus its associated sea quarks and gluons due to the dressing processes in QCD. In a bound-state problem those processes are virtual and a good approximation for the problem is to regard a valon as one integral unit whose internal structure cannot be resolved. Thus it may be identified as a constituent quark. In a scattering problem, such as a proton going through a nucleus, there is no reason to believe that a constituent quark in the projectile would traverse a certain distance in the nuclear matter without interaction and then makes a scattering with another constituent quark in the target, followed by perhaps another rescattering in the nucleus before emerging with an altered momentum, but remaining intact as a constituent quark. The point is that calling it a constituent quark leads one to a picture that it is an indivisible, point-like object which undergoes collision as an integral unit. In a scattering situation the virtual partons in a valon can be excited and be put on mass shell. It is then more appropriate to think of a valon as a cluster of partons with some momentum distribution. As a valon propagates through the nuclear medium (or any hadron) its identity (viz. the momentum profile of its parton) cannot be maintained since the fast and slow partons interact differently with the target owing to the short-range correlation of the partons. Thus in a collision problem a valon should not be thought of as a constituent quark, particularly since what happens to the latter under scattering is unknown.

The model that we are describing provides a way of unifying the static properties of a hadron with what have been revealed in deep inelastic probing. The proton, say, has three valons which, on the one hand, interact with one another in a way that is characterized by the valon wave function, and which, on the other hand, respond independently in an inclusive hard collision with a Q^2 dependence that can be calculated in QCD at high Q^2. In the former role the valons are the constituents of the bound-state problem; in the latter they are the valence quark clusters whose internal structures are probed with high resolution. One involves the confinement problem at large distance, the other the short-distance problem of current operators. It is not yet possible to make the connection precise in QCD, but models have been constructed to link them in various ways, all sharing the same basic idea.[22,26-29]

Whereas the Q^2 dependence of the parton distribution in a valon is calculable at high Q^2, the distribution itself at low Q^2 (call it the cluster distribution P) is hard to determine. As we have learned from electroproduction at SLAC, the quark distribution in a proton already exhibits precocious scaling for Q^2 in a range as low as 1-3 GeV^2. That is, Q^2 evolution has already run its course and further increase in Q^2 would only result in logarithm correction to scaling. Thus even at low Q^2 the parton distribution may be represented by a diagram shown in Fig. 2.1 where v denotes valons and q denotes quark. G(y) is the valon distribution in hadron and P(z) is the parton distribution in a valon, which is the low-Q^2 limit of a Q^2-dependent distribution $p(Q^2,z)$. The convo-

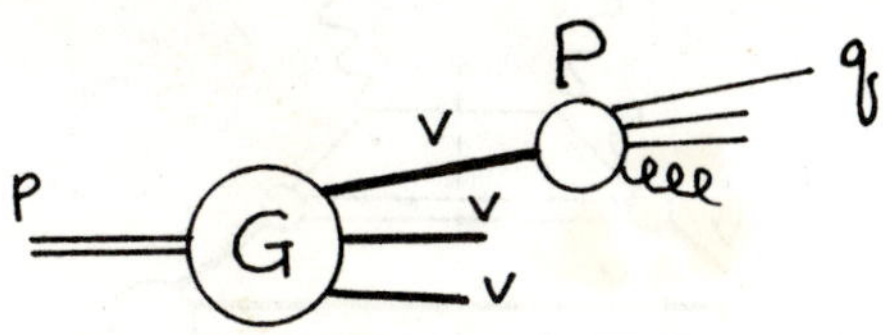

Fig. 2.1 Parton distribution in a proton is the convolution of valon distribution G(y) in proton with parton distribution P(z) in valon.

luted result is the parton distribution in a hadron. Note that even at high Q^2 the valon distribution does not depend on Q^2. (See next section.) Thus we describe the hadron wave function in the valon representation for all Q^2; it is intimately connected with the solution of the bound-state problem. This description is different from the Fock-space description used extensively for high-Q^2 processes.[30] There, the different sectors of the Fock space $|qqq\rangle$, $|qqq,g\rangle$, ... $|qqq,q\bar{q}\rangle$ ... contribute with varying degree of importance depending on Q^2.

2.3 Valon Distribution

If we let Ψ_h denote the wave function of a hadron h in terms of valon momenta, then $|\psi_h|^2$ is related to the valon distribution $G_v^h(y)$. Since ψ_h is basic to all hadronic problems, it enters in all hadronic reactions, notably those involving form factors (exclusive) and structure functions (inclusive). They are illustrated in Fig. 2.2: (a) diagram for nucleon form factor at low Q^2, and (b) deep inelastic scattering at high Q^2 with the dotted line signifying sum over final state. The blobs connected to the external hadron lines in both cases represent ψ_h (hence, different from Fig. 2.1, in which the big blob represents $|\psi_h|^2$). Evidently, if we know ψ_h from some confinement model, we can calculate both quantities. On the other hand, we can also use the experimental data on those quantities to extract $|\psi_p|^2$ phenomenologically. Since we have no prejudices about models of confinement, our approach will be the latter. What we obtain for ψ_h is probably more realistic and therefore more reliable, compared to a theoretical result based on some assumed confinement mechanism.

To facilitate phenomenological analyses it is useful to adopt a simple parametrization for the valon distribution in a hadron. Let y_i denote the momentum fraction of the $\underline{i}$th valon. If, for simplicity's sake for the moment, we assume flavor independence of the valons in a nucleon, then a simple form would be[23]

$$G_v^p(y_1,y_2,y_3) = \alpha_p (y_1 y_2 y_3)^a \, \delta(y_1+y_2+y_3-1) \tag{2.1}$$

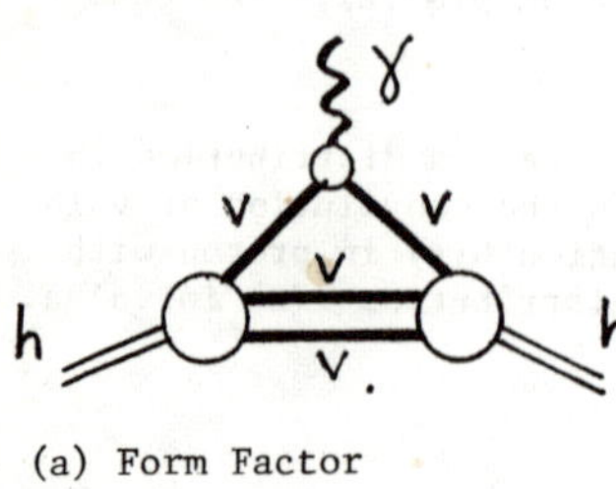

(a) Form Factor

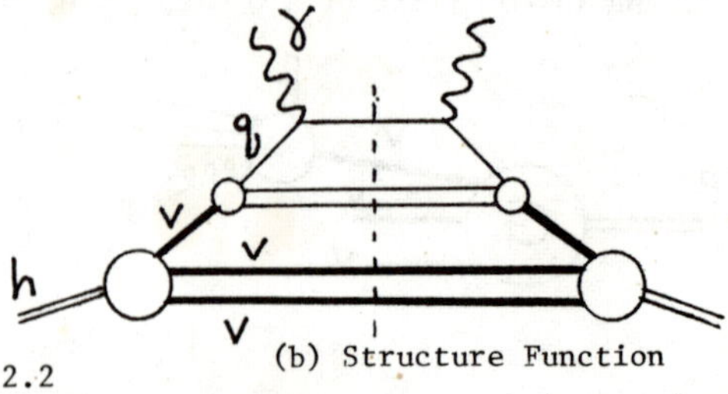

(b) Structure Function

Fig. 2.2

which exhibits symmetry among the three valons. Integrating over y_2 and y_3 yields the one-valon inclusive distribution

$$G_v^p(y) = [B(a+1, 2a+2)]^{-1} y^a (1-y)^{2a+1} \tag{2.2}$$

where the parameter α_p has been determined by the normalization condition $\int_0^1 G_v^p(y)dy = 1$. Similarly, for pion we have[23]

$$G_v^\pi(y_1, y_2) = \alpha_\pi (y_1 y_2)^b \, \delta(y_1 + y_2 - 1) \tag{2.3}$$

$$G_v^\pi(y) = [B(b+1, b+1)]^{-1} [y(1-y)]^b \tag{2.4}$$

In this case α_π is just $[B(b+1,b+1)]^{-1}$. For these two basic hadrons, the only parameters are a and b. Of course, one could choose to use more complicated forms involving more parameters. But it is more important to recognize the physics of the input: the valons (3 for p, 2 for π) exhaust the hadron mementum (i.e. no extra momentum carried by the glue) and the effect of binding is taken into account by the y dependence of $G_v^h(y)$.

On the latter point about binding it is instructive to consider the difference between deuteron and pion. In both cases there are two low-Q^2 constituents: d has two nucleons and π has two valons. Because d is loosely bound, the nucleons do not exchange much momentum in the bound system, so the nucleon momentum distribution $G_N^d(y)$, where y is a fraction of the deuteron momentum, is a function sharply peaked around y = 1/2. That is, each nucleon carries roughly half the deuteron momentum. But for pion it is a tightly bound system of the two valons. We have found that $G_v^\pi(y)$ is nearly constant in y,[23,28] indicating that the valons readily exchange momentum. Since the wave functions are drastically different in the two cases, the form factors will have very different Q^2 dependences. However, in both cases the constituents have internal structures that can be further resolved in higher Q^2 experiments.

As we have discussed in the beginning of this subsection, the valon distribution can be determined from either the structure function or the form factor. We shall do both in the following.

2.4 Structure Function

Fig. 2.2(b) suggests that the structure function of a hadron involves a convolution of two distributions, i.e. valon in hadron and quark in valon:[22,23]

$$F^h(x, Q^2) = \sum_v \int_x^1 dy\, G_v^h(y) F^v\left(\frac{x}{y}, Q^2\right) \tag{2.5}$$

where F^v is the structure function of the probed valon, and is calculable in perturbative QCD to a certain degree of approximation. If Q^2 is small enough, we may at some point identify $F^v(z,Q^2)$ as $\delta(z-1)$ for the reason that we are unable to discern internal structure of a valon at that Q^2, call it Q_v^2. As Q^2 is increased from Q_v^2, one expects $F^v(z,Q^2)$ to develop a tail in the $0<z<1$ region due to momentum leakage through gluon radiation (or figuratively due to finer probing) although the mathematical features of the tail are not precisely calculable. When Q^2 is sufficiently high, then the Q^2-dependence is calculable reliably in the leading-log approximation; let it be for $Q^2>Q_1^2$, say. Between Q_v^2 and Q_1^2 non-leading and higher-twist terms are either so complicated or so poorly understood that no definitive prediction can be made, but that is also the region where the most important part of the Q^2 evolution takes place.

Given this unsatisfactory theoretical situation a sensible phenomenological approach is to adopt the leading-log approximation for $Q^2>Q_o^2$ where Q_o is an adjustable parameter lying somewhere between Q_v and Q_1. Using (2.5) to fit the data on structure function at high Q^2, we can extract $G_v^h(y)$ which should then be regarded as the effective valon distribution defined at Q_o^2. It would not be the actual distribution at Q_v^2, which will be discussed in the next subsection and in Sec. 6. The actual distribution would not be of any use to us in calculating $F^h(x,Q^2)$ because the ignorance barrier concerning $F^v(z,Q^2)$ that stands between Q_v and Q_1 prevents us from using (2.5) for that purpose. On the other hand, the effective valon distribution determined at Q_o is perfectly useful for calculating the quark and gluon distributions at high Q^2 using leading log approximation only.

Using the parametrization (2.2) for the flavor-independent valon distribution $G_v^p(y)$, and using neutrino deep inelastic scattering data,[31] we obtained the value $a = 1/2$, hence[22,23]

$$G_v^p(y) = \frac{105}{16}\, y^{1/2}(1-y)^2 \tag{2.6}$$

This is to be compared to the usual description of valence quark distribution, say the u quark,

$$q_u^p(x) \propto x^{-1/2}(1-x)^3 \tag{2.7}$$

at moderate Q^2. They are related by

$$q^p(x) = \sum_v \int_x^1 \frac{dy}{y}\, G_v^p(y)\, p^v\left(\frac{x}{y}\right) \tag{2.8}$$

where $p^v(z)$ is the momentum distribution of a quark in a valon. Thus, the small x behavior in (2.7) is a direct consequence of a similar small z behavior in $p^v(z)$. As for the behavior in the $x \to 1$ limit, because $p^v(z) \to$ constant as $z \to 1$, we obtain

$$q_v(x) \propto \int_x^1 dy\, G(y) \propto (1-x)^3, \quad x \to 1 \tag{2.9}$$

as desired. The vanishing of G(y) as $y \to 0$ is mathematically connected to the $y \to 1$ behavior by the symmetry property contained in (2.1), and physically a reflection of the unlikehood for a valon (which is a cluster of valence quark, sea quarks and gluons) to have zero total momentum. It should be evident by now that the difference between a valon and a valence quark is enormous.

A more detailed analysis of the structure functions of proton and neutron can give the U- and D-valon distributions separately. This has been done in Ref. 32 using the muon scattering data[33] at $Q^2 = 22.5\ \text{GeV}^2$. The results are

$$G_U^p(y) = [B(a+1,3)]^{-1} y^a (1-y)^2 \tag{2.10a}$$

$$G_D^p(y) = [B(2-a,2a+2)]^{-1} y^{1-a} (1-y)^{2a+1} \tag{2.10b}$$

$$a = 0.65$$

On the basis of these distributions, calculations have been made for xF_3 as well as F_2 at very high Q^2 and the results agree well with the data of CDHS[34] and EMC[35] respectively. Moreover, it is possible to determine the quark and gluon distributions for any high Q^2. A convenient set of parametrizations is provided in Ref. 32, which, we believe, is better than the standard Buras-Gaemers parameterization[36] since more recent and more accurate data have been incorporated. At high Q^2 a sharp peak at very small x is predicted for each of those distributions, which is a characteristics of soft gluon emission.[37,38]

For application later, it is important to determine the valon distribution in a pion, for which we unfortunately do not have deep inelastic data. However, from lepton-pair production[39] the quark distribution in a pion has been found to behave as $(1-x)^1$ at large x. Thus from the first half of (2.9) we get immediately that $G_v^\pi(y)$ should approach a constant as $y \to 1$. It then follows from (2.4) that[23]

$$G_v^\pi(y) = 1 \tag{2.11}$$

This is the result alluded to earlier that reveals the strong binding of the pion.[28]

It is of interest to remark that the valon distribution of the proton expressed in either (2.6) or (2.10) has been compared to the result of a calculation of the proton wave function based on a covarient harmonic oscillator model.[40,41] The general shapes agree surprisingly well at least for $y > 0.4$.[42]

The forms of the valon distributions for kaons and hyperons with unequal valon masses taken into account are discussed in Ref. 23. They can best be determined phenomenologically by low-p_T reactions. A careful study in the valon model remains to be done. See Sec. 4.5 below.

2.5 Form Factors

We mention briefly here the valon distributions that can be extracted from form factors, leaving the more detailed discussions to Sec. 6. By considering form factor at low Q^2, which is the Fourier transform (in the transverse coordinate) of quark distributions in a hadron, it is possible to determine in the framework of the valon model the "actual" valon distribution as distinct from the "effective" ones extracted from structure functions. We call it actual because no leading-log approximation is needed to extrapolate from high Q^2 to where the valons are identified as constituent quarks. The result for proton turns out to be[43]

$$G_U^p(y) = 10.8\, y^{0.9}(1-y)^2 \tag{2.12a}$$

$$G_D^p(y) = 4.6\, y^{0.1}(1-y)^{2.8} \tag{2.12b}$$

Both the proton and neutron form factors have been used to obtain these distributions.

Although (2.12a) and (2.12b) differ significantly from each other, their differences from (2.10a) and (2.10b), respectively, are not great. In fact, it is remarkable that the effective distributions are so close to the actual ones. It dissolves any pressing need to improve upon the leading-log approximation used in conjunction with the structure functions since the change that would result on the effective distributions would not be drastic enough to justify the considerable increase in labor and uncertainties that non-leading and higher-twist terms would entail.

The pion form factor at low Q^2 has also been used to determine[43]

$$G_v^\pi(y) = 1.8\,[y(1-y)]^{0.3} \tag{2.13}$$

This is actually very close to (2.11) so again the actual and effective distributions are very similar. These are all indications that the valon model has captured the essence of the hadron structure since it is in the framework of that model that we can understand simultaneously structure functions at high Q^2 and form factors at low Q^2.

2.6 Recombination Function

Having determined the valon distribution of a hadron, we know immediately by time reversal the probability for recombination. However, it is important to understand what is recombining. For a meson one usually says that it is a quark-antiquark recombination. But if it is a quark in the sense of a parton, then what about gluons? A simple recombination function that takes into account of the gluons is possible only if we consider the recombination of valons. One may then wonder how one would calculate the availability of valons in a reaction for the purpose of recombining them. It is here that we need to recognize that a quark can dress itself up to become a valon with the momentum of the valon being the same as that of the original quark. This is depicted in Fig. 2.3. Any exchange of gluons can be included in the bubble representing the interaction between the valons that form the meson. Thus effectively one can calculate the $q\bar{q}$ distribution in a reaction as in (1.1), and then recombine them using a recombination function appropriate for valons. The question of how the gluons in the original collection of partons are converted into hadrons will be dealt with in Sec. 3.2.

For hadronization into a meson we start with the valon distribution in that meson; the general form for the case of equal valon mass is given by (2.3) and (2.4). The distribution with invariant normalization is

$$F_V^M(y_1,y_2) = y_1 y_2 G_V^M(y_1,y_2) = \alpha_M(b)\,(y_1 y_2)^{1+b}\,\delta(y_1+y_2-1). \tag{2.14}$$

Recombination function for the formation of such a meson at x from two valons at x_1 and x_2 is then

$$R^M(x_1,x_2,x) = \alpha_M(b)\left(\frac{x_1 x_2}{x^2}\right)^{1+b}\delta\left(\frac{x_1}{x}+\frac{x_2}{x}-1\right). \tag{2.15}$$

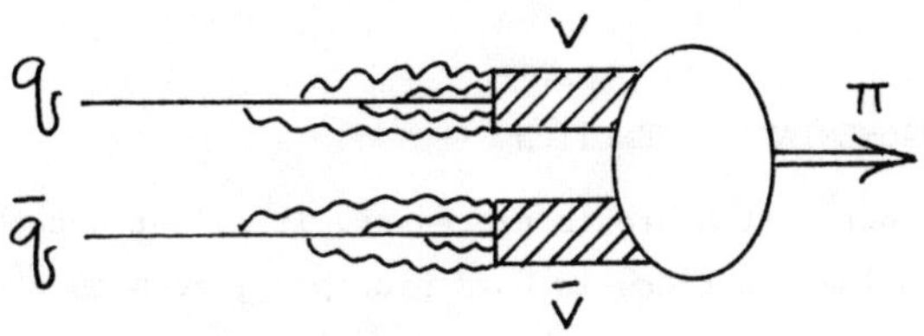

Fig. 2.3
Dressing of quarks before recombination

For pions we have from (2.11) or (2.13) a value for b ranging between 0 and 0.3. The original value used in Ref. 3 was b = 0, but $\alpha_\pi(b)$ was uncertain [cf. Eq. (1.3)]. Now in the valon model we know that the corresponding value is $\alpha_\pi(b=0) = 1$. So we have

$$R^\pi(x_1, x_2, x) = \frac{x_1 x_2}{x^2}\,\delta\left(\frac{x_1}{x} + \frac{x_2}{x} - 1\right) \tag{2.16}$$

which is what we shall use hereafter for pion recombination. In the case of resonances (e.g. ρ, f, etc.), we expect the value of b to be larger because a more loosely bound system results in a distribution that is more sharply peaked at y = 1/2.

Although there is no way as yet to determine the valon distributions for meson resonances, there are reasons to believe that the inclusive distributions of produced mesons are rather insensitive to the value of b (for pseudoscalar or vector particles). This is so provided that the antiquark distribution $F_{\bar q}(x_2)$ is rapidly damped, like $(1-x_2)^7$. To see this qualitatively, let us return to the simple recombination model (1.1) and (1.2), and calculate

$$\int \frac{dx_2}{x_2} F_{\bar q}(x_2)\left(\frac{x_1 x_2}{x^2}\right)^{1+b} \delta\left(\frac{x_1}{x} + \frac{x_2}{x} - 1\right) \tag{2.17}$$

For fixed x this is sharply peaked in x_1 near x, with very little dependence on the precise value of b. Hence, the next integration over x_1 in (1.1) yields a distribution in x that is very close to $F_q(x_1)$, which corresponds to the phenomenological observation of Ochs.[44] As will be discussed in Sec. 4, the insensitivity on b is a phenomenological feature in resonance production.

It is of interest to remark that (2.15) with $b \geq 0$ corresponds to a recombination mechanism that prefers x_1 and x_2 to be near each other. Indeed, the vanishing of R as $x_i \to 0$ is crucial in suppressing the wee partons from participating in the hadronization process in the fragmentation region. It has been shown in Ref. 10 that the recombination function for b = 0 corresponds to a correlation length of two units of rapidity.

The recombination functions for the production of kaon, nucleon and hyperon can all be related to the corresponding valon distributions, and are given in Ref. 23.

3. FRAGMENTATION FUNCTION

Although this School is about soft hadronic processes, it is not out of place to discuss here fragmentation functions in hard processes, such as e^+e^-

annihilation, because the hadronization problem is the same. Moreover, for a quark jet at high Q^2 the first part of the problem involving quark and gluon production can be calculated in perturbative QCD. Thus the recombination mechanism can be more cleanly tested when compared to experiment. The agreement with data (with no arbitrary parameters) not only gives support to the recombination model but suggests that our hadronization mechanism is likely to be more physical than a purely phenomenological approach (such as the cascade model[45]) whose input is adjusted to fit the data.

3.1 Quark-antiquark Distribution

The quark (or gluon) fragmentation function $D_i^\pi(x,Q^2)$ for the decay of a parton i into a pion can be calculated in the recombination model according to[46]

$$x D_i^\pi(x,Q^2) = \int \frac{dx_1}{x_1}\frac{dx_2}{x_2} F_{i\to k\ell}(x_1,x_2,Q^2)\, R_{k\ell}^\pi(x_1,x_2,x) \tag{3.1}$$

This is very similar in structure to (2.5) except for the reversal in the in- and out-states and for the two-valon intermediate state linking F and R. It is depicted in Fig. 3.1(a), in which the shaded blob represents $F_{i\to k\ell}$, the inclusive $q_k\bar{q}_\ell$ distribution in a parton i with all other produced partons not exhibited explicitly in the figure. We have already discussed in Sec. 2.6 in connection with Fig. 2.3 how the partons k and ℓ become valons before recombination. Consistent with the procedure used in determining the valon distribution from structure function (Sec. 2.4), we also use perturbative QCD here to calculate $F(x_1,x_2,Q^2)$ in the leading-log approximation when Q^2 is large. Regarding R as known (Sec. 2.6), we then evaluate (3.1) to obtain $D(x,Q^2)$. There are therefore no adjustable parameters in the calculation.

The most general diagram for the shaded blob in Fig. 3.1(a) is the one shown in Fig. 3.1(b),[47] where each of the three blobs represents single-parton inclusive distribution. Let the first one be denoted by $G_{i\to j}(y,Q^2,k^2)$, signifying the probability of finding a parton of type j in parton i with momentum fraction y as Q^2 is degraded to k^2 through gluon bremsstrahlung and pair-cre-

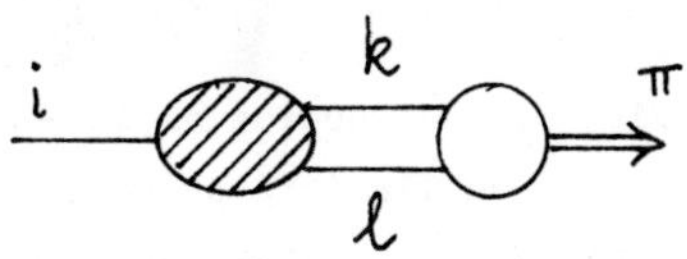

(a) Quark decay in recombination model

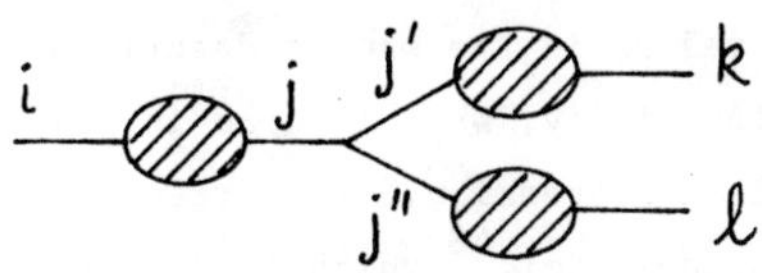

(b) Two-parton distribution

Fig. 3.1

ation processes. In leading-log approximation this is completely known. The bifurcation vertex $j\to j'+j''$ is described by $P_{j\to j'}(z/y)$ where z is the momentum fraction associated with parton j'. Its form is also well-known. Putting the various pieces together in accordance to Fig. 3.1(b), we have

$$F_{i\to k\ell}(x_1,x_2,Q^2)=\sum_{j,j',j''}\int_{Q_0^2}^{K^2}\frac{dk^2}{k^2}\frac{\alpha(k^2)}{2\pi}\int_{x_1+x_2}^{1}dy\,G_{i\to j}(y,Q^2,k^2)$$
$$\cdot\int_{x_1}^{y-x_2}\frac{dz}{y}P_{j\to j'}\left(\frac{z}{y}\right)\frac{x_1}{z}G_{j'\to k}\left(\frac{x_1}{z},k^2,Q_0^2\right)\frac{x_2}{y-z}G_{j''\to\ell}\left(\frac{x_2}{y-z},k^2,Q_0^2\right)\tag{3.2}$$

Q_0 is a parameter determined by structure function analysis[23] (Q_o=0.8GeV) and K^2 = min (Q^2, Q_1^2) where Q_1^2 is some cutoff taken for definiteness to be 30 GeV^2 because of finite range in k_T^2 for recombination; the final answer is insensitive to its precise value.[46] The summation in (3.2) extends over all parton types: quark, antiquark and gluon.

3.2 Gluon Conversion

Eqs. (3.1) and (3.2) contain all the necessary components to determine the fragmentation function, each component being completely calculable. But there is still a missing piece of physics. Consider a quark decaying into a π^+ plus anyting. It seems that we need only determine the $u\bar{d}$ distribution from (3.2). But then, what happens to all the gluons in the quark jet? They carry a sizable fraction of the jet momentum, and they must hadronize also. Now, our hadronization mechanism is recombination from valons to hadrons. Thus our procedure for hadronization of the gluons is, as it is for quarks, to convert first from the quark-gluon representation to the valon representation, whereupon the recombination mechanism can be applied. It means that when the Q^2 degradation in the jet reaches Q_o^2, we insert a complete set of valon states between the parton and hadron states:

$$\sum_{v_1,v_2\cdots}\langle qq\cdots\bar{q}\bar{q}\cdots ggg|v_1v_2\cdots\rangle\langle v_1v_2\cdots|\pi+X\rangle$$

We implement this by attaching a $g\to q\bar{q}$ conversion process to every gluon at Q_o^2 and then let all the resultant quarks and antiquarks dress themselves up to become valons in the manner described in Sec. 2.6.

Pictorially, what is done is illustrated in Fig. 3.2(a) showing explicitly gluon conversion. It corresponds in some sense to an implicit process in the cascade model,[48] which is depicted in Fig. 3.2(b) where the blobs are some unknown pion emission function to be determined by fitting data.[45] What is implied is that some gluons are converted to $q\bar{q}$ pairs as shown in Fig. 3.2(c), and that recombination rather than fragmentation is the more basic subprocess.

We use the lowest-order perturbation result for the conversion:

$$\bar{P}_{g\to q\bar{q}}(z) = \frac{3}{2f}\left[z^2 + (1-z)^2\right] \tag{3.3}$$

where f is the number of flavors. The normalization is adjusted to satisfy

$$\sum_{j=1}^{f} \int_0^1 dz\, z\, \bar{P}_{g\to q_j\bar{q}_j}(z) = \frac{1}{2} \tag{3.4}$$

so that on the average each quark carries half the gluon momentum. The precise form in (3.3) is not crucial since the gluon momentum is to be integrated over the whole range.

3.3 Pion Distribution in a Quark Jet

The computation of $D_q^{\pi}(x,Q^2)$ in accordance to (3.1) and (3.2) involves going to the moments and then inverting. It is tedious but straightforward.[46] The result can be presented with four components corresponding to the four parts of $F_{q\to k\ell}$:

$$F = F^{(0)} + F^{(a)} + F^{(b)} + F^{(c)} \tag{3.5}$$

Each component is shown in Fig. 3.3, the last three being contributions from gluon conversion. From the calculated results on $D_q^{\pi\pm}$, q = u,d, and s, we can obtain

$$N_{e^+e^-}^{\pi^++\pi^-} = (2\sigma_T)^{-1}\frac{d\sigma^{\pi^++\pi^-}}{dx} = \frac{5}{6}D_u^{\pi^+} + \frac{7}{6}D_u^{\pi^-} \tag{3.6}$$

For Q = 5 GeV, the result is shown as dashed line in Fig. 3.4 with the contributions from the different components of (3.5) exhibited in dash-dot lines. When resonances are taken into account,[12] the total inclusive distribution of

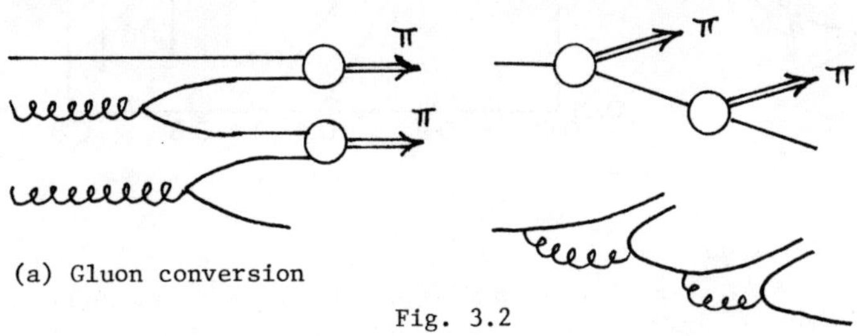

(a) Gluon conversion

(b) Cascade emission of pions

(c) Blobs of (b) exposed

Fig. 3.2

charge pions is increased to what is shown by the solid line. The agreement with data[49] is good in both normalization and shape. Note that the contribution from component (a) is important at large x and that all components are important at small x.

We stress that the result is essentially free of adjustable parameters. If we believe that the quark and gluon proliferation process in a jet is reliably specified by perturbative QCD calculations, then the good agreement with data indicates that the hadronization process is well described by the recombination model.

Hadron distribution in <u>gluon</u> jet can also be calculated,[46] but data with identified particles are not yet available for comparison.

3.4 Glueball Production

The U(1) problem (η' being too massive to be the ninth pseudoscalar Goldstone boson) is resolved by associating the mass of η' to a strong coupling of η' to two gluons.[50] That is, $\langle 0|\frac{3\alpha_s}{4\pi}G^a_{\mu\nu}\tilde{G}^{\mu\nu}_a|\eta'\rangle$ is large where α_s is the QCD

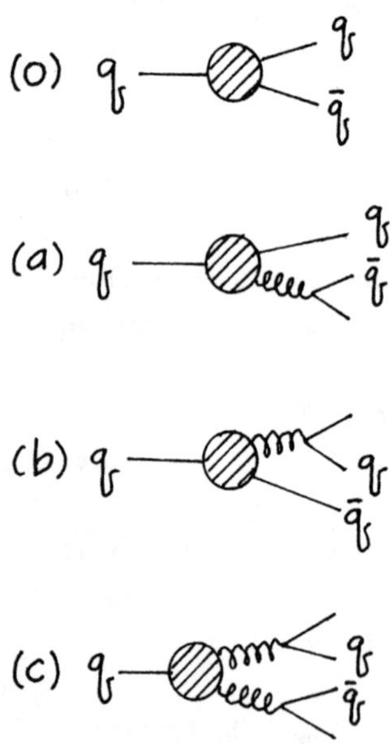

Fig. 3.3
$F_{i\to k\ell}(x_1,x_2,Q^2)$

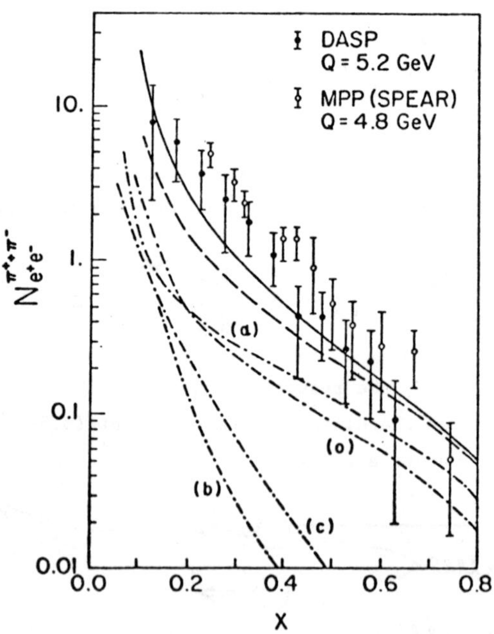

Fig. 3.4 Pion distribution in e^+e^- annihilation

coupling constant, $G^a_{\mu\nu}$ is the gluon field tensor and $\tilde{G}^{\mu\nu}_a = \frac{1}{2}\epsilon^{\mu\nu\alpha\beta}G^a_{\alpha\beta}$. In the following we shall identify η' as a bound state of two glue-valons, and regard it as a glueball.

While the meson distribution in a jet reveals the quark (and antiquark) content in the jet, the glueball distribution should then reveal the glue content. Furthermore, if triple-gluon coupling is significant, the glue content in a gluon-initiated jet would be different from a quark-initiated jet. The detection of glueballs should then provide a way of differentiating gluon jet from quark jet. With these possibilities in mind it is of interest to calculate the fragmentation functions for η' production using the method already described in this section.[51]

The recombination function of two gluons hadronizing into η' requires the knowledge of the valon distribution for η'. Since that is unavailable, we proceed in the following way. We shall provisionally use (2.16) as the recombination function with the understanding that b = 0 in (2.15) is not necessarily reliable. To minimize the error made in connection with this uncertainty, we calculate the ratio of fragmentation functions in gluon and quark jets both referring to η' production. The ratio should be less sensitive to b than would be the individual D functions, which, according to the discussion in Sec. 2.6, are themselves likely to be rather insensitive to b.

The calculation of $D^{\eta'}_g$ and $D^{\eta'}_q$ follows closely the procedure of previous subsections. There is, however, one tacit assumption made, namely: the final-state hadrons do not compete in the hadronization of quarks and gluons. That is, it is assumed that there are enough open channels in the final state so the formation of η' is independent of that of π. Obviously, the assumption is made mainly for the sake of simplicity.

The result[51] is shown in Fig. 3.5 for (a) Q^2 = 25 GeV^2 and (b) Q^2 = 1000 GeV^2. Clearly, it is more likely to find η' in a gluon than a quark jet, as compared to the production of π^o. This property is made more explicit in Fig. 3.5 (c) where we show the ratio of ratios. The figure infers that the η'/π^o ratio is greater in a gluon jet than in a quark jet. That it is always greater than 1 is a consequence of the existence of triple-gluon coupling. Hence, the η'/π^o ratio is a good signature for the type of parton that initiates the jet. Measuring it around Υ should show a rise on resonance.

Extension of the ideas here to soft processes obviously suggests that the measurement of η' at low p_T in the fragmentation region should reveal the gluon content of the incident hadron. Such measurements are therefore strongly urged.

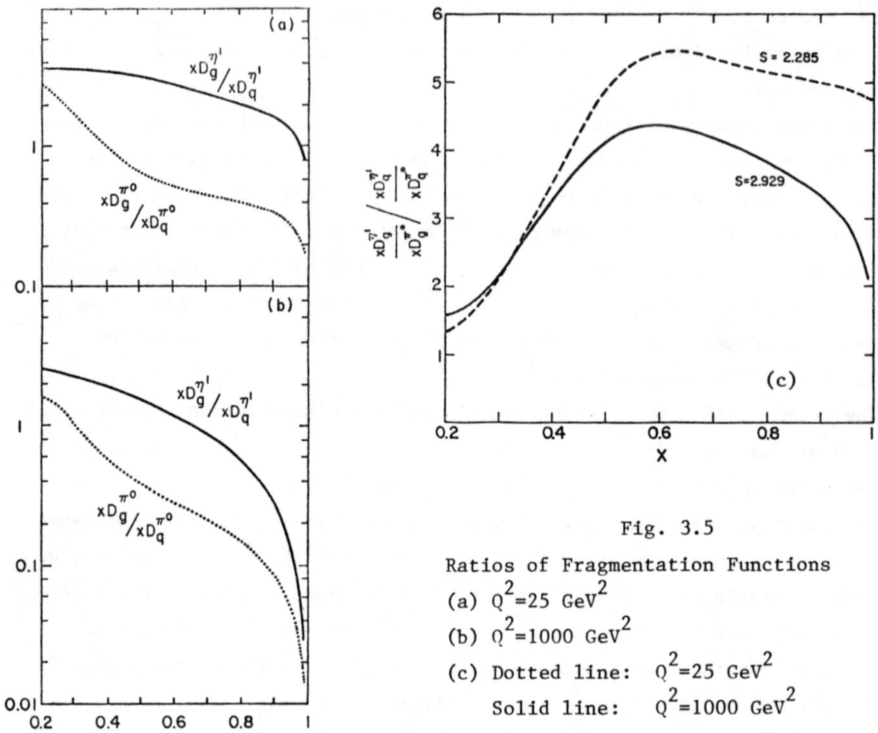

Fig. 3.5

Ratios of Fragmentation Functions

(a) Q^2=25 GeV^2

(b) Q^2=1000 GeV^2

(c) Dotted line: Q^2=25 GeV^2

Solid line: Q^2=1000 GeV^2

4. LOW p_T INCLUSIVE REACTIONS

The conventional wisdom about low-p_T reactions is that particles produced in the projectile fragmentation region is independent of the target. While we shall examine the degree of that independence quantitatively in Sec. 7, let us adopt that view as a working hypothesis (which will be shown to be extremely good). Then, as we have already motivated in Sec. 1, the parton distributions in an incident hadron are what characterize the inclusive distributions of the produced hadrons. In Sec. 2 we have considered the hadron structure in terms of valons. Toward the end of that section and in Sec. 3 the hadronization of partons is discussed in the recombination model. What remains to be specified for the completion of the description of soft processes in the fragmentation region is the link between the valons of the incident hadron and the quarks (and antiquarks) that recombine. The study of that problem and its application to various reactions at low p_T is the subject of this section.

4.1 Cluster Distribution

In collision processes where a valon is to be regarded as a cluster of partons, the momentum distribution of the partons will be called cluster distribution $P_i^v(z)$, where z is the fraction of valon momentum. The convolution of $P_i^v(z)$ with the valon distribution $G_v^h(y)$ gives the parton distribution in a hadron

$$q_i^h(x) = \sum_v \int_x^1 \frac{dy}{y} G_v^h(y) P_i^v\left(\frac{x}{y}\right) \tag{4.1}$$

as shown in Fig. 2.1. During collision the fast and slow (wee) partons interact differently with the opposite-going partons of the other hadron because the interaction is short-ranged in rapidity. Thus the x distribution of the partons, in principle, does not remain invariant in time during the course of collision, especially in nucleus. But for hadron-hadron collision, we may regard the x distribution as unaltered (according to our working hypothesis) except for wee partons at or near x=0. Note that if $P_i^v(x)$ changes significantly, as would be expected in a hadron-nucleus collision, it would be inappropriate to regard a valon as a constituent quark since a change of momentum profile would suggest the cluster property of a valon rather than the indivisible nature of a (constituent) quark.

It is important to point out here the conceptual difference between our approach and one that emphasized color separation. We regard the picture of a hadron being a collection of partons as being meaningful as soon as the confinement condition imposed on a hadron bag is broken by collision. The projectile partons then separately and independently traverse the target hadron and recombine downstream. This is different from a picture in which a proton has three "quarks", one of which (a color 3) is separated from a "diquark" (a color $\bar{3}$) of the other incident proton, forming a chain of $q\bar{q}$'s due to color separation.[52]

Now, let us focus on the cluster quark distribution $P_q^v(z)$. It is not calculable by present methods in QCD since it is defined at a low Q^2 relevant to low-p_T reactions. It may be treated as the boundary value for evolution to higher values of Q^2. However, that part of the evolution is the easier part since it is scaling violation of order log Q^2. The major part of evolution occurs at low Q^2 as evidenced by the fact that even in early SLAC experiments ($Q^2 \sim$ 1-3 GeV^2) nearly half the nucleon momentum already resides in the neutral gluon sector. Since this part of the evolution is theoretically untractable, the determination of $P_q^v(z)$ is per force phenomenological. Our formulation of the low-p_T physics is such that a knowledge of $P_q^v(z)$ would facilitate the prediction for a large variety of soft processes. Thus, in principle one can use one or two standard reactions, such as $p + p \to \pi^{\pm} + X$, to determine phenomenologically $P_q^v(z)$. Using that result, one can then predict the distribuitons for other reactions, such as $\pi^{\mp} + p \to \pi^{\pm} + X$, recognizing that a change of the incident hadron affects only the valon distribution $G_v^h(y)$ which is known. That is, $P_q^v(z)$ is the same for all valons due to the universality of QCD effects independent of valon flavor or hadron structure. This is an approach that has not yet been undertaken by anyone but should surely be done. It justifies doing experiments at low-p_T.[98]

Another approach which was used in Ref. 23 is more direct, but involves an assumption: the quark distribution $q_i^h(x)$ is to be identified with that probed in electroproduction at low Q^2. Then by virtue of (4.1), $P_i^v(z)$ can be extracted from electroproduction data. The advantage of this approach is that, as far as low-p_T reactions are concerned, absolute prediction of inclusive distributions is possible without adjustable parameters. The approach should however, be regarded as provisional, since the validity of the above assumption is not quantitatively guaranteed, although Ochs' observation[44] is qualitatively suggestive. This approach will be described in the following.

There are two parts to P_q^v. The favored part K(z) is the probability of finding a quark at z having the same flavor as the valon. The unfavored part is L(z) which is the distribution of a quark or antiquark having a different flavor from the parent valon. We may write

$$K = K_{NS} + L \tag{4.2}$$

where K_{NS} refers to the so-called "valence" quark and L the "sea" quark. We use the invariant phase space for the normalization of K and L. Thus, with the appropriate charges of the quarks taken into account, the structure function $\nu W_2(x)$ and sea quark distribution $xS(x)$ of the nucleon can be obtained, with the help of (4.1),

$$\nu W_2^N(x) = \int_x^1 dy\, G^N(y)\left[K_{NS}\left(\frac{x}{y}\right) + 4L\left(\frac{x}{y}\right)\right] \tag{4.3}$$

$$x S^N(x) = 3\int_x^1 dy\, G^N(y)\, L\left(\frac{x}{y}\right) \tag{4.4}$$

We have for simplicity ignored flavor dependence of $G^N(y)$. A more careful analysis using (2.10) instead of (2.6) can be done, but the improvement on K and L would not be significant.

Now, to determine K_{NS} and L we need phenomenological input for νW_2 and S. For the former we use the early but classic SLAC data at low Q^2 parametrized by[53]

$$\nu W_2(x) = (1-x)^3 \left[1.274 + 0.599(1-x) - 1.675(1-x)^2\right]. \tag{4.5}$$

For the latter there is a lot more uncertainty. In Ref. 23 we use the formula

$$x S(x) = 0.17(1-x)^9 \tag{4.6}$$

which is obtained from a variety of parametrizations based partly on dimuon production data[54] and partly on guesses.[55] In retrospect we believe that the exponent in (4.6) is too high; some value near 6 or 7 would be more appropriate, and would lead to a result with better agreement with data. Nevertheless, we proceed with (4.5) and (4.6) in our review here.

To extract $K_{NS}(z)$ and L(z) from (4.3)-(4.6) we adopt some simple forms involving the least number of parameters. For K_{NS} we use a one-parameter form for its moments

$$K_{NS}(n) = \int_0^1 dz\, z^{n-2} K_{NS}(z) \tag{4.7}$$

$$K_{NS}(n) = \exp(-d_n^{NS}\zeta) \tag{4.8}$$

There is nothing deep and significant with (4.8) except that when ζ is large, it is a valid QCD form with the anomalous dimensions d_n^{NS} completely known. Here we use it merely as a convenient parametrization and let ζ take whatever value the fitting of νW_2 demands. As for L(z) we use the canonical form

$$L(z) = a(1-z)^c \tag{4.9}$$

Thus we have three free parameters to fit $\nu W_2(x)$ and S(x) and obtain[23]

$$\zeta = 2, \qquad a = 0.08, \qquad c = 3.5 \tag{4.10}$$

The value ζ=2 turns out to correspond to $Q \sim 1.6$ GeV, a rather reasonable number for the values of Q at which the data were taken. As a log log function $\zeta = 2$ is actually a rather large number for an evolution parameter. It signifies that even at $Q \sim 1.6$ GeV the evolution has been substantial, a circumstance which is not surprising from the point of view of precocious scaling and large momentum fraction in the gluon sector.

If we invert the moments $K_{NS}(n)$, we get to a good approximation the z distribution

$$K_{NS}(z) = 1.2\, z^{1.1} (1-z)^{0.16} \tag{4.11}$$

This can be used for direct application to the calculation of inclusive spectra.

4.2 $\underline{F(x_1,x_2)}$

Having determined the cluster distributions K and L, computation of the two-parton distribution $F(x_1,x_2)$ should be straightforward. The two partons can belong to the same valon or to two separate ones, as shown in Fig. 4.1 for the case of a nucleon. In either case there is at least one spectator valon. This feature has some similarity to the DTU approach[52] where a "diquark" of one nucleon stretches a flux tube with a held-back (spectator) quark in the other hadron. The "diquark" is assumed to fragment and contributes to secondary hadrons in the fragmentation region. It is in this parallelism that one might find a germ of hope for the unification of recombination and fragmentation models. However, we investigate more thoroughly the problem of hadronization, while in the DTU approach the D function for fragmentation is adopted from jet data on the assumption that the $\bar{3}$ for a "diquark" behaves the same way as a $\bar{3}$ for an antiquark.

Prior to a description of $F(x_1,x_2)$ it is necessary first to take care of the problem of gluon conversion, as has been done in Sec. 3.2, before recombination can be applied. It means that the sea must be saturated[8] because the conversion from gluons to $q\bar{q}$ raises the normalization of L(z) until the momentum of a valon is completely shared by the valence and sea quarks. We find

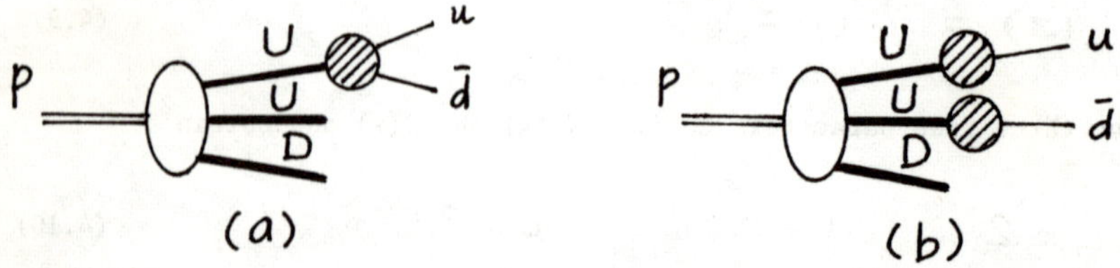

Fig. 4.1 Two partons from (a) the same valon, (b) different valons

that[23]

$$L(z) \rightarrow \tilde{L}(z) = \tilde{a}\,(1-z)^{c} \tag{4.12}$$

where $\tilde{a} = 0.41$. It is an assumption that the exponent c is unaltered by gluon conversion, which is reasonable. Since $K_{NS}(z)$ is unaffected, we also have $K(z) \rightarrow \tilde{K}(z) = K_{NS}(z) + \tilde{L}(z)$.

Now, for the case of $q\bar{q}$ originating from the same valon [Fig. 4.1(a)], if there is one favored quark, the saturated two-parton cluster distribution is

$$\tilde{P}^{v_i}_{q_i\bar{q}_k}(z_1, z_2) = \frac{1}{2}\left[\tilde{K}(z_1)\tilde{L}\left(\frac{z_2}{1-z_1}\right) + \tilde{K}\left(\frac{z_1}{1-z_2}\right)\tilde{L}(z_2)\right] \tag{4.13}$$

while if neither partons are favored, one has

$$\tilde{P}^{v_i}_{q_j\bar{q}_k}(z_1, z_2) = \frac{1}{2}\left[\tilde{L}(z_1)\tilde{L}\left(\frac{z_2}{1-z_1}\right) + \tilde{L}\left(\frac{z_1}{1-z_2}\right)\tilde{L}(z_2)\right] = \tilde{a}\tilde{L}(z_1+z_2) \tag{4.14}$$

The second equality follows from the form of (4.12). Thus, for $p \rightarrow \pi^{\pm}$, the $u\bar{d}$ and $d\bar{u}$ distributions are

$$F_{\pi^{\pm}}(x_1, x_2) = F^{(1)}_{\left\{u\bar{d} \atop d\bar{u}\right\}}(x_1, x_2) + F^{(2)}_{\left\{u\bar{d} \atop d\bar{u}\right\}}(x_1, x_2) \tag{4.15}$$

$$F^{(1)}_{\left\{u\bar{d} \atop d\bar{u}\right\}}(x_1, x_2) = \int_{x_1+x_2}^{1} dy \left[2G^p_U(y)\tilde{P}^U_{\left\{u\bar{d} \atop d\bar{u}\right\}}\left(\frac{x_1}{y}, \frac{x_2}{y}\right) + G^p_D(y)\tilde{P}^D_{\left\{u\bar{d} \atop d\bar{u}\right\}}\left(\frac{x_1}{y}, \frac{x_2}{y}\right)\right]$$

$$F^{(2)}_{\left\{u\bar{d} \atop d\bar{u}\right\}}(x_1, x_2) = \int_{x_1}^{1} dy_1 \int_{x_2}^{1-y_1} dy_2\, G^p_{UU}(y_1, y_2)\, 2\left\{\tilde{K}(x_1/y_1) \atop \tilde{L}(x_1/y_1)\right\} \tilde{L}\left(\frac{x_2}{y_2}\right)$$

$$+ \int_{x_1}^{1} dy_1 \int_{x_2}^{1-y_1} dy_2\, 2G^p_{UD}(y_1, y_2)\left[\left\{\tilde{K}(x_1/y_1)\tilde{L}(x_2/y_2) \atop \tilde{L}(x_1/y_1)\tilde{K}(x_2/y_2)\right\} + \tilde{L}\left(\frac{x_1}{y_1}\right)\tilde{L}\left(\frac{x_2}{y_2}\right)\right].$$

Clearly, $F^{(1)}$ and $F^{(2)}$ correspond to diagrams of the types illustrated in Fig. 4.1(a) and (b), respectively.

Note that as $x_1+x_2 \rightarrow 1$, we have $y \rightarrow 1$ in $F^{(1)}$ and $y_1+y_2 \rightarrow 1$ in $F^{(2)}$. In both cases the integrals vanish because $G(y) \sim (1-y)^2$ and $G(y_1, y_2) \sim (1-y_1-y_2)^{\frac{1}{2}}$. Hence, $F(x_1, x_2)$ vanishes at the kinematical boundary without the ad hoc phase-space factor $\rho(x_1, x_2)$ in (1.2). This is the primary reason why the valon model provides a more proper formulation of the low p_T problem.

4.3 Inclusive Hadron Distribution

Having obtained the two-parton distribution $F(x_1,x_2)$ in (4.13)-(4.15), we are ready to calculate the inclusive hadronic distribution in the recombination model using (1.1). It is a straightforward multiple integral to be evaluated since all terms in the integrand are already specified in previous subsections with R being given by (2.16). Note that there is no color factor in any of the equations even though one would at first thought conclude that the formation of a color singlet from two triplets should have such a factor. But we cannot keep track of color for $F(x_1,x_2)$ because it is always possible for the valons to emit a very soft gluon to change the color structure at x_1 and x_2 without changing the joint momentum distribution of q and $\bar{q}$. Thus both $F(x_1,x_2)$ and $R(x_1,x_2,x)$ are color averaged, and the question of color factor should not enter into the problem.

In Ref. 23 the flavor dependence of G is ignored so (2.6) is used instead of (2.10). The moment method adopted there is merely a way to reduce the number of integrals, but is not a necessary procedure. The predicted result for $p\to\pi^+$ can be well approximated by the analytic formula

$$\frac{x}{\sigma_T}\frac{d\sigma}{dx} = 0.6\times(1-x)^{3.5} + 0.9\,(1-x)^5 \tag{4.16}$$

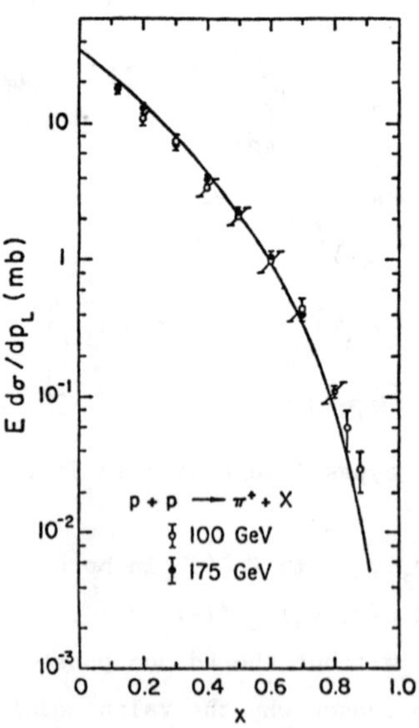

Fig. 4.2 Inclusive Cross Section

Comparison with data[56] on inclusive cross section with σ_T taken to be 38.5 mb is shown in Fig. 4.2. The agreement is remarkably good over nearly three orders of magnitude with no adjustable parameter.

The striking agreement with data gives us the confidence to raise questions now on secondary effects in order to fine tune the result. First of all, the flavor-dependent G functions in (2.10) should be used and the π^+/π^- ratio in the proton fragmentation region should be calculated and compared with data. Secondly, if one looks closely at Fig. 4.2, the predicted curve falls slightly more steeply than the data. Could that be due to our assumption about $\nu W_2(x)$ and $xS(x)$ from which K(z) and L(z) are determined? A more slowly falling function for xS(x) than that given in (4.6) would certainly decrease the value of c in (4.9) and (4.10) and result in an even better fit of the data. This observation naturally leads to the question of whether the cluster distributions $\tilde{K}$ and $\tilde{L}$ should not be determined directly from the $p + p \to \pi^{\pm} + X$ data, since the reliance on $\nu W_2(x)$ and $xS(x)$ is after all an assumption. This approach has already been discussed in Sec. 4.1, and deserves serious consideration.

In Ref. 23 it is found that of the two components in (4.15) the one with larger magnitude throughout most part of the x range is $F^{(2)}$, which corresponds to Fig. 4.1(b). If we express its contribution to $(x/\sigma)d\sigma/dx$ by

$$H^{(2)}(x) = \sum_{ij} \int dy_1 dy_2 \, G^p_{v_i v_j}(y_i, y_j) \, J(y_i, y_j, x) \tag{4.17}$$

where J is a convolution of $\tilde{K}$, $\tilde{L}$, and R, then (4.17) can be given the interpretation of divalon fragmentation. It is to be compared to the corresponding integral in the fragmentation model[52]

$$\int dy_3 \, \text{"}G(y_3)\text{"} (1-y_3) D_{\text{"diquark"} \to \pi}(1-y_3, x) \tag{4.18}$$

where "$G(y_3)$" is the distribution of the "held-back quark". If it refers to our third spectator valon, the corresponding distribution would be $y_3^{\frac{1}{2}}(1-y_3)^2$. Thus it cannot be so since it would offer no held-back effect without divergence as $y_3 \to 0$. It is not even the valence quark with the canonical distribution $y_3^{-\frac{1}{2}}(1-y_3)^3$ because the large y_3 behavior is not what is desired.[52] The requirements on it from Regge theory make it into a mathematical object, called boundary-line "quark," with a momentum distribution quite unrecognizable in the parton model.

4.4 Resonances

So far in this section we have treated pion production without considering resonances. Of course, pions can arise from the decay of resonances. In the simple recombination model[3] the effect of resonance production and subsequent decay on the pion distribution has been studied,[12] and it is found that the change on the x distribution is not more than 25% for $x > 0.5$ assuming certain ρ/π ratio in the hadronization rate. But for $x < 0.5$ the effect is large so a prediction of the pion spectrum is not realistic without the resonances taken into account seriously. Since the dynamics of resonance versus "stable" meson production is not yet well understood, one could take the view that resonances are intermediate states in the hadronization from partons to stable mesons. In that case one would consider either the resonances or the stable mesons, but not both to avoid double counting. This view is needed to justify the acceptance of the good agreement between theory and experiment exhibited in Fig. 4.2.

One approach that is rather safe from the uncertainties mentioned above is to study the resonance spectra themselves. It is reasonable to assume that ρ, f, g, h particles are not decay products of other higher resonances. Inclusive distributions of those resonances should then be a reliable indication of the initial hadron structure without the complication of cascading. The drawbacks are, however, numerous. Firstly, the quality of data is poor; the statistics is low with limited ranges in x and p_T. Secondly, theoretical understanding of the mass and spin effects on the recombination function is crude, so any prediction of the resonance distribution can only be tentative. Despite these shortcomings some progress has been made in studying resonance production.[57]

It is first confirmed by computation that the x distribution is very insensitive to the parameter b in the recombination function (2.15) as we have argued in Sec. 2.6. Thus by relinguishing the possibility of checking the normalization of the inclusive cross sections due to uncertainties in the recombination functions, the shapes of the x distributions can still be examined in the recombination framework. An impressive empirical fact is that those shapes are universal at least for x up to 0.4 where resonance data exist.[58] This is shown in Fig. 4.3 for p → ρ, f, g, h as well as $\pi^\pm$ for comparison. The implication of this universality is that resonance productions (whatever their recombination functions may be) reveal directly the same parton

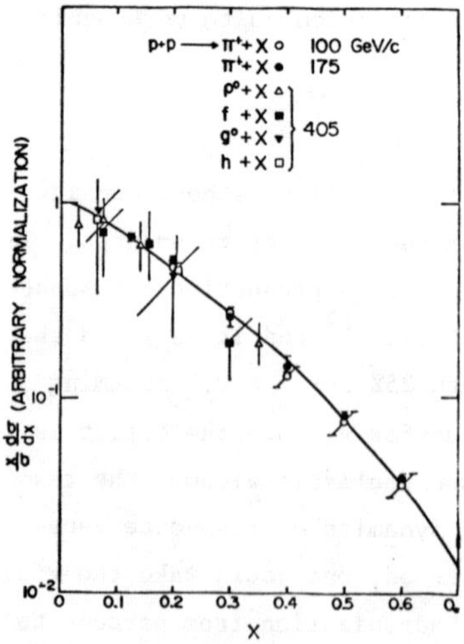

Fig. 4.3 Inclusive distribution for $\pi^\pm$ and resonance production.[58] The normalizations are arbitrarily adjusted to agree with one another. The solid line shows the universal theoretical result obtained in the recombination model.[57] The universality is not expected to include such resonances as K^{*-}.

distribution in the incident proton. It is another indication of the validity of the recombination model. It also supports the thesis that soft hadronic processes provide a direct access to hadron structure.

From the universal x distribution at low x (<0.4) it is possible to determine the shape of $\tilde{L}(z)$, assuming $K_{NS}(z)$ to be given by (4.11). Xie[57] found the value 1.5 for c in (4.12); $\tilde{a}$ can be fixed as before. This lower value of c was anticipated in Sec. 4.1 because the exponent of (1-x) in (4.6) is high compared to the usual parametrization of the sea distribution. This is an example of an investigation where low-p_T data are used to determine the characteristics of hadron structure.

Proceeding from the new value of c, Xie calculated π^+/π^- ratio in the proton fragmentation region for all x. Although the agreement with data is acceptable, the theoretical result is weakened by the fact that the flavor dependence of the valon distributions in (2.10) is not taken into account, thus neglecting a feature of the proton that is important for the π^+/π^- ratio.

Data on K^{*+}/ρ^+ would be extremely useful from which to learn about the strange sea-quark distribution in the nucleon. Based on the fact that the strange sea-quark mass is not much greater than that of non-strange quark (unlike the case of strange vs. non-strange valons), one would not expect a significant difference in their x distributions. It then would follow that the K^{*+}/ρ^+ ratio should be approximately constant in x. This expectation is generally borne out by recent experiments.[59] The observed x dependence of K^+/π^+ ratio is then due to the asymmetric decay of $K^{*+} \to K^+\pi^o$ favoring K^+ at higher momentum. Detailed analysis of this type would be very fruitful when more high quality data at high energy become available.

4.5 Proton, Antiproton, Photon and Strange Particles

We have already commented on flavor dependence immediately above. To make the discussion systematic, we should distinguish two types of flavor dependence. One is for fixed beam particle; the behaviors of different types of detected particles are investigated. The other is to study the dependence on beam particles for the same detected particle. The former probes the question of flavor dependence of the sea in the incident hadron, while the latter reflects the flavor dependence of the valon distributions. They are different because the valons have constituent quark masses which vary significantly between strange and non-strange valons, but current quarks have much lower masses, usually neglected (for u, d, s) in the parton model.

Duke and Taylor[8] have investigated the π^+/π^-, π^+/K^+, K^+/K^-, and π^-/K^- ratios in the proton fragmentation region in the framework of the simple recombination model.[3] The same should be done with the implementation of the valon model. However, as noted in the previous subsection, the contamination of the pseudoscalar mesons due to vector meson decay may render the derived sea-quark distribution unreliable. Replacing those ratios by their vector equivalents, e.g. ρ^+/ρ^-, ρ^+/K^{*+}, etc. would be far better for analysis, but data are not as accurate. The issue is whether the unfavored cluster distribution $\tilde{L}(z)$ is dependent on strangeness. This question cannot yet be accurately settled. As mentioned at the end of Sec. 4.4 there are hints to suggest independence.

An interesting problem in the category of ratio of detected particles is the $p/\bar{p}$ ratio in the π^+ fragmentation region. The data show that this ratio is significantly greater than 1. This is not a charge effect since for both p and $\bar{p}$ two sea quarks need be picked up. Migneron and Robinson[14] attributed the large ratio to some peculiar sea quark distributions for the two cases. However, in the valon model the excess of proton over antiproton can be traced easily to the presence of extra diagrams for the proton. This is illustrated in Fig. 4.4 where in (a) the interchange of the two unfavored quarks results in a new diagram, while in (b) the interchange of the two $\bar{u}$'s does not. If all three quarks (antiquarks) originate from the same valon (antivalon), then statistics applies equally to both and there would not be any extra terms for p. The sum of all diagrams clearly yields a $p/\bar{p}$ ratio that is greater than one, and is x dependent. This problem is worthy of a careful calculation in the valon model with no free parameters needed. The inclusive cross sections for the production of p, $\bar{p}$, and $\bar{\Lambda}^\circ$ from p and $\pi^\pm$ beams have actually been calculated already by Takasugi and Tata[60] in the context of Kuti-Weisskopf model[21] as supplement to the recombination model. With no adjustable parameters the results agree well with data. The physical content for the large $p/\bar{p}$ ratio is similar to the mechanism described above in the valon model.

A very interesting consequence of the recombination model that has recently been discovered by DeGrand and Miettinen[61] is the polarization of the produced hyperon.[62] They suggest that the polarization arises via Thomas preces-

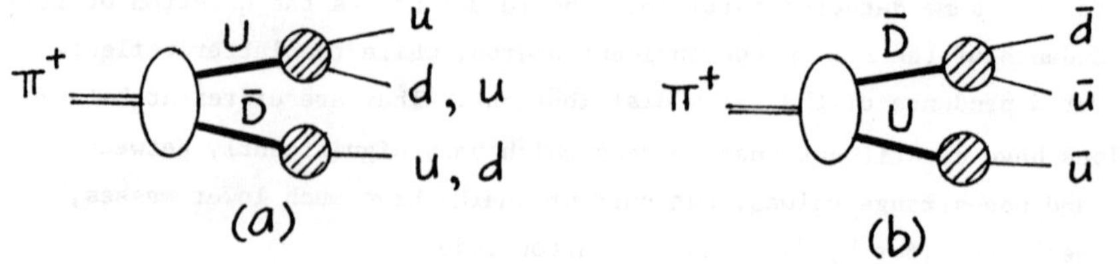

Fig. 4.4 Valon-parton diagrams for (a)$\pi^+\to p$, (b)$\pi^+\to\bar{p}$

sion of the quarks' spins in the recombination process. They have shown that the mechanism accounts for all of the qualitative features of the baryon and antibaryon polarization data.

Turning to flavor dependence of the incident particles, the only non-leptonic beams we have at high energies are photon, proton, pion and kaon. Photon is included here not merely because it can provide vector-dominated hadrons, but also because it can have point-like interaction with $q\bar{q}$ pairs. In some region the latter dominates and one can calculate the photon structure function without hadronic complication.[63] Inclusive pion distribution at large x in the photon fragmentation region can then be determined in the recombination model.[64] The result compares favorably with the very meager data available, an experimental status that should certainly be improved upon. The aim of that problem is to avoid the unknowns about the incident particle insofar as possible, but to learn about hadrons we must focus on hadronic beams.

We have discussed the valon distribution of a pion in Secs. 2.3 and 2.4 which has been used as basis for the pion recombination function. It is symmetrical in y_1 and y_2 because a pion has two equal-mass valons. For a kaon the strange and non-strange valon masses are significantly different so the valon momentum distribution is asymmetric:

$$G_U^K(y) = [B(a'+1, b'+1)]^{-1} y^{a'} (1-y)^{b'} \tag{4.19a}$$

$$G_S^K(y) = [B(a'+1, b'+1)]^{-1} y^{b'} (1-y)^{a'} \tag{4.19b}$$

where a' and b' are constrained by the requirement that the average momentum of a valon is proportional to its mass[23]

$$\frac{a'+1}{b'+1} = \frac{m_U}{m_S} \simeq \frac{2}{3} \tag{4.20}$$

Although kaon-initiated reactions have not yet been analyzed in the valon model, (4.19) and (4.20) have been used to constrain the recombination function for kaons.[23] It is then applied by Aitkenhead et al.[18] to the analysis of a complete set of reactions involving $\pi^{\pm}$ and $K^{\pm}$ for both beam and detected particles using the simple recombination model.[3] The results are $b+1 = 1.0 \pm 0.1$ for pion [see (2.4) for definition of b] and $b'+1 = 2.5 \pm 0.6$ for kaon. While the former agrees with other findings of the pion structure function, the latter seems a bit high compared to the results of other experiments.[65] A definitive determination of b' must await an accurate measurement of $K^+ \to \pi^+$ in the large x region and a detailed phenomenological study of the data using the valon-recombination model.

4.6 Two-Particle Correlation

While recombination and other models all claim success in explaining single-particle inclusive cross sections, not all models would have the same predictions for two-particle correlations. Indeed, correlation data give stringent tests on models. That is another reason why soft hadronic experiments should be pursued vigorously.

An outstanding example of what one can learn from two-particle inclusive distributions in the proton fragmentation region is the observation[66] of $\pi^+\pi^-/\pi^-\pi^-$ ratio rising sharply with increasing x similar to the K^+/K^- ratio in support of the recombination model. Another discriminating experiment is the measurement by ACCMOR collaboration of $(K^- \to \pi^-\pi^-)/(\pi^- \to \pi^-\pi^-)$. The decrease of this ratio with increasing x (sum of momentum fractions of the two pions) can be readily understood in the recombination model, but in the fragmentation model one would naively predict a constant ratio of roughly 1/2.[65,67]

The most detailed theoretical work done to date on particle correlations is by Takasugi and Tata[68] in the framework of recombination model supplemented by modified Kuti-Weisskopf quark distributions. They use single-particle distributions as input to fix the quark distributions of various beam hadrons. Then without further adjustable parameters they predict numerous two-particle distributions all of which are in good agreement with data[69] in absolute normalizations as well as in x dependences. The results are so impressive that they enable the recombination model to pass a definitive test to which other models have not yet been submitted.

It remains to be seen whether the valon model can provide the multiparton distribution in such a way as to give good fits to the correlation data also. There is no arbitrariness in the model; it is only a matter of computation which awaits being carried out.

5. HADRON FRAGMENTATION IN HARD PROCESSES

Recent experiments in deep inelastic scattering have progressed to semi-inclusive measurement of distributions of produced hadrons not just in the current fragmentation region, but also in the target fragmentation region. They are therefore correlation experiments relating hard and soft processes, and have long been suggested as a means of testing hadronization models.[9] Lepton-pair production can also be used as trigger for the soft production process. These are all reactions that can be treated by the recombination model in conjunction with perturbative QCD. We describe briefly here some of the problems that are anticipated.

5.1 Punctured Hadron

First of all, I would like to dispel a misnomer. For $x_{bj} = Q^2/2M\nu > 0.3$, it is reasonable to identify the struck quark as being the valence quark in the target proton. The residual system has commonly been called "diquark", which results from a confusion between current and constituent quarks, as mentioned in Sec. 2.1. A proton being composed of three valons, it is not even appropriate to call the residual system a divalon since only a current quark is knocked out. Recall that the latter is only a part of a valon. A more suitable (physical as well as figurative) name for the system is "punctured hadron" which implies that it retains all the complications of the initial hadron, if not more. Inclusive hadronization of a punctured proton is illustrated by the schematic diagram in Fig. 5.1. All three valons contribute to the production of secondary hadrons in the target fragmentation region. The valon model combined with recombination mechanism is ideally suited to describe this process.

There are two x variables in the problem, one being x_{bj} already defined above, the other being the longitudinal fraction of the initial hadron momentum. Momentum conservation puts obvious constraint on them. The flavor dependence is also easy to see, especially if the hard process is a Drell-Yan reaction,[70] e.g., $\pi^+ + p \rightarrow (\mu^+\mu^-) + h + X$. At high x_F of the dilepton the d valence quark in p is annihilated by the $\bar{d}$ in π^+; the detected h if it is a pion should then have a large π^+/π^- ratio at large x. Qualitative observations such as this can be made on the basis of simple recombination model,[9] but quantitative prediction with Q^2 dependence taken into account must rely on perturbative QCD calculations.

5.2 Evolution Up and Down

Consider the usual deep inelastic scattering process in perturbative QCD. Leaving out the hadronic complication that can be factorized,[71] (the same com-

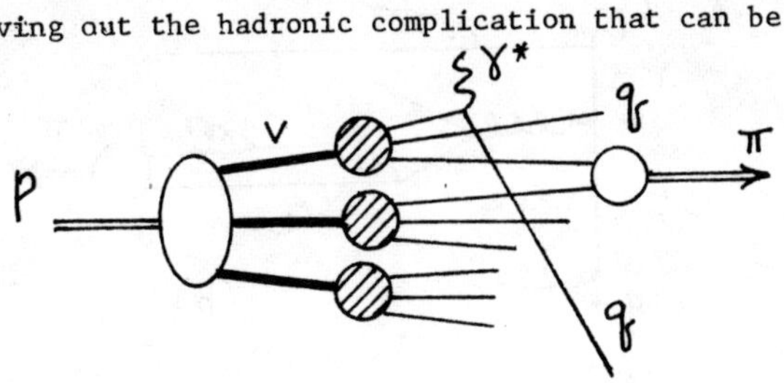

Fig. 5.1
Valon-parton diagram for hadronization of punctured proton

plication that is represented by the valon distribution), the diagram for γ^*-quark scattering is shown in Fig. 5.2(a), where gluon emission before and after the interaction gives rise to scaling violation. The nested integral is of the form

$$\int^{Q^2} \frac{dk_1^2}{k_1^2} \frac{\alpha(k_1^2)}{2\pi} \int^{k_1^2} \frac{dk_2^2}{k_2^2} \frac{\alpha(k_2^2)}{2\pi} \cdots$$

where $Q^2 \gtrsim k_1^2 \gtrsim k_2^2 \ldots$. Evolution is up as one goes up the ladder in the sense that the gluons on the top rungs are most far off their mass shells. Quark pairs that are created by those top-rung gluons cannot hadronize immediately. In order to recombine the quarks must evolve down (or degrade) as in quark and gluon jets discussed in Sec. 3. This is depicted in Fig. 5.2(b).

In general, quarks that recombine may come from

(A) a high off-shell rung in the struck valon and then degrade to Q_o^2,

(B) a low off-shell rung in the struck valon already near Q_o^2, or

(C) a different valon insensitive to Q^2.

We can represent the various processes leading to recombination by using the following notations: x_{bj} labels the quark line struck by virtual vector boson, x_1 and x_2 denote either quark or antiquark that eventually recombine, ⊘ signifies cluster distribution of partons (q, $\bar{q}$ or g) in a valon, and ⇡ (⇣) implies evolution function up (down) in off-shell-ness. We then have six generic types of processes in giving rise to a $q\bar{q}$ pair for recombination, as shown in Fig. 5.3, with obvious labelling in terms of combinations of (A), (B), (C). The last two diagrams belong to the same type CC which is what one would normally expect in a naive quark picture. They are particularly relevant if x_{bj} is large and $x/(1-x_{bj})$ is also large, where $x = x_1+x_2$ is the pion momentum fraction. Since the q and $\bar{q}$ in the CC-type diagrams are from valons that are not struck by the high-Q^2 probe, their contribution to the fragmentation of the puncture nucleon is independent of Q^2. The diagrams involving evolution up or down are all in principle Q^2 dependent. Interesting features of the fragmentation process reside in both the soft (BB-CC) and hard parts (AA-AC).

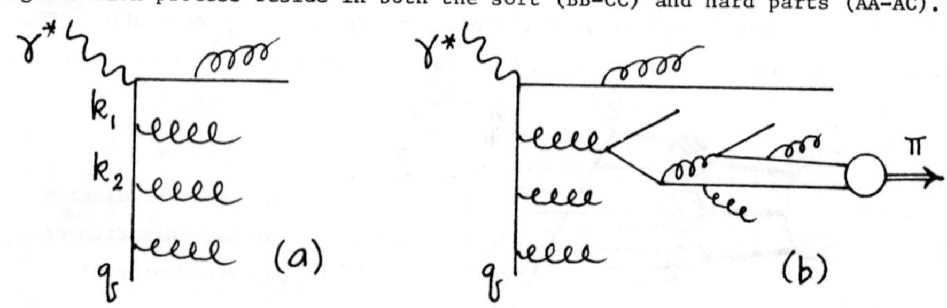

Fig. 5.2 Deep inelastic scattering (a) structure function (b) hadronization

Only by measuring the full dependence on all three variables, x_{bj}, x, and Q^2, can the detailed properties of a punctured hadron be obtained.

The above theoretical description is in the valon model. Again, there are no free parameters. Except for computational complexity there is no difficulty in principle to carry out the calculation. Predictions will serve as a good test of the valon model. The result is expected to differ from that of cascade-type model[72] which has no Q^2 dependence.

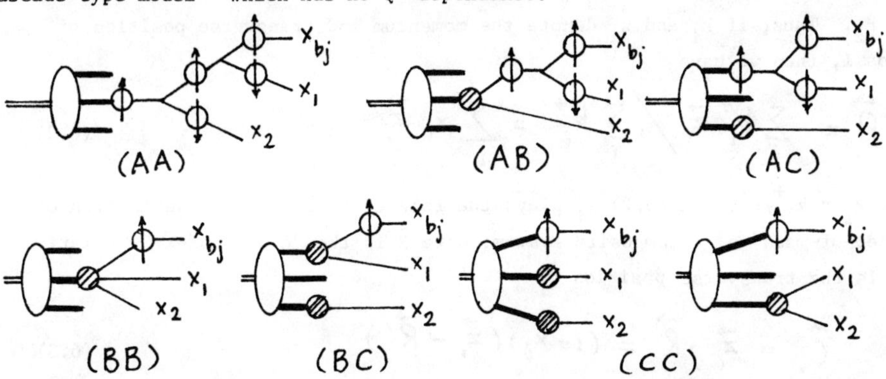

Fig. 5.3 Diagrams for hadronization of punctured nucleon

6. FORM FACTORS

Form factors of hadrons are traditionally regarded as indicators of hadron structure. With deep inelastic scattering processes being given the prominent role for probing hadron structure during the past ten years, no dramatic progress has been made recently in the experimental study of form factors. Theoretical work has been done on the subject but the aim has been to study the high-Q^2 behavior as consequence of QCD.[73,74] However, information on the hadron structure is contained mainly in the low-Q^2 behavior of a form factor. One need only be reminded that the charge radius is related to the derivative of the form factor at Q^2=0. Perturbative QCD is, of course, useless at low Q^2, while non-perturbative QCD is not at a stage ready for phenomenological interpretation. The valon model is the ideal framework to describe hadron structures and their form factors. We discuss below how the valon wave functions are to be extracted from form factors and show how the result obtained for pion can lead to a very sensible determination of the pion decay constant.[43]

6.1 Transverse Coordinates

The charge form factor $F_h(Q^2)$ can be related to the quark distribution q_i^h in hadron h by[75]

$$F_h(Q^2) = \sum_i e_i \int_0^1 dx \int d^2r\, e^{i\vec{Q}\cdot\vec{r}}\, q_i^h(x,\vec{r}) \tag{6.1}$$

where $\vec{r}$ is the transverse distance of quark i from the center $\vec{R}$ of P^+ where P^+ is the + component of the light-cone momentum of the hadron, i.e. $P^+ = P_0 + P_z$. Thus, if k_i and $\vec{z}_i$ denote the momentum and transverse position of parton i, then we have

$$\vec{R} = \sum_i k_i^+ \vec{z}_i \Big/ \sum_i k_i^+ = \sum_i x_i \vec{z}_i \tag{6.2}$$

where $x_i = k_i^+/P^+$. In (6.2) x_i plays the role of "mass" in the definition of "center-of-mass" of a composite system; here $\vec{R}$ is the "center-of-x." Relative to $\vec{R}$ is the transverse position $\vec{r}_i$

$$\vec{r}_i = \vec{z}_i - \vec{R} = (1-x_i)(\vec{z}_i - \vec{R}') \tag{6.3}$$

where $\vec{R}'$ is the center-of-x of the residual system without quark i. Defining $\vec{y}_i = \vec{z}_i - \vec{R}'$ with the interpretation that it is the distance between quark i and the center of the residual system, then one gets the generic relation[75]

$$\vec{r} = (1-x)\vec{y} \tag{6.4}$$

referring to a quark at x.

It is reasonable to expect the quark distribution to be factorizable in the variables x and $\vec{y}$ rather than in x and $\vec{r}$ since the definition of $\vec{r}$ depends on x. It has been shown by Soper[75] that

$$q_i^h(x,\vec{r})\,d\vec{r} = |\phi_i^h(x,\vec{y})|^2\, d\vec{y} \tag{6.5}$$

where ϕ_i^h is the wave function of quark i in hadron h. The difference between $\vec{r}$ and $\vec{y}$ is important as we shall see below.

Note that (6.1) is simple because the quark distribution is specified in terms of a mixed representation of longitudinal momentum fraction and transverse coordinate. The Fourier transform in $\vec{r}$ makes explicit the usual notion that the form factor describes the size of a hadron.

6.2 Form Factor at Low and High Q^2

The relationship between quark and valon distributions is a convolution

$$q_i^h(x,Q^2) = \sum_j \int_x^1 \frac{dx'}{x'} G_j^h(x') p_i^j\left(\frac{x}{x'}, Q^2\right) \tag{6.6}$$

from which (2.5) can be derived. p_i^j is the distribution of quark i in valon j. The origin of (6.6) can be traced to the operator product expansion. We generalize it to an expression that includes the $\vec{r}$ dependence[43]

$$q_i^h(x,\vec{r},Q^2) = \sum_j \int_x^1 \frac{dx'}{x'} \int d^2r' \, V_j^h(x',\vec{r}\,') p_i^j\left(\frac{x}{x'}, \vec{r}-\vec{r}\,', Q^2\right) \tag{6.7}$$

It is a reasonable but model-dependent description of the quark distribution, assumed to be valid over a range of low values of Q^2, and is of a form from which (6.6) can be recovered upon integration over $\vec{r}$ with

$$G_j^h(x') = \int d^2r' \, V_j^h(x',\vec{r}\,') \tag{6.8}$$

Eq. (6.7) is not expected to be valid for $Q^2 \to \infty$ because in perturbative QCD terms not included in it become important at large Q^2 corresponding to exchange of hard gluons between quarks. Eq. (6.7) itself is, of course, not derivable from perturbative QCD which is invalid at low Q^2. Where is the transition between the two regions of Q^2? The answer is a priori not known, but the question is a matter of phenomenology which we can and shall examine in the next subsection.

To understand theoretically the difference in physics at high and low Q^2, we first note from (6.1) that the dominant contribution to $F_h(Q^2)$ at high Q^2 comes from the $\vec{r} \to 0$ region. From (6.4) it implies either $x \to 1$ for any $\vec{y}$, or $\vec{y} \to 0$ for any x, or both. Which kinematical region is important depends on how fast $q_i^h(x,\vec{r},Q^2)$ vanishes as $x \to 1$. If it does not vanish very fast, then the dominance of the $x \to 1$ region leads to the Drell-Yan-West relation[76]

$$F_h(Q^2) \sim (Q^2)^{-p}, \qquad \nu W_2(x) \sim (1-x)^{2p-1} \tag{6.9}$$

Otherwise, (6.9) is invalid.[73,74]

To see more clearly the issues involved in terms of the wave functions, consider the general formula (valid for all Q^2) that follows from (6.1), after using (6.4) and (6.5),

$$F_h(Q^2) = \sum_i e_i \int_0^1 dx \int \frac{d^2k}{(2\pi)^2} \tilde{\phi}_i^h(x,\vec{k}) \tilde{\phi}_i^{h*}(x, \vec{k}+\vec{Q}(1-x)) \tag{6.10}$$

where $\tilde{\phi}_i^h(x,\vec{k})$ is defined according to

$$\phi_i^h(x,\vec{y}) = \int \frac{d^2k}{(2\pi)^2} e^{i\vec{k}\cdot\vec{y}} \tilde{\phi}_i^h(x,\vec{k}) \tag{6.11}$$

Note that $\vec{k}$ is the transverse momentum of a quark relative to the incident hadron with $\vec{P}=\vec{0}$, but in the same frame a quark with momentum $(xP,\vec{k}+\vec{Q})$ has a transverse momentum $\vec{k}+\vec{Q}(1-x)$ relative to the outgoing hadron with $\vec{P}'=\vec{Q}$. At low Q^2 all parts of the x-integration in (6.10) are important. Moreover, the photon having poor resolution interacts with a valon as a whole, so the wave functions in (6.10) refer to valons. In the case of pion this is represented by the diagram in Fig. 6.1(a) where the valon form factor is very nearly one (see next subsection). The valon wave function for a pion is approximately constant, so $\tilde{\phi}(x,\vec{k}+\vec{Q}(1-x))$ is important in the $x\to 1$ region even when Q^2 becomes moderately large.

On the other extreme, as $Q^2\to\infty$, the photon is coupled to a quark whose wave function may be suppressed in the $x\to 1$ region. If so, then to achieve $\vec{r}\to 0$ in (6.1) for $x \not\approx 1$ it is necessary for $\vec{y}\to 0$. Short (transverse) distance separation means large momentum transfer mediated by the exchange of a gluon as examplified by the diagram in Fig. 6.1(b). Such processes are precisely what have been studied in perturbative QCD to obtain the large-Q^2 behaviors of form factors.[73,74] The quark wave function $\phi(x,Q_o^2)$ at some low Q_o^2 has been taken to be as sharp as $\delta(x-\frac{1}{2})$, but by iterating gluon exchange for high Q^2 the wave function can be broadened to $\sim x_1x_2$. Thus, as $Q^2\to\infty$, one has $|\phi(x,Q^2)|^2 \sim [x(1-x)]^2$, which is still considerably damped as $x\to 1$ compared to the valon distribution. To achieve collinearity of the recombining quark and antiquark, a gluon of transverse momentum $\vec{Q}/2$ must be exchanged. Terms in the wave function that behave as $f(x)/Q^2$ disappear in the $Q^2\to\infty$ limit, but can be important at low Q^2. The result of the investigation in the large Q^2 region determines the Q^2 dependence of $F_h(Q^2)$, but uncovers very little about the transverse structure of the hadron. The latter is revealed in

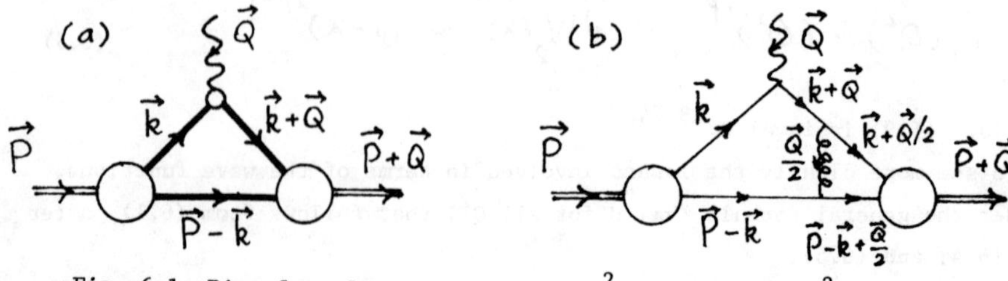

Fig. 6.1 Pion form factor at (a) low Q^2 and (b) high Q^2

the low-Q^2 region where $\vec{y}$ is allowed to be non-zero, consequently sampling the entire spatial extent of the hadron.

At intermediate values of Q^2 the quarks in a valon can be resolved so the valons acquire form factors themselves. The relevant diagram is shown in Fig. 6.2(a) where the upper triangle refers to the valon form factor. There is, of course, also competition from the gluon exchange diagrams such as the one shown in Fig. 6.2(b). Without phenomenology it is not possible to predict the value of Q^2 at which the two types of diagrams are of comparable importance. We investigate the former only at low and intermediate Q^2 and compare with existing data for $Q^2 \lesssim 10$ GeV^2.

We end this subsection with a physical description of the difference between large and small Q^2 regions. At short distance the potential between quarks is of the singular form 1/r because of massless gluon exchange. Hence, it increases with Q as $V \sim Q$ apart from log Q factors. Thus impulse approximation is not good (unless the log Q factors come to the aid as in asymptotic freedom--but very slowly). This is the reason why in using perturbative QCD to treat the large Q^2 regime[73,74] it is necessary to consider diagrams with hard gluon exchanges, the physical explanation of which is that the spectator quarks are not independent from the struck quark when the potential is strong enough to invalidate impulse approximation. But at low Q^2 the potential between valons is not strong; the poor resolution in 1/Q prevents any sensitivity to the 1/r potential. The impulse approximation is therefore good. The virtual photon sees each valon one at a time in a way that depends on the valon wave function. This is the same physical approximation found valid in molecular, atomic and nuclear physics. The valon model extends it to hadron physics. It stops there because at the next level (quark physics at high Q^2) the potential becomes singular. Where the transition takes place is what we proceed now to investigate.[43]

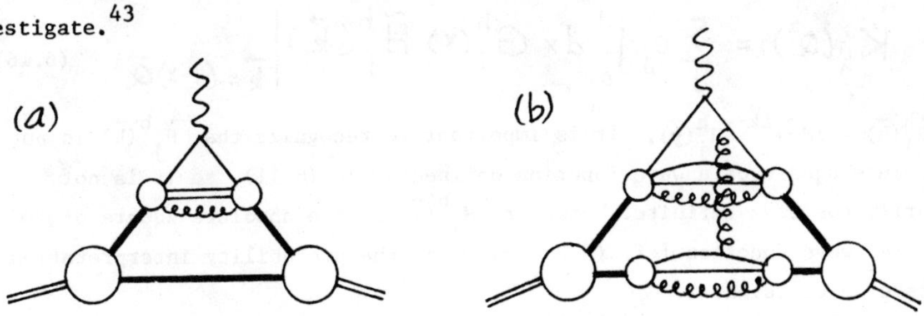

Fig. 6.2 Schematic diagrams for form factor at intermediate values of Q^2 (a) without or (b) with gluon exchange between struck and spectator quarks.

6.3 Valon Distributions from Form Factors at Low Q^2

If we substitute (6.7) into (6.1), we get

$$F_h(Q^2) = K_h(Q^2)\, F_v(Q^2) \tag{6.12}$$

$$K_h(Q^2) = \sum_j e_j \int_0^1 dx \int d^2r\, e^{i\vec{Q}\cdot\vec{r}}\, V_j^h(x,\vec{r}) \tag{6.13}$$

$$F_v(Q^2) = \sum_i \frac{e_i}{e_j} \int_0^1 dz \int d^2\rho\, e^{i\vec{Q}\cdot\vec{\rho}}\, p_i^j(z,\vec{\rho},Q^2) \tag{6.14}$$

$F_v(Q^2)$ is independent of flavor and color because apart from the over-all charge e_j the valons have universal structure due to QCD virtual processes. Normalizations of the probabilities imply $F_v(0) = 1$ and $K_h(0) = \sum_j e_j = e_h$. Clearly, $K_h(Q^2)$ is the contribution to $F_h(Q^2)$ due to the valon wave function describing the spatial extent of h. $F_v(Q^2)$ is the valon form factor that gives further damping at higher Q^2. When Q^2 is sufficiently small, say < 1 GeV^2, the valon structure cannot be resolved. Then we set $F_v(Q^2)=1$ and identify $F_h(Q^2)=K_h(Q^2)$.

We assume that $V_j^h(x,\vec{r})$ is factorizable in the variables x and y, as discussed earlier. By virtue of (6.8) we then have

$$V_j^h(x,\vec{r}) = \frac{G_j^h(x)}{(1-x)^2}\, H_j^h(\vec{y}) \tag{6.15}$$

where $H_j^h(\vec{y})$ is normalized by $\int H_j^h(\vec{y})\,d\vec{y} = 1$. It then follows from (6.13) that

$$K_h(Q^2) = \sum_j e_j \int_0^1 dx\, G_j^h(x)\, \tilde{H}_j^h(\vec{k})\Big|_{\vec{k}=(1-x)\vec{Q}} \tag{6.16}$$

where $\tilde{H}_j^h(\vec{k}) = \int d\vec{y}\, e^{i\vec{k}\cdot\vec{y}} H_j^h(\vec{y})$. It is important to recognize that $\tilde{H}_j^h(\vec{k})$ is not the absolute square of a wave function defined as in (6.11), so it is not necessarily positive definite. However, $H_j^h(\vec{y})$ is the absolute square of the spatial wave function [cf. (6.25)] and has the probability interpretation in transverse coordinates.

Using (6.16) to fit the low-Q^2 data of $F_h(Q^2)$ for h = p, n, π can readily be carried out, provided that we adopt some simple parametrization for G_j^h and $\tilde{H}_j^h$. For G_v^p we use (2.10) while for G_v^π we use (2.4) with a and b being free parameters. For $\tilde{H}_j^h(\vec{k})$ we use Gaussian form (not power-law behavior which is sometimes mistakenly interpreted as the suggestion of QCD) because the hadron size has a well-accepted notion related to diffraction peak[77] and p_T distribution in soft hadronic processes, both of which suggest Gaussian behavior for $\vec{k}$ not too large. We thus write

$$\tilde{H}_j^h(k^2) = \exp\left(-\beta_j^h k^2\right) \tag{6.17}$$

whereby three more parameters β_U^p, β_D^p and β^π are introduced.

Fitting all three form factors (using charge symmetry for neutron) for $Q^2 < 1\ \mathrm{GeV}^2$ yields

$$a = 0.9\ , \quad \beta_U^p = 6.1\ \mathrm{GeV}^{-2}, \quad \beta_D^p = 3.0\ \mathrm{GeV}^{-2} \tag{6.18}$$

$$b = 0.3\ , \quad \beta^\pi = 6\ \mathrm{GeV}^{-2}. \tag{6.19}$$

There is very little freedom in adjusting these parameters to give the best fit to all three $F_h(Q^2)$. The results of the fits are shown as solid lines in Fig. 6.3. The deviations from data for $Q^2 > 1\ \mathrm{GeV}^2$ are to be accounted for by $F_v(Q^2)$. Thus we have extracted the valon distributions from low-Q^2 form factors. The corresponding G_j^h functions have already been exhibited in (2.12) and (2.13), which are not too far from the expressions determined from structure functions. For comments see Sec. 2.5.

F_v being universal, the factorizability of (6.12) implies

$$\frac{F_\pi(Q^2)}{F_p(Q^2)} = \frac{K_\pi(Q^2)}{K_p(Q^2)} \tag{6.20}$$

for all Q^2 where (6.7) is dominant. This can be tested since the r.h.s. can be calculated from (6.16) for $Q^2 > 1\ \mathrm{GeV}^2$ and compared with the data on l.h.s. Instead of taking the ratio of data, we approximate $F_p(Q^2)$ by the dipole formula $F_p^D(Q^2) = (1+Q^2/.71)^{-2}$ as shown in Fig. 6.3. Then we calculate $(K_\pi/K_p)F_p^D$ and compare it to the data on F_π. From Fig. 6.3 it is clear that the agreement is excellent

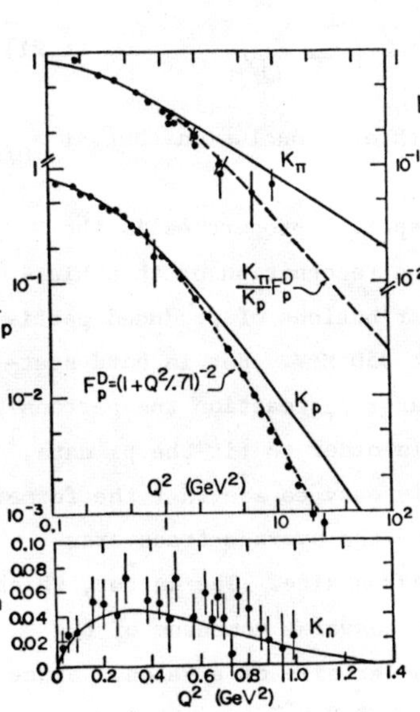

Fig. 6.3 Form factors for p, n, π.

for the existing data. It suggests that the component of $F_h(Q^2)$ involving hard gluon exchange not included in (6.7) is not dominant for $Q^2 < 10\ \mathrm{GeV}^2$. Thus the transition region between low and high Q^2 questioned in the previous subsection appears to lie above 10 GeV^2 where no data for F_π exist.

6.4 Spatial Properties of Hadrons and Valons

It is highly significant that the U and D valon distributions in a proton are such that, according to (6.18), $\beta_U^p \approx 2\beta_D^p$. In two-dimensional transverse space one has $\beta_j^p = \langle k_T^2\rangle_{j/p}^{-1} \propto \langle r^2\rangle_{j/p}$. That means $\langle r^2\rangle_{U/p} \sim 2\langle r^2\rangle_{D/p}$, i.e. the U's are on the average on the outside while the D is on the inside. In the quark model the identical constituents are spin parallel and repel; non-identical ones are spin antiparallel and attract--in agreement with our result. Thus for neutron the average spatial arrangement, if written out in a linear array, would spell DUD. The same conclusion has been reached by others on the basis of other arguments.[78,79]

It is interesting to note that $\beta_U^p > 2\beta_D^p$ is also compatible with an experiment on the p_T distributions of the detected pions in the target fragmentation region of $\pi^- + n \to \pi^\pm + X$.[80] It is found that the ratio π^+/π^- decreases with increasing p_T. Although we have not discussed p_T distribution in these lectures, it is not difficult to arrive at the relation (for neutron)

$$\langle p_T^2\rangle_{\pi^-} - \langle p_T^2\rangle_{\pi^+} = \langle k_T^2\rangle_{D/n} - \langle k_T^2\rangle_{U/n} \tag{6.21}$$

Experimentally, this is negative, in agreement with our conclusion that $\langle r^2\rangle_{D/n} > \langle r^2\rangle_{U/n}$.

Another perhaps more basic character of the spatial properties is the hierarchy of sizes: hadron, valon and parton. The recognition of this hierarchy immediately resolves a puzzle in the p_T distributions of produced particles. The average p_T in low-p_T reactions is about 350 MeV. But in hard scattering processes such as dilepton production or large-p_T reaction the partons are given a primordial $\langle k_T\rangle$ of about 600-800 MeV in order to fit the p_T data.[1] Why are there two scales? In the valon model it is easy to see that the former, being associated with soft processes, is related to the average transverse momentum of the valons, which characterizes the hadron size. The latter, which is associated with hard processes, is due to the transverse momentum of the partons in valons, and is characterized by the smaller size of a valon. Since primordial $\langle k_T\rangle \approx 0.8$ GeV would correspond to $\beta^v \approx 1.2\ \mathrm{GeV}^{-2}$, we obtain $\langle r^2\rangle_{\text{valon}} / \langle r^2\rangle_{\text{hadron}} = \beta^v/\beta_v^h \approx 1/5$. Thus constituent quarks are far from being point-like; the hadrons are not very "spacious" (like the atoms) nor very "crowded" (like the nuclei).

6.5 Pion Decay Constant

The pion is not only a bound state of two valons but also a Goldstone boson, i.e. it would have been the zero-mass boson if chiral symmetry were exact. To understand these dual roles has been a problem of theoretical interest.[81-84] A test of that understanding is the determination of the pion decay constant f_π, since on the one hand it depends on the wave function of its constituents, while on the other it is defined in accordance to the Goldstone character of the pion by

$$\langle 0| A^i_\mu(0) | \pi^j(P) \rangle = \delta_{ij}\, i f_\pi P_\mu \tag{6.22}$$

where A_μ is the axial-vector current and f_π is experimentally 93 MeV.

One may first ask what sets the scale for f_π. Is it the mass? That cannot be since $f_K/f_\pi \sim 1.2$ while $m_K/m_\pi \sim 3.5$. The only other characteristic scale of a hadron is its size, and that seems plausible since the sizes of pion and kaon are comparable. We shall indicate below how that connection can be established quantitatively.

Eq. (6.22) implies that f_π is related to the wave function of the current quarks at the origin. Indeed, it can be shown that[43]

$$f_\pi = i \frac{3}{\sqrt{\pi}} \int_0^1 dx\, \phi(x, \vec{0}) \tag{6.23}$$

where $\phi(x,\vec{y})$ is the quark wave function and the factor 3 is due to color. The integration over the momentum fraction x is equivalent to evaluating the spatial wave function at the origin. In a model where a π^+ consists of only $u\bar{d}$ quarks, Soper[75] has shown

$$q(x, \vec{r}) = \frac{3}{(1-x)^2} \left| \phi(x, \vec{y}) \right|^2 \tag{6.24}$$

which differs from (6.5) only by a factor of 3, inserted here on account of color. But a model with only $u\bar{d}$ quarks (which is a simple idealization at the quark level) is precisely our $U\bar{D}$ valon model (which is a realistic description of the pion with known wave function). Hence, instead of discussing quark distribution $q(x,\vec{r})$, we should consider the valon distribution $V(x,\vec{r})$, and (6.24) is to be replaced by

$$V(x, \vec{r}) = \frac{3}{(1-x)^2} \left| \Phi(x, \vec{y}) \right|^2 \tag{6.25}$$

where $\Phi(x,\vec{y})$ is the valon wave function averaged over color. It then follows from (6.15) that

$$|\Phi(x,\vec{0})| = \frac{1}{\sqrt{3}\,2\pi}\left[G_v^\pi(x)\int d^2k\,\tilde{H}_v^\pi(k^2)\right]^{1/2} \tag{6.26}$$

where the integral over $\vec{k}$ yields π/β^π on account of (6.17).

If it were reasonable to approximate the quark wave function $\phi(x,\vec{0})$ called for in (6.23) by the valon wave function in (6.26), then we would have

$$f_\pi = \frac{\sqrt{3}}{2\pi}(\beta^\pi)^{-1/2}\int_0^1 dx\,[G_v^\pi(x)]^{1/2} \tag{6.27}$$

The integral over x can be evaluated using (2.4) and (6.19), and one gets 0.994. It is close to one because G_v^π is normalized by $\int_0^1 dx G_v^\pi(x) = 1$. Now, the difference between $\phi(x,\vec{0})$ and $\Phi(x,\vec{0})$ is not great because the relevant Q^2 is low for the present problem ($Q^2=m_\pi^2$). Any minor modification on the x dependence would not change the integral in (6.27) significantly, so we shall approximate the x integral by one. As for the $\vec{k}$ integral in (6.26) one sees from Fig. 6.4 that the W boson is coupled to the quarks which in turn are linked exclusively to the two valons. Thus the k distribution of the ingegrand in (6.26) which refers to the valons should be modified by a multiplicative factor that describes the k distribution of a quark in a valon. We do not know the latter very well, but for $Q^2=m_\pi^2$ we may take it to be the primordial $\langle k_T\rangle$ distribution, characterized by a Gaussian with $\beta^v \sim 1.2\ \text{GeV}^{-2}$. Hence, the integral is changed from π/β^π to $\pi/(\beta^\pi+\beta^v)$. The result for f_π is therefore[43]

$$f_\pi = \frac{\sqrt{3}}{2\pi}(\beta^\pi+\beta^v)^{-1/2} \tag{6.28}$$

Putting in the value of 6 GeV^{-2} for β^π from (6.19) and 1.2 GeV^{-2} for β^v (which is a low, conservative estimate), we obtain f_π = 102 MeV, missing the experimental value of 93 MeV by less than 10%. If one wants to fit the experimental number by varying β^v, one would get $\beta^v = 2.79\ \text{GeV}^{-2}$, corresponding to a pri-

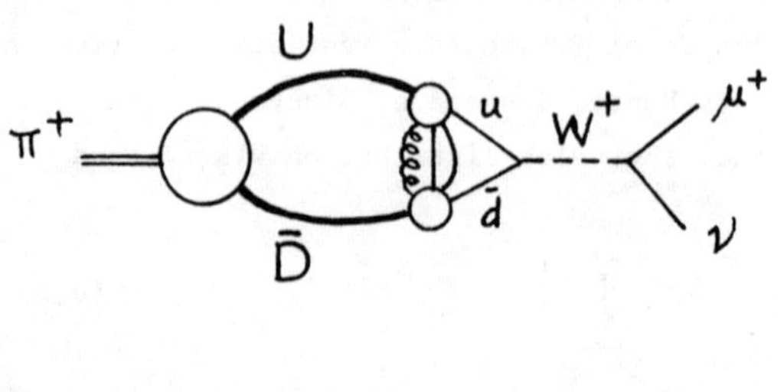

Fig. 6.4
Valon-parton diagram for pion decay.

mordial $\langle k_T \rangle$= 0.6 GeV, which has also been suggested as a reasonable empirical value for certain experiments.[1] I tried to avoid using this number at the outset for fear of causing disbelief that the theory can predict to an accuracy of 1%. Indeed, it cannot. The notion of primordial $\langle k_T \rangle$ is not precise and its empirical value equally approximate. However, the theory clearly indicates that it is the hadronic size (parametrized by β^{π}) that dominates the determination of f_{π} in (6.28) and that the valon size adds a small correction factor which is somewhat fuzzy quantitatively, but which is qualitatively a definite signature of the heierachy of sizes discussed in Sec. 6.4.

7. RADIATION LENGTH OF FAST QUARK IN NUCLEAR MATTER

We now bring in nucleus into our consideration. It is well known that in nuclear matter an incident hadron provides free quark beams (since hadronization takes place outside the nucleus) and that the nucleons in the nucleus serve not only as scatterers but also as detectors of quarks. Thus to understand quark interaction it is important to examine nuclear effects either as clues to the basic interaction or as tests of theoretical predictions.

As a start a reader familiar with general facts about strong interaction is invited to make a guess on what the radiation length of a very fast quark is in traversing a nuclear medium taken hypothetically to be arbitrarily large, but of the usual density. Would it be 1, 10, 10^2 or 10^3 fm? A number of people I asked attempted to answer by starting with total σ_{pp} as a guide. It is a wrong track based on the confusion between quarks and valons, as discussed in Sec. 2.1. The elucidation of these problems is the subject of this and the following section.

7.1 Factorization and Nuclear Attenuation

A basic property of multiparticle production at high energy is factorization which states that particles produced in the beam fragmentation region is independent of the target, and vice-versa. Another property is short-range correlation in rapidity, which is, of course, not unrelated to factorization. Suppose we take factorization to be exactly true. Interpreted at the constituent level it means that the fast quarks in a beam particle that finally become constituents of a detected particle are unimpeded by the target particle. If so, then it would remain to be true even with nuclei as targets. That would mean that a fast quark can penetrate nuclear matter with infinite radiation length. Surely, that cannot be true. If not, how does one give a quantitative estimate of the radiation length?

If the quark momentum is degraded as it traverses a nucleus, then the most noticable effect should be the pion inclusive distribution in the projectile fragmentation region. Any dependence of that distribution on the nucleon number A is a signal for non-factorizability.[85] Indeed, experimentally such a dependence has been found,[86,87] and the cross section has been parametrized by

$$\left(x \frac{d\sigma}{dx}\right)_A = \left(x \frac{d\sigma}{dx}\right)_0 A^{\alpha(x)} \tag{7.1}$$

This formula has no theoretical basis and does not give the best fit to the data.[87] However, that is how data have been summarized in terms of $\alpha(x)$, and it suffices for our purpose.

For $p+A\to\pi(\text{or}K)+X$, the exponent $\alpha(x)$ can be approximated by a straight line as shown in Fig. 7.1, where the data points[86] have been taken from a compilation in Ref. 88. Writing

$$\alpha(x) = \alpha(0) + \alpha' x \tag{7.2}$$

we get for the best fit

$$\alpha(0) = 0.62 , \qquad \alpha' = -0.17 \tag{7.3}$$

An explanation for the average value of $\alpha(x)$ being less than 2/3 has been advanced by Dar and Takagi[88] on the basis of "quark attennuation and recombination." However, the physics of that has more to do with valons than with quarks, as we shall explain in the next section. In our view it is α' (ignored in Ref. 88) that provides the crucial empirical information about the radiation length of a fast quark.

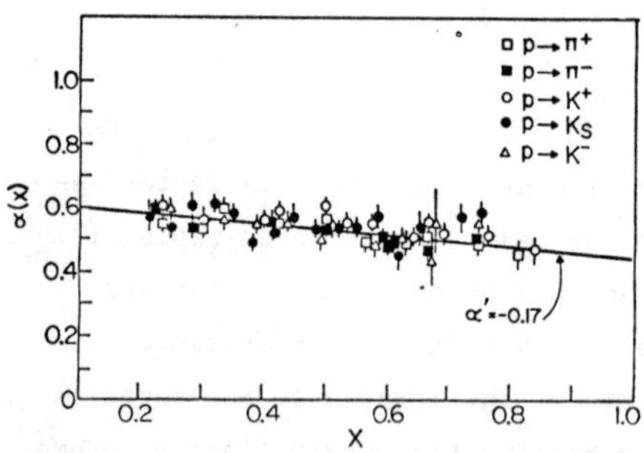

Fig. 7.1 Data on $\alpha(x)$ for the production of $\pi^{\pm}$ and $K^{\pm}$ at 24 GeV proton beam momentum and K_S at 300 GeV. The straight-line is an eyeball fit.

Obviously, what is needed to discover the implication of the value of α' is a reliable method that can relate the inclusive distribution of the detected hadron to the momentum distribution of quarks traversing the nucleus. That is, of course, precisely what we have in the recombination model. In Sec. 4 we have seen how $(x/\sigma)(d\sigma/dx)$ can be calculated for a proton projectile assuming that the partons are unaffected by the nucleon target. Now, the nuclear-target effect must be taken into account in order to reproduce the observed α'.[89]

7.2 Momentum Degradation of Fast Partons

The physics of why α' is negative can be understood as follows. Factorization in the form exhibited in (7.1) need not be right, but given that form adopted by the experimentalists one may associate the significance of $(d\sigma/dx)_0$ with the projectile and $A^{\alpha(x)}$ with the target. Relative to $\sigma_{inel} \sim A^{2/3}$, $A^{\alpha(x)}$ is a damping factor for the detected hadron, which becomes more severe at larger x for negative α'. But at the parton level the incremental loss of momentum per nucleon may still be scaling. The dependence of the damping factor on x is to be attributed to the fact that given an interval between x and x+dx partons can not only leave the interval due to momentum loss in a scattering but also enter it from a higher x-interval due to the same mechanism. Evidently, what we are about to suggest is an evolution equation for the parton distribution as a function of the number of nucleons that a parton traverses.

Let P(x) be the probability that a fast parton has a fraction x of the incident momentum after traversing a nucleon. Let it be written in the form

$$P(x) = \delta(x-1) + A(x) \tag{7.4}$$

so that A(x) has the interpretation of a momentum-degradation function. We adopt the normalization $\int_0^1 dx P(x) = 1$ which can possibly be relaxed to account for flux attenuation, a problem presumably related to the value $\alpha(0)$ but one that will not be treated here. Then we have

$$\int_0^1 A(x)\, dx = 0 . \tag{7.5}$$

We shall for simplicity ignore the differences between quarks and gluons and consider only average partons. The justification is mainly that we do not have precise knowledge about A(x) and that only a crude average information can be extracted from one number, viz. α'.

It is crucial to recognize that we are dealing with partons, not valons. The valons include wee partons and interact with nucleons in totally different ways from fast partons. By short-range correlation a fast parton would not interact effectively with a target nucleon so A(x) should be sharply peaked at x=1. But what is its x distribution for x<1? We expect the canonical radiation tail. In perturbative QCD if one considers the loss of momentum of a quark due to gluon radiation, one would get in lowest order $A(x)=\frac{4}{3}(1+x^2)/(1-x)$ for x<1. The denominator is the ω^{-1} tail characteristic of emission of a massless boson. We keep that as the only essential part relevant for our problem and write a phenomenological formula

$$A(x) = \frac{k}{(1-x)_+} \tag{7.6}$$

where k is a small constant and the + symbol signifies regularization at x=1 according to

$$\frac{1}{(1-x)_+} = \frac{1}{1-x} - \delta(x-1)\int_0^1 \frac{dx}{1-x} \tag{7.7}$$

Note that (7.5) is automatically satisfied. If k is sufficiently small, A(x) is important only for x very near 1 so that any complicated dependence on x that could have appeared in the numerator of (7.6) can be well approximated by a constant in that extreme region. Thus (7.6) is a reasonable formula and k is our only unknown in the problem. Our aim is to relate k to α' which can be achieved only if the model makes phenomenological sense.

Let the momentum distribution of a parton after penetrating N nucleons be q(x,N). The change on q(x,N) as a result of traversing one more nucleon is then,[89] owing to (7.4),

$$\frac{d}{dN} q(x,N) = \int_x^1 \frac{dx'}{x'} q(x',N)\, A\left(\frac{x}{x'}\right). \tag{7.8}$$

This is an evolution equation similar to that of Altarelli and Parisi[90] except that N plays the role log Q^2 in scaling violation. . Thus the evolution of the x distribution as N increases is of a familiar nature: q(x,N) becomes more suppressed in most of the x region except at very small x where it is enhanced. Eq. (7.8) can trivally be solved in terms of the moments $q_n(N)$, yielding

$$q_n(N) = \exp\left(kNM_n\right) \tag{7.9}$$

where M_n are the moments of (7.7). In (7.9) we have used the boundary condition that $q_n(0) = 1$ for all n, i.e. $q(x,0) = \delta(x-1)$.

Having obtained the N dependence due to nuclear degradation, or equivalently the A dependence if we set $N=A^{1/3}$, we can now apply it to the recombination model and calculate $(x/\sigma)(d\sigma/dx)_A$. The only difference from the procedure followed in Sec. 4.3 is that $F(x_1,x_2)$ should first be convoluted with $q(x_1',A)$ and $q(x_2',A)$ before hadronization. The physics is that the parton distribution of the projectile, previously assumed to be unaffected by the target hadron, is now modified by the target nucleus before recombination. The question is whether the resultant inclusive distribution can be approximated by (7.1) in which the A-dependent part of the x dependence is exp $(\alpha' x \ell n A)$. As it turns out, due to the very fortunate circumstance that $A^{1/3} \simeq \ell n A$ for $5<A<200$, an acceptable correspondence can be established, provided that[89]

$$k = -0.12 \quad \alpha' = 0.02 \, . \tag{7.10}$$

We have therefore achieved not only in getting a constraint between k and α' that renders our model phenomenologically sensible, but also in obtaining a value of k which is small enough to make the model self-consistent.

7.3 Radiation Length

Having determined the value of k it is now straightforward to calculate the radiation length. Combining (7.6) and (7.10) with (7.4) we obtain for the average momentum fraction $\langle x \rangle$ of a parton after emerging from a nucleon

$$\langle x \rangle = \int_0^1 x \, P(x) \, dx = 1 - k = 0.98 \tag{7.11}$$

If t denotes the average thickness of a nucleon and L the radiation length of a fast parton in nuclear matter, then by the exponential law of attenuation we have

$$\langle x \rangle = \exp(-t/L) \simeq 1 - t/L \tag{7.12}$$

which implies

$$L = \frac{t}{k} \simeq 50 \text{ fm} \tag{7.13}$$

where we have taken $t \simeq 1$ fm as a reasonable approximation. Even if we give a generous margin of error of about 100% to the estimation of $L \sim 50$ fm due to the uncertainties related to α' and (7.6), a range from 20 to 100 fm is nevertheless the first quantitative answer to a question for which there is very little intuitive guidance for an educated guess. The most important implication that the result underlines is that a fast quark behaves drastically differently from a fast valon.

With (7.11) we now see that a fast quark loses only ∿2% of its initial momentum in a hadron-hadron collision. Consequently, factorization is a very good approximation. This justifies the procedure used in Sec. 4.3 where the effects of hadron target on the partons in the projectile fragmentation region are totally ignored. But in hadron-nucleus eollision the successive losses of 2% to each nucleon along the trajectory through a nucleus lead to a noticable change in the inclusive distribution, and factorization ceases to be valid even approximately.

8. CENTRAL REGION

We have deferred discussion about the central region because we know least about it. In the fragmentation region we have made use of the transparency of a hadron (at a level of 2% error concluded in the last section) to ignore parton-hadron interaction and calculated the inclusive distribution using recombination. In the central region hadronization is still through the recombination mechanism, but the interaction between wee partons from the two incident hadrons cannot be ignored. It is precisely that interaction which is complicated and difficult to describe. On the other hand, if the cross section has a scaling limit as $s\to\infty$, there is not as much to calculate. Unlike the situation in the fragmentation region where we can examine the x dependence of the hadrons produced, in the central region it is only the height of the rapidity plateau that is to be determined. It maps onto one number at the point $x=0$ in the x distribution in the scaling limit. But even that one number is hard to predict. Other more complicated features in the central region such as correlation and s dependence (without Regge input) are problems for the future. Obviously, those are important problems that must be understood in the parton picture if strong interaction is to be described at the constituent level.

What will be offered in this section is only a modest step in the right direction. It is important to make contact between the physics in the fragmentation region and that in the central region within the same framework of valons and partons. It is necessary to answer such questions as: if a hadron is so transparent to a fast quark, why is the cross section so large? Our qualitative answer in the next subsection will be developed into a quantitative description about hadron-nucleus cross sections in the following subsection.

8.1 Valon Picture

Dominant contribution to the multiplicity of produced particles and to total cross section in hadron-hadron collision comes from the central region. By definition the central region is where the wee partons hadronize. Since the wee partons are parts of the valons, it is the valon-valon interaction that is responsible for the total cross section. Unfortunately, at this stage of the development of the valon model we can say very little precisely about the valon-valon interaction. It is certainly totally different from quark-quark interaction. A valon has the flavor and average color of its valence quark. But when two valons interact at large relative momentum the nature of that interaction is rather insensitive to their flavors and colors because those quantum numbers are carried by the valence quarks which have high probability of also having large relative momentum, hence weakly interacting. The strong interaction exists among the multitude of wee partons (of all flavors and colors) independent of the quantum numbers of the valence quarks in the fragmentation region. Thus the first general statement that can be made about valon-valon interaction is its universality on the average, including valon-antivalon interaction. We remark parenthetically that this universality is quite contrary to the rules of "quark-diquark chain formation" in the fragmentation model.[52] That is, we see no compelling reason for the notion of $3-\bar{3}$ color separation as the basic mechanism for particle production.

It should be clear from the above discussion that a valon once interacted will not emerge as the same valon. The momentum profile of the partons is altered, and dressing of the partons would take place farther downstream. In this respect regarding the valons as constituent quarks can be very misleading.

On the basis of the universality of valon-valon interaction, we now want to derive additivity in total cross sections, i.e. $\sigma_{pp}/\sigma_{\pi p} \simeq 3/2$. It is, however, necessary first to know how likely the valons will interact in a hadron-hadron collision. Consider the transverse plane in pp collision. We assume that two valons (one from each proton) must spatially overlap in order to give rise to any significant valon-valon interaction through their wee partons. Since a proton is not densely packed with valons as is a nucleus with nucleons, the probability for two pairs of valons to overlap is considerably lower than for just one pair. Then in a single pp collision, the most probable occurence is for just one pair of valons to interact. The same is true with πp collision. Since the total number of possible pairings in pp is 9, while in πp it is only 6 (this being where the notion of additivity arises), we conclude that $\sigma_{pp}/\sigma_{\pi p} \simeq 3/2$.

The above argument is, of course, very crude. At finite energies the quantum numbers are relevant. But it does show that in the valon model the additivity property depends on the universality of the valon-valon interaction and on the assumption that it is short-ranged in transverse distance. To see whether the assumption is necessary, not just sufficient, we need to test the idea in other problems, such as in hadron-nucleus collisions.

8.2 Hadron-Nucleus Cross Sections

In principle one can learn about the central region by studying particle multiplicity as well as cross section in hadron-nucleus collisions. Unfortunately, at present energies the cascade effect in nuclear fragmentation region contaminates the average multiplicity so much that no reliable information about central production can be extracted from it. Data on absorption cross sections do not have the same defect.

Now in a nucleon-nucleon collision if the probability is high that at most only one valon from each nucleon interacts, then in a nucleon-nucleus collision the spectator valons in the projectile can interact with other nucleons in the target downstream. This is very similar to the idea of "wounded quarks" of Bialas.[85] In the valon model every valon contributes in the same way to the particles produced in the fragmentation region whether or not it participates in central interaction with other valons--because it is the fast quarks (which are decoupled from the wee partons) which hadronize in the fragmentation region. The hadron-nucleus cross section σ_{hA} is a measure not only of the valon-valon central interaction but also of the number of valons in h for further interaction downstream. The fact that $\sigma_{hh'}$ is additive does not imply that σ_{hA} is additive. Indeed, $\sigma_{hh'}$ is the infinitesimal component of σ_{hA}; the "transparency" of hh' interaction exponentiates into "opaqueness" for hA interaction, losing additivity as a consequence.

The exponentiation property of nuclear attenuation is not new.[91,92] We want to derive here a remarkably simple formula that works extremely well. The basic input is consistent with the valon model although not limited to it. Let us first define a geometrical cross section.

$$\sigma_{hA}^{geom} = \sigma_{hN} A^{2/3} \tag{8.1}$$

which has the interpretation of what the incident hadron "sees" if there is just one layer of $A^{2/3}$ nucleons facing the incident flux. One may think of the layer as the frontal view of a cubical nucleus. The details of nuclear size and density distribution are not important. Now, σ_{hA}^{geom} is, of course, not the absorption cross section σ_{hA}^{abs} since valons that do not interact with the nucleons in the first layer would go on and interact with others downstream. Thus σ_{hA}^{abs} is enhanced over σ_{hA}^{geom} by a factor proportional to 1-exp $[-\sigma_{hN} T(A)]$ where T(A) is the effective nuclear thickness, related to the nuclear density $\rho(z,b)$ by

$$T(A) = \int dz\, \rho(z,b) \tag{8.2}$$

The exponential term is what survives after nuclear absorption. We normalize the enhancement factor and define

$$E_h(A) = \frac{1 - \exp[-\sigma_{hN} T(A)]}{1 - \exp[-\sigma_{hN} T(1)]} \tag{8.3}$$

so that $E_h(1) = 1$. We then have

$$\sigma_{hA}^{abs} = \sigma_{hA}^{geom} E_h(A) \tag{8.4}$$

In both (8.1) and (8.3) σ_{hN} refers to the absorptive (inelastic) hN cross section. Consistent with our simple picture of the nucleus we assume that the nuclear density is constant. Then, since $\int d^2b\,T(A) = A$ and $\int d^2b = \sigma_{NA}^{geom}$, we have

$$T(A) = A/\sigma_{NA}^{geom} = A^{1/3}/\sigma_{NN} \tag{8.5}$$

Consequently, $\sigma_{hN} T(A) = r_h A^{1/3}$ where

$$r_h = \sigma_{hN}/\sigma_{NN} = \begin{cases} 1 \\ 2/3 \end{cases} \text{for } h = \begin{cases} N \\ \pi \end{cases} \tag{8.6}$$

The additivity property discussed in the previous subsection has been used in (8.6). Thus, we have from (8.3) and (8.4)

$$E_h(A) = \frac{1 - \exp(-r_h A^{1/3})}{1 - \exp(-r_h)} \tag{8.7}$$

and

$$\sigma_{hA}^{abs} = \sigma_{NN}\, r_h A^{2/3} E_h(A) \tag{8.8}$$

This is a one-parameter formula for all A and for h = N or π. The parameter is the overall normalization σ_{NN}, the inelastic N-N cross section, which is 31.6 mb for beam momentum in the 200-300 GeV range. It corresponds to valon-valon absorptive cross section being 3.5 mb. The predictions of (8.8) are shown as solid lines in Fig. 8.1 and evidently agree with data[92] marvelously well.

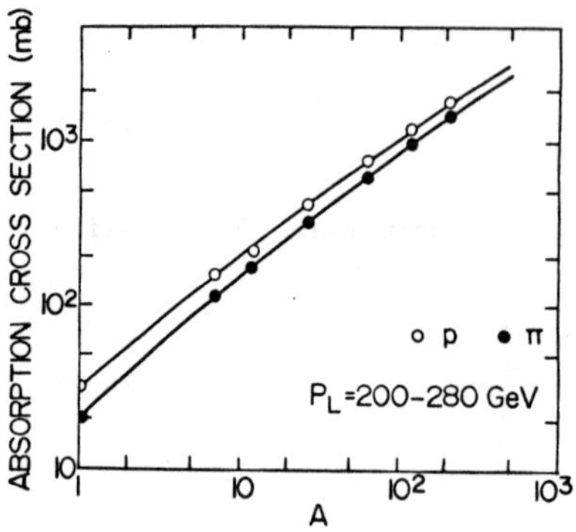

Fig. 8.1

Hadron-Nucleus Cross Sections

The physics of the result is utterly simple and is insensitive to the nuclear-structure details. The essence of (8.4) is the enhancement over the geometrical cross section due to nuclear attenuation. The enhancement factor (8.7) depends on the nuclear thickness (proportional to $A^{1/3}$) and the number of valons (related to r_h) in the incident hadron. Although nuclear attenuation can be discussed without valons, the valon model is strengthened by the possibility to incorporate a successful interpretation of the hadron-nucleus collision phenomena in the central region, when it is already shown (in the last section) to be successful in the fragmentation region.

9. CONCLUSION

The organizer of this School has urged the speakers to make comparisons between models. Indeed, it should be very useful to point out similarities and differences and to explore possible areas for unification. I shall mainly be concerned here with comparing the recombination model (RM) with the fragmentation model (FM) since they appear to be on opposite poles. But theorists can only express taste. Ultimately, it is the experimenters who provide the hard data that weed out unrealistic views.

9.1 Comparison with Fragmentation Model

The FM is fragmented into different schools of approaches.[93-95] The Lund group[93] emphasizes color dynamics, while the Saclay group[94] uses dual topological unitarization (DTU)[96] as the basis. The Orsay group[95] incorporates both and has splinter factions with varying emphases.[97] All groups attempt to make contact with quarks and partons. It is very difficult for an outsider (and in frank circumstances even an insider) to ascertain just what kind of quarks are referred to in the FM. Various adjectives have been used: constituent, current, bare, dressed, held-back, boundary-line, valence, diquark, etc. It would help if terminology about quarks is first defined and then diagrams drawn, rather than the other way around as is done in the DTU approach.

It should be recognized that DTU is a valid approach based on S-matrix theory and Regge behavior. It has a rich historical background with successes that cannot be ignored. To get Regge behavior as an output in a constituent model based on partons is still in the realm of fond hope for the future. However, in approaching the problem in the opposite direction by starting from DTU and incorporating parton ideas so as to give substance to the FM, ambiguities are encountered in the interpretation of the lines of the dual diagrams. For example, Capella[52] has indicated that the "quark" lines have a momentum distribution $x^{-1/2}(1-x)^{3/2}$ in a proton and $x^{-1/2}(1-x)^{-1/2}$ in a pion. These are, of course, rather unfamiliar quark distributions in the parton model. If they are not the familiar quarks, how does one justify using the fragmentation function for the familiar quarks determined in e^+e^- annihilation experiments?

There is another difficulty related to diquark fragmentation in soft processes. The "held-back quark" is assigned a color triplet, so the "diquark" is regarded as a color antitriplet whose fragmentation function is assumed to be the same as the D function determined phenomenologically in e^+e^- annihilation--the cornerstone of FM in the DTU approach. But unlike an antiquark in a hard process for which successive gluon emission does not change the $\bar{3}$ color nature of the $\bar{q}$, a "diquark" is a composite object (see Sec. 5), which need not remain as a $\bar{3}$ when gluons (with color 8) are emitted.

I have seriously attempted to explore areas of unification between FM and RM. The closest one can come to a duality of the two models is shown in Fig. 9.1. Cutting at the parton lines one emphasizes the recombination mechanism, while cutting at the valon lines one speaks of fragmentation. But even this qualitative observation cannot be translated into quantitative compatibility since the "quark" lines in FM have Regge properties built in by hand, while the valon and quark lines in RM have properties derived from structure functions and form factors. We have already mentioned their conflict at the

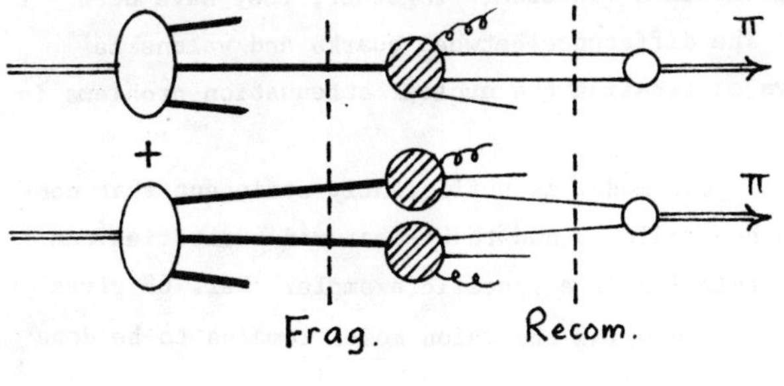

Fig. 9.1 Valon-parton diagrams for hadronization

end of Sec. 4.3.

If DTU is a well-honored feature of strong interaction, then it is a burden of the valon-parton model to derive it as a consequence. In considering the central region I have tried to obtain the two-chain structure from the valon picture, but without success. Short-range interactions among the wee partons have no way of knowing the quantum numbers of the valence quarks. Instead, we have universality in valon-valon interaction as discussed in Sec. 8.1.

The most outstanding problem with the valon model is that Regge behavior does not emerge naturally. It is an important direction to investigate for the future. For DTU-FM, on the other hand, the most serious problem is that it does not help us to understand better the hadron structure. It would be hard to win sympathy for doing low-p_T experiments if the justification is not placed heavily on the probing of the constituent structure of hadrons.

9.2 Retrospection and Outlook

Despite the fact that we cannot, as yet, derive Regge behavior or treat exclusive scattering processes, we have, in retrospect, successfully dealt with a large portion of strong-interaction physics. We have used the valon model to describe hadron structure and the recombination model to describe hadronization. They are the entry and exit windows of a black box containing hot quark soup. As we have argued, the soup is actually quite clear viewed in the forward window (fragmentation region) but rather murky in the side windows (large-p_T region). If the aim is to learn about hadron structure, then one should study hadron production in the fragmentation region in soft and hard processes. If one wants to learn about quark interaction dynamics, then one should look at the soup and do experiments at large angles or in the central region.

We have tested our entry and exit windows separately. For hadron structure, we have examined form factors in the valon model and calculated the pion decay constant. For hadronization, we have applied the recombination model to the calculation of quark fragmentation function. Together, they have been applied to low-p_T reactions. The difference between quarks and valons is stressed in our different ways of treating the nuclear attenuation problems in Sec. 7 and 8, respectively.

In the fragmentation region our model is sufficiently stringent that confrontation between certain to-be-predicted and to-be-measured quantities can be decisive. Two-particle correlation is a specific example. Ref. 68 gives hope for optimism, but a calculation using the valon model remains to be done and experiments at higher energies with independent momenta measurements are yet to be performed.

There are many more areas either to be refined or to be understood for the first time. Some aspects of polarization have already begun to be clarified.[61] Resonance productions must be understood better; more accurate data on their inclusive cross sections are sorely needed. The p_T distribution at low p_T, especially its thermodynamic characteristics, has not yet been seriously considered at the parton level. Valon-valon interaction is still barely understood. Why is σ_{vv} around 3.5 mb? Ultimately, the list of questions must include soft two-particle reactions and Regge behavior. When we have found the answers to those questions on a fundamental basis, we will have come a long way toward understanding strong interaction physics.

ACKNOWLEDGMENT

I am grateful to many collegues for discussions, critical comments and other forms of help which have collectively contributed toward developing the materials described in these lectures. An incomplete list would include V. Chang, N. G. Deshpande, G. Eilam, C. S. Lam, R. G. Roberts, D. Soper, E. Takasugi, M. Teper, and M. S. Zahir. Then there are also others who have helped me either to understand alternative theoretical ideas or to be acquainted with relevant experimental facts; acknowledgments go to D. Barton, G. Brandenburg, A. Capella, C. B. Chiu, P. Garbincius, W. Kittel, and U. Sukhatme. I am particularly thankful to R. T. Van de Walle for inviting me to give these lectures at Erice and for creating a friendly atmosphere in which I can learn as well as disseminate. The support of EPS and U.S. DOE is gratefully acknowledged.

REFERENCES

1. See various articles contained in the Proceedings of Symposium on Jets in High Energy Collisions, published in Physica Scripta 19, No. 2, 1979.
2. R. P. Feynman, Photon-Hadron Interactions (W. A. Benjamin, New York, 1972).
3. K. P. Das and R. C. Hwa, Phys. Lett. 68B, 459 (1977).
4. R. C. Hwa, Proc. of IX International Symposium on High Energy Multiparticle Dynamics, Tábor, Czechoslovakia, 1978.
5. U. Sukhatme, Proc. of XIV Rencontre de Moriond, ed. J. Tran Thanh Van (1979); J. Królikowski, T. Hofmokl, and A. Jachołkowska, Acta Phys. Polonica B10, 1067 (1979).
6. See, for example, R. D. Field and R. P. Feynman, Phys. Rev. D 15, 2590

(1977); R. C. Hwa, A. J. Spiessbach, and M. J. Teper, Phys. Rev. Lett. 36, 1418 (1976); D. Sivers, S. J. Brodsky, and R. Blankenbecler, Phys. Rep. 23C, 1 (1976).

7. M. J. Teper, Rutherford Lab Report RL-78-022 (1978); L. Van Hove, Acta Physica Austriaca, Suppl. XXI, 621 (1979).
8. D. W. Duke and F. E. Taylor, Phys. Rev. D 17, 1788 (1978).
9. T. A. DeGrand and H. I. Miettinen, Phys. Rev. Lett. 40, 612 (1978); T. A. DeGrand, Phys. Rev. D 19, 1398 (1979).
10. E. Takasugi, Z. Tata, C. B. Chiu, and R. Kaul, Phys. Rev. D 20, 211 (1979).
11. E. Takasugi and X. Tata, Phys. Rev. D 21, 1838 (1980).
12. R. G. Roberts, R. C. Hwa, and S. Matsuda, J. Phys. G 5, 1043 (1979).
13. R. C. Hwa and R. G. Roberts, Z. Phys. C 1, 81 (1979).
14. R. Migneron and J. L. Robinson, Fermilab-Pub-80/72-THY (1980).
15. B. Buschbeck et al., Z. Phys. C 3, 97 (1979); B. Buschbeck, Proc of XI International Symposium on Multiparticle Dynamics, Bruges (1980); M. Markytan, Proc. of XX International Conference on High Energy Physics, Madison (1980).
16. F. C. Erne and J. C. Sens, CERN preprint (1978).
17. D. Cutts et al., Phys. Rev. Lett. 43, 319 (1979).
18. W. Aitkenhead et al., Phys. Rev. Lett. 45, 157 (1980).
19. E. Lehman et al., Phys Rev. D 18, 3353 (1978); N. N. Biswas et al., Phys. Rev. D 19, 1960 (1979); W. D. Shephard, Proc. of Tábor Symposium (1978).
20. J. Kalinowski, S. Pokorski, and L. Van Hove, Z. Phys. C 2, 85 (1979).
21. J. Kuti and V. F. Weisskopf, Phys. Rev. D 4, 3418 (1971).
22. R. C. Hwa, Phys. Rev. D 22, 759 (1980).
23. R. C. Hwa, Phys. Rev. D 22, 1593 (1980).
24. R. C. Hwa, Proc. of Bruges Symposium (1980).
25. A. Chodos et al., Phys. Rev. D 9, 3471 (1974); T. deGrand et al., Phys. Rev. D 12, 2060 (1975).
26. G. Altarelli, N. Cabibbo, L. Maiani and R. Petronzio, Nucl. Phys. B 69, 531 (1974); N. Cabibbo and R. Petronzio, Nucl. Phys. B 137, 395 (1978).
27. V. V. Anisovich, Phys. Lett. 57B, 87 (1975); A. Bialas, W. Czyz and W. Furmanski, Acta Phys. Polonica B8, 585 (1977).
28. A. De Rújula and F. Martin, Phys. Rev. D 22, 1787 (1980).
29. R. L. Jaffe and G. G. Ross, Phys. Lett. 93B, 313 (1980). R. L. Jaffe, Proc. of Madison Conference (1980).
30. See, for example, S. J. Brodsky and G. P. Lepage, SLAC-PUB-2447, Proc. of SLAC Summer Institute (1979).

31. P. C. Bosetti et al., Nucl. Phys. B142, 1(1978); J. G. H. deGroot et al., Phys. Lett. 82B, 292 (1979); 82B, 456 (1979).
32. R. C. Hwa and M. S. Zahir, Phys. Rev. D 23, (1981).
33. B. A. Gordon et al., Phys. Rev. D 20, 2645 (1979); D. Duke and R. G. Roberts, Nucl. Phys. B166, 243 (1980).
34. J. G. H. deGroot et al., Z. Phys. C 1, 143 (1979).
35. E. Gabathuler, Proc. of International Conference on High Energy Physics, Geneva, 1979; P. R. Norton, Proc. of Madison Conference (1980).
36. A. J. Buras and K. J. F. Gaemars, Nucl. Phys. B132, 249 (1978).
37. M. Glück and E. Reya, Nucl. Phys. B130, 76 (1977).
38. F. Martin, Phys. Rev. D 19, 1382 (1979).
39. C. B. Newman et al., Phys. Rev. Lett. 42, 951 (1979).
40. R. P. Feynman, M. Kislinger, and F. Ravndal, Phys. Rev. D 3, 2706 (1971).
41. Y. S. Kim and Noz, Phys. Rev. D 15, 335 (1977); Prog. Theor. Phys. 60, 801 (1978).
42. P. E. Hussar, Maryland preprint 81-169 (1981).
43. R. C. Hwa and C. S. Lam, OITS-158 (1980).
44. W. Ochs, Nucl. Phys. B118, 397 (1977).
45. R. D. Field and R. P. Feynman, Nucl. Phys. B136, 1 (1978).
46. V. Chang and R. C. Hwa, Phys. Rev. Lett. 44, 439 (1979); Phys. Rev. D 23, 728 (1981).
47. K. Konishi, A. Ukawa, and G. Veneziano, Phys. Lett. 78B, 243 (1978); 80B, 259 (1978); Nucl. Phys. B157, 45 (1979).
48. See the lectures of U. Sukhatme in this volume.
49. R. Brandelik et al., Nucl. Phys. B148, 189 (1979); H. F. W. Sadrozinski, Proc. of International Symposium on Lepton and Photon Interactions at High Energy, Hamburg, 1977.
50. S. Weinberg, Phys. Rev. D 11, 3583 (1975); H. Fritzsch, M. Gell-Mann and H. Leutwyler, Phys. Lett. 47B, 365 (1973); G. ´t Hooft, Phys. Rev. Lett. 37, 8 (1976); Phys. Rev. D 14, 3432 (1975).
51. V. Chang, G. Eilam and R. C. Hwa, OITS-156 (1980), to be published in Phys. Rev.
52. See the review by A. Capella in this lecture series.
53. G. Miller et al., Phys. Rev. D 5, 528 (1972).
54. E. L. Berger, Proc. of the Third Vanderbilt Conference, ed. R. S. Panvini and S. E. Csorna (AIP, New York, 1978).
55. R. D. Field and R. P. Feynman, Phys. Rev. D 15, 2590 (1977).
56. Fermilab Experiment No. E118 (private communication from G. W. Brandenburg,

P. Garbincius, and V.A. Polychronakos).

57. Q.B. Xie, OITS-164 (1981).

58. A. Suzuki, Nucl. Phys. B172, 327 (1980).

59. For a summary of ρ^{o} and K^{*} data, see E.A. de Wolf, Bruger Symposium (1980).

60. E. Takasugi and X. Tata, Phys. Rev. (to be published).
E. Takasugi, Bruges Symposium (1980).

61. T. DeGrand and H.I. Miettinen, Phys. Rev. D23, 1227 (1981).

62. G. Bunce et al., Phys. Rev. Lett. 36, 1113 (1976); K. Heller et al., Phys. Lett. 68B, 480 (1977); Phys. Rev. Lett. 41, 607 (1978); F. Lomano et al., Phys. Rev. Lett. 43, 1905 (1979).

63. E. Witten, Nucl. Phys. B120, 189 (1977); W.R. Frazer and J.F. Gunion, Phys. Rev. D20, 147 (1979); C.H. Llewellyn Smith, Phys. Lett. 79B, 83 (1978); R.J. DeWitt et al., Phys. Rev. D19, 2046 (1979).

64. V. Chang and R.C. Hwa, Phys. Lett. 85B, 285 (1979).

65. See the review by W. Kittel in this lecture series.

66. W. Lockman et al., Phys. Rev. Lett. 41, 680 (1978).

67. R.C. Hwa, talk given at EPS conference, Geneva (1979).

68. E. Takasugi and X. Tata, Osaka and Texas preprint (1981); see also E. Takasugi, Bruges Symposium (1980).

69. E. Lehman et al., Phys. Rev. D18, 3353 (1978); H. Abramowicz et al., Cracow Univ. preprint (1979); E.A. de Wolf et al. (to be published).

70. S.D. Drell and T.M. Yan, Phys. Rev. Lett. 25, 316 (1970).

71. D. Amati, R. Petronzio, and G. Veneziano, Nucl. Phys. B140, 154 (1978); B146, 29 (1978); R. Ellis et al., Phys. Lett. 78B, 281 (1978).

72. U.P. Sukhatme, K.E. Lassila, and R. Orava, Fermilab-Pub-81/20-THY (1981).

73. G.P. Lepage and S.J. Brodsky, Phys. Rev. Lett. 87B, 359 (1979); Phys. Rev. Lett. 43, 545 (1979); Phys. Rev. D22, 2157 (1980); S.J. Brodsky, T. Huang, and G.P. Lepage, SLAC-PUB-2540 (1980)

74. A. Duncan and A.H. Mueller, Phys. Lett. 90B, 159 (1980); Phys. Rev. D21, 1636 (1980); A.V. Efremov and A.V. Radyushkin, Phys. Lett. 94B, 245 (1980).

75. D.E. Soper, Phys. Rev. D15, 1141 (1977).

76. S.D. Drell and T.M. Yan, Phys. Rev. Lett. 24, 181 (1970); G.B. West, ibid. 24, 1206 (1970).

77. T.T. Chou and C.N. Yang. Phys. Rev. 170, 1591 (1968); T.T. Chou ibid. D19, 3327 (1979).

78. R. Carlitz, S.D. Ellis, and R. Savit, Phys. Lett. 64B, 85 (1977).

79. N. Isgur, G. Karl, and D.W.L. Sprung, RL-80-053 (1980).

80. Fermilab-Krakow-Warsaw-Washington Collaboration (private communication from

A. Zieminski).

81. J. Donoghue and K. Johnson, Phys. Rev. D21, 1975 (1980).

82. H. Pagels and S. Stokar, Phys. Rev. D20, 2947 (1979).

83. J.M. Cornwall, Phys. Rev. D22, 1452 (1980).

84. M. Scadron, Report on Progress in Physics (to be published).

85. A. Bialas and E. Bialas, Phys. Rev. D20, 2854 (1979); A. Bialas, Tabor Symposium (1978); Workshop on Ultra-relativistic Nuclear Collisions, LBL-8957 (1979); Bruges Symposium (1980).

86. T. Eichten et al., Nucl. Phys. B44, 333 (1972); S.A. Azimov et al., Phys. Lett. 73B, 500 (1978); P.L. Skubic, Ph.D. Thesis, Univ. of Michigan Report No. HE-77-32, 1977.

87. C. Halliwell (to be published); D.S. Barton, Bruges Symposium (1980).

88. A. Dar and F. Takagi, Phys. Rev. Lett, 44, 768 (1980).

89. R.C. Hwa, OITS-163 (1981).

90. G. Altarelli and G. Parisi, Nucl. Phys. B126, 298 (1977).

91. V.V. Anisovich, Yu M. Shabelsky, and V.M. Shekhter, Nucl. Phys. B133, 477 (1978).

92. For a review see L. Bergström, S. Fredriksson, G. Berlad, and G. Eilam, Phys. Report (to be published).

93. B. Andersson, G. Gustavson, and C. Peterson, Phys. Lett. 69B, 231; 71B, 227 (1977). See also the lecture by G. Gustavson in this volume.

94. G. Cohen-Tannoudji et al., Phys. Rev. D17, 2930 (1978); D19, 3398 (1979); D21, 2689 (1980); R. Peschanski, Bruges Symposium (1980).

95. A. Capella, U. Sukhatme, C.I. Tan, and J. Tran Thanh Van, Phys. Lett. 81B, 68 (1979); A. Capella, U. Sukhatme, and J. Tran Thanh Van, Z. Phys. C3, 329 (1980).

96. G. Veneziano, Nucl. Phys. B74, 365 (1974); B117, 519 (1976); H.M. Chan et al., ibid. B86, 479 (1975); B92, 13 (1975); G.F. Chew and C. Rosenzweig, ibid. B104, 290 (1976); Phys. Rep. 41C, 263 (1978).

97. Contrast the emphases of V. Sukhatme (Ref. 48) and A. Capella (Ref. 52) in these lectures.

98. After the completion of this manuscript I received a preliminary report from Nijmegen, in which an investigation of the type mentioned has been carried out.

Chapter V: DUAL FRAGMENTATION MODELS OF SOFT HADRON-HADRON AND HADRON-NUCLEUS INTERACTIONS

A. Capella

L.P.Th.P.E. - Orsay

Abstract:

A fragmentation model for hadron-hadron interactions is formulated in the S-matrix framework of dual models and Regge field theory. The starting point is derived from the topological approach to strong interactions and the 1/N expansion. The model provides a comprehensive description of the mechanism of multi-hadron production both in the central and fragmentation region. Extension to soft hadron-nucleus collisions is straightforward and does not involve any new concepts or assumptions.

1. Introduction

In these lectures I am going to review the fragmentation models which have been formulated in the framework of dual models.

The aim of these models (as well as of the recombination models), is to describe low-p_T physics in terms of the hadronic constituents. The latter have played an important role, first in the additive quark model and later on in the duality diagrams. Impressive evidence on the importance of the hadron constituents in low-p_T physics came when Ochs and the Lund group discovered two remarkable regularities of the low-p_T hadronic spectra at large x [1].

Ochs [2] remarked that there is a striking similarity between low-p_T hadronic spectra at large x and the structure function of pointlike constituents, namely

$$x \frac{d\sigma^{pp\to\pi^+}}{dx}(x) \sim u_p(x) \sim (1-x)^3 \qquad (1.1)$$

$$(x > 0.4 \div 0.5)$$

$$x \frac{d\sigma^{pp\to\pi^-}}{dx}(x) \sim d_p(x) \sim (1-x)^4 \qquad (1.2)$$

Note that the valence quark u is common to p and π^+, whereas the valence quark d is common to p and π^-.

More recently it has been shown [3] that a similar relationship is also true for $\pi^\pm \to \pi^\mp$, $K^\pm$,....

The above similarities are at the basis of the so-called recombination models [4]. In these models one assumes that the glue and sea of the colliding hadrons (which is concentrated at very low x) is responsible for multihadron production in the central region. The valence quarks, on the other hand, will pick up (recombine with) a parton from the sea to produce a hadron at large x. Since the sea parton has very small x, the momentum of the hadron will be essentially the same as that of the valence quark, hence eqs. (1.1) and (1.2).

The observation of the Lund group [5] is that there is an equally striking similarity (both in shape and absolute normalization) between low-p_T hadronic spectra in the fragmentation regions ($x_F > 0.2$), and the fragmentation functions measured in hard processes, namely

$$\frac{1}{\sigma}\frac{d\sigma^{\pi^+p\to\pi^+}}{dx}(x) = \tfrac{1}{2}[D^{u\to\pi^+}(x) + D^{\bar{d}\to\pi^+}(x)] = D^{u\to\pi^+}(x) \tag{1.3}$$

$$\frac{1}{\sigma}\frac{d\sigma^{\pi^+p\to\pi^-}}{dx}(x) = \tfrac{1}{2}[D^{u\to\pi^-}(x) + D^{\bar{d}\to\pi^-}(x)] = D^{u\to\pi^-}(x) \tag{1.4}$$

Eqs. (1.3) and (1.4) suggest that, as a result of the interaction, the q and the $\bar{q}$ in a meson are in a very asymmetric configuration, the q being very slow and the $\bar{q}$ very fast or vice-versa. This is indeed assumed to be the case in the first version of the fragmentation model proposed by the Lund group. In this model, the slow constituent is in fact supposed to be stopped (at x = 0) by the effect of the interaction. The hadronization of the fast constituent is governed by the same fragmentation functions as measured in hard processes [5].

For proton initiated reactions one has (see the lectures by H. Lubatti)

$$\frac{1}{\sigma}\frac{d\sigma^{pp\to\pi^+}}{dx}(x) = D^{uu\to\pi^+}(x) + D^{uu\to\pi^+}(x) \tag{1.5}$$

$$\frac{1}{\sigma}\frac{d\sigma^{pp\to\pi^-}}{dx}(x) = D^{uu\to\pi^-}(x) + D^{ud\to\pi^-}(x) \tag{1.6}$$

Eqs. (1.5) and (1.6), which seem to be true at a 10 ÷ 20% level of accuracy, suggest that a diquark in the proton (i.e.: two quarks acting in a coherent way) carries essentially all the momentum of the hadron - the odd quark being slowed down by the interaction.

Obviously, if eqs. (1.1) to (1.6) (and the corresponding ones for other processes) are well verified (as seems to be the case) it becomes a challenge to theorists to show why this is so, and in particular, why the r.h.s. of eqs. (1.3) - (1.4) and (1.5) - (1.6) have the same x-dependence at large x.

In these lectures I am going to show how a fragmentation model can be formulated in the S-matrix framework of dual models and Reggeon field theories. The model provides a comprehensive description of the mechanism of multihadron production, both in the central region and fragmentation region. Moreover, it can be extended in a rather straightforward way and without any new concepts or assumptions, to soft high energy collisions involving nuclei.

A related fragmentation model [6], of a non-dual type, is described in U. Sukhatme lectures.

I shall start with a brief review of the topological approach to strong interactions and the 1/N expansion. This provides the general framework of the dual fragmentation models and suggests a specific mechanism of multihadron production. Subsequently I will describe in detail the dual fragmentation model and its applications to hadron-hadron and hadron-nucleus interactions.

2. Topological Approach: The 1/N Expansion

The topological approach is very general in scope. It is based on the idea that the strong coupling constant is big and therefore one has to try to find a different expansion parameter allowing a perturbative treatment [7]. In dual models an expansion in powers of $1/N_f$ (where N_f is the number of flavors) was introduced in order to implement unitarity. The approach is known as Dual Topology Unitarization (D.T.U.) [8]. The physical arguments is favour of the reliability of such an expansion are the following. More than 20 years of experimental and phenomenological work in strong interactions has revealed beyond doubt that the final multi-hadron state is not produced in a random way, but with 'order'. The manifestations of such order are well known: dominance of short range correlations (short-range-order), limited p_T, small diffractive dissociation, jet-like final state, etc. Moreover, the resonance produced have rather narrow widths and are located on Regge-trajectories which are approximately linear. Also, the O.Z.I. rule is known to be approximately verified. Since order is closely related to topology one gets the idea of a topological approach in which the graphs with the simpler topologies are assumed to dominate. It can indeed be shown that dominance of the graphs with the simplest topology yields the properties of the multi-hadron final state mentioned above.

One could ask at this point why these properties are not exact and how one is led to a 1/N expansion. The answer is: because of unitarity. Indeed, once the graphs with the simplest topologies are introduced, the unitarity condition forces more complicated topologies to appear. The more the topology of a graph is complicated the more the order of the produced final state is violated. However, it turns out that these more and more complicated topologies are depressed by higher and higer powers of $1/N_f$ - hence the relevance of an expansion in powers of this parameter.

It can be shown from simple topological arguments that a general S-matrix element for a process with n external mesons can be written as

$$G^{(n)} \sim g^{n-2} \, (g^2N)^{\omega} \, (g^2)^{2h+b-1} \, \tilde{G}^{(n)} \qquad (2.1)$$

where g is the coupling constant, N the number of flavors, ω the number of quark loops and b and h numbers which characterize the topology of the graph. The complexity of the topology increases with increasing b and/or h. The quantity b represent the minimum number of quark loops to which one can attach the external pions; it is easy to show that a graph with a given b violates the O.Z.I. rule (n-1) times (see fig. 1). The quantity h characterizes the least complicated surface in which one can embed a given graph: a sphere (h=0), a sphere with one handle or torus (h=1), a sphere with n-handles (h=n). Thus the number h gives the J-plane singularity of the graph. Higher values of h correspond to multi-Regge cuts and the corresponding graphs violate short-range-order. The origin of the factor $(g^2N)^\omega$ in eq. (1) is obvious: every quark loop introduces a factor g^2N (fig. 2). Since the presence of quark loops does not change the topology of a graph, one should keep the product g^2N_F fixed and expand in powers of the $1/N_f$ expansion.

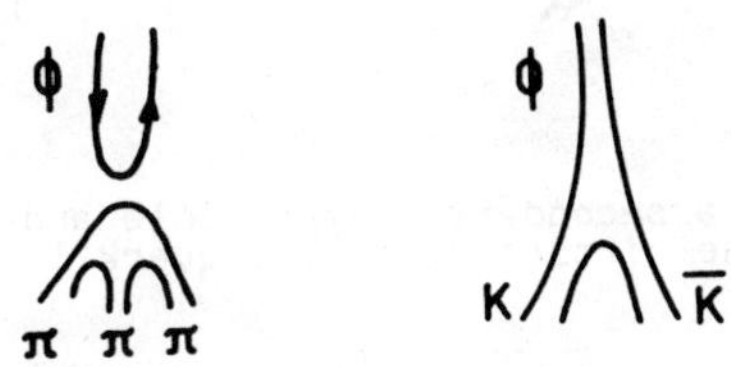

Fig. 1 Examples of planar (b=1) and non-planar (b=2) graphs. The second one violates the O.Z.I. rule

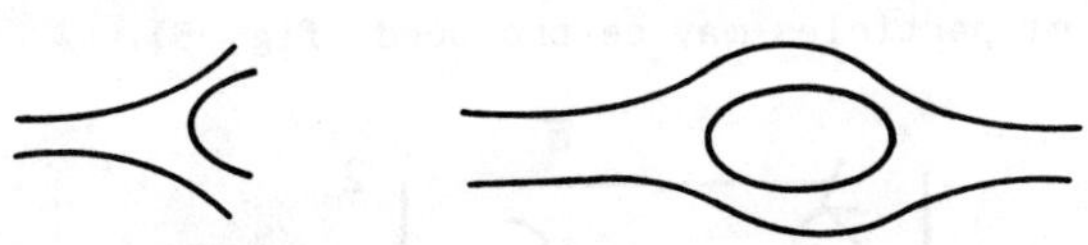

Fig. 2: Graphs representing the coupling constant g and a quark loop.

Let us consider the case n = 4. One gets from eq. (1)

$$G^{(4)} \sim (g^2)^{b+2h} \sim (\frac{1}{N})^{b+2h} \qquad (2.2)$$

In this case the graphs with the simplest topologies are:

<u>b = 1, h = 0:</u> order 1/N, corresponds to a planar graph, representing a sec-

ondary Regge pole (fig. 3).

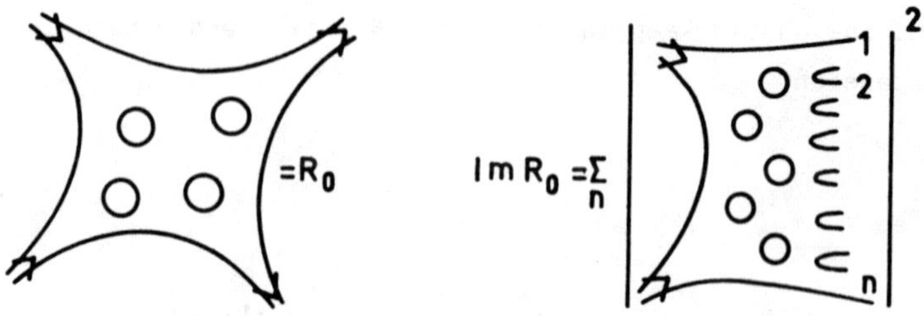

Fig. 3: Graphs for a secondary Regge pole and its s-channel discontinuity. The circles represent quark loops.

The s-channel discontinuity contains one chain or string of produced hadrons (two back-to-back jets).

b = 2, h = 0: order $(1/N)^2$, corresponds to the simplest non-planar topology (sphere or cylinder) and represents a Pomeron pole (fig. 4). The s-channel discontinuity contains two chains of produced particles

b = 1, h = 1: order $(1/N)^3$, represents an $R \otimes P$ cut

b = 2, h = 2: order $(1/N)^4$, corresponds to the topology of a cylinder with a handle or torus and represents a Pomeron-Pomeron ($P \otimes P$) cut. In this case four chains of particles may be produced (fig. 5).

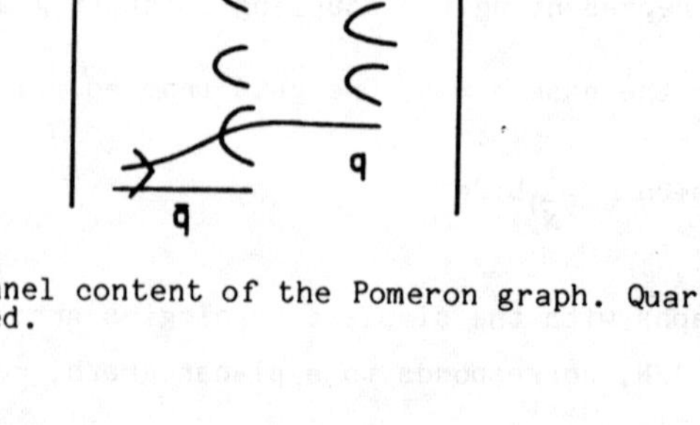

Fig. 4: s-channel content of the Pomeron graph. Quark loops have been omitted.

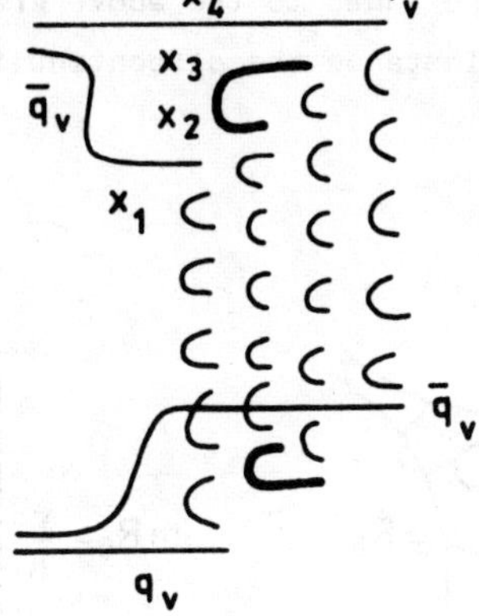

Fig. 5: Example of discontinuity of a graph representing a P⊗P cut. Quark loops have been omitted. Two of the chains are stretched between $\bar{q}$'s and $\bar{q}$'s from the sea.

The following remarks are in order:

a. In order to obtain the dominant graphs at higher energies one has to take into account both the order in the 1/N expansion and the energy dependence. The Reggeon graph corresponds to the lowest order (1/N) but is not dominant because it behaves like $(s)^{-1/2}$.

b. By comparing with the same s-dependence, it can be seen that the true expansion parameter is $(1/N)^2$. For instance P⊗P/P $\sim (1/N)^2$; R⊗P/R $\sim (1/N)^2$, etc.

c. The $1/N_f$ expansion provides a link between dual resonance model, and Reggeon field theory.

d. So far only quark loops are considered (no gluons, no color).

Obviously the motivation for the $1/N_f$ expansion (namely the necessity of finding a suitable expansion parameters when the coupling constant is large) is also relevant for Q.C.D. In fact 't Hooft [9] has suggested to use in this case an expansion in powers of $1/N_c$ where N_c is the number of colors (with $N_c g^2$ fixed). This $1/N_c$ expansion adds gluon loops to the naive parton models and provides a link between dual resonance models and gauge theories. However, it has in common with the former the lack of unitarity. For this reason, Veneziano [7],[8] has introduced an expansion in powers of 1/N with N_f/N_c fixed and, of course, Ng^2 fixed. In this way unitarity corrections (cuts) are introduced (see the discussion above) and a conceptual link between gauge theories and Reggeon field theory is obtained. Obviously in this way gluons are added to the above graphs (which so far contained only quark loops). For instance the discontinuity of the Reggeon graph is now the one of fig. 6.

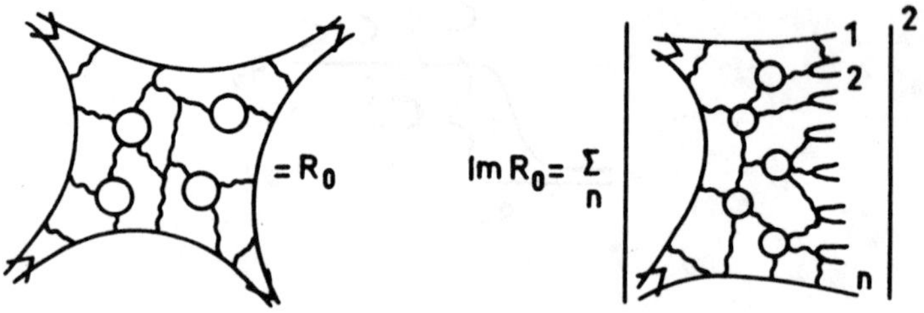

Fig. 6: Same as fig. 3 with gluons added.

With our present knowledge it is not possible to compute the contributions at the various orders of the 1/N expansion- although interesting attempts are now under way putting the theory on a lattice. Moreover, confinement has to be assumed, even at the planat level. However, in this way, we obtain a unified description of lepton and hadron processes, and a Q.C.D. inspired (dual string) picture of hadrons, satisfying the general requirements of an S-matrix theory. In the case of $e^+e^- \to$ hadrons and $\ell p \to \ell X$, the picture is well known (fig. 7).

The corresponding picture for purely hadronic processes, such as $h_1 + h_2 \to h_3 + X$ is rather different from other approaches. The main contributionat high energies is the Pomeron with a production mechanism as depicted in figs. 4 and 8.

Note that the flavor quantum numbers of the projectile divide themselves over the two chains (this is necessary for the Pomeron component since b = 2: see also the end of this section). For this reason all valence quarks of a colliding hadron cannot go into same produced particle - except in diffractive events which are not included in the Pomeron component (see

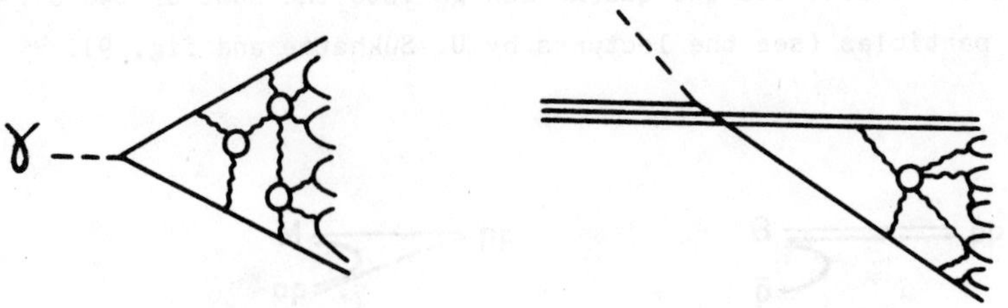

Fig. 7: Q.C.D. graphs for $e^+e^- \rightarrow$ hadrons and $\ell p \rightarrow \ell X$ (for the case in which the hard current hits a valence quark of the proton).

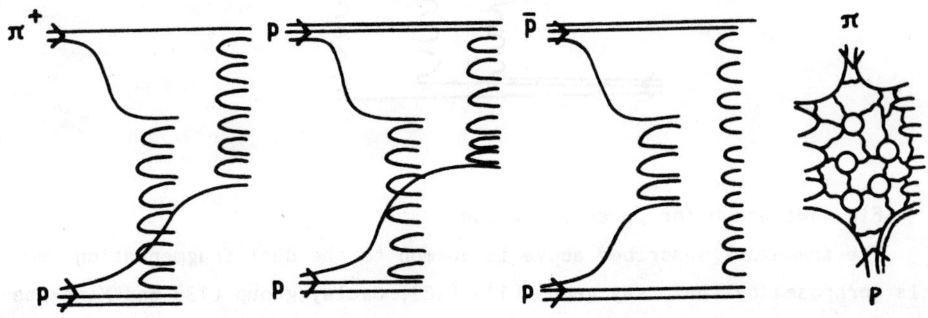

Fig. 8: The dominant production graphs for meson-baryon and baryon-baryon interactions. The last graph is the same as the first one, but includes the quark loops and gluons - which are always present but often omitted to simplify the figure.

also the discussion at the end of section 3). In addition, although the diquark belongs in one chain, its two quarks can go into the same or two different produced particles (see the lectures by U. Sukhatme and fig. 9).

Fig. 9: Diquark fragmentation

Note also that when more than two chains of hadrons are produced, the corresponding contribution is a non-dominant one. For instance in $\bar{p}p$ annihilation 3 chains are produced (fig. 10).

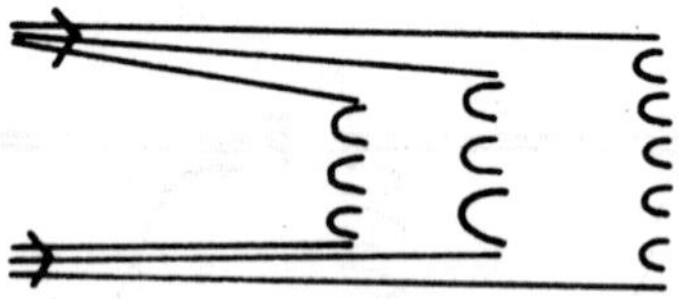

Fig. 10: Graph for $\bar{p}p$ annihilation

The framework described above is common to the dual fragmentation models proposed by the Orsay group [11]-[12], Saclay group [13] and Minekata [14]. Differences in the detailed formulations of these three models will be discussed in sections 4 and 5.

Before closing this section let me remind you of the physical arguments which are at the origin the two chain picture described above - i.e. the picture in which the valence quarks of the colliding hadrons divide themselves over the two chains. This is a specific feature of the dual picture, distinct from the pictures introduced by Low [15] and Nussinov [16] in which there is color exchange between the two colliding hadrons (in fact a one gluon exchange) and it is possible to form a single chain and to have all the valence quarks of the colliding hadrons sticking together. The physical origin of these two chains is just the Reggeized multiperipheral model, with clusters (resonances) produced in the final state (fig. 11).

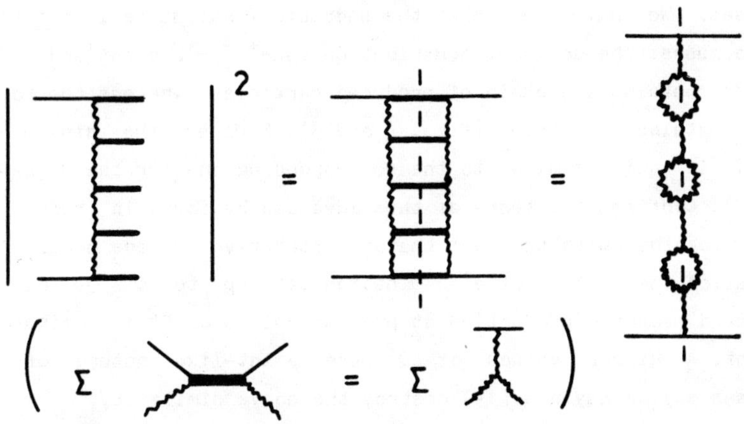

Fig. 11: Reggeized multiperipheral graph for cluster production. The wavy lines represent secondary Regge exchanges, and the thick lines are clusters (resonances). Implicit sums over the number of produced clusters and their masses are assumed. The last equality uses the finite mass sum rules shown in parenthesis.

We see from the last graph in fig. 11, that the Pomeron is coupled to the external particles via a secondary Regge trajectory (f-dominant Pomeron). This means that at the two ends of the Pomeron there will be a planar topology (corresponding to the secondary Regge pole) with only one chain. The average extension in rapidity of each of these two planar pieces is given by the size of the rapidity gap produced by a Regge pole, which is roughly equal to $1/\alpha_R \sim 2$ units. In order to have a secondary Regge pole exchange in these two-planar-rapidity-gap, it is necessary that one valence quark, from the neighboring hadron, travels through each of them. This produces a splitting of each colliding hadron into two colored systems, which will fragment forming two chains as described above.

In order to have a complete formulation of the dual fragmentation model, one has to define the momentum distribution of the valence quarks (border lines of the various chains) and the hadron spectra of each chain.

Before doing this in detail we desribe some experimental verifications of the two-chain picture.

3. Test of the Topological Picture for Lepton Hadron Induced Processes.

The first question one can ask is: are jets the same in soft and hard processes? We have seen that the hadronic final state is not the same in the two cases: the dominant contribution to $e^+e^- \to$ hadrons and $\ell p \to \ell X$ (fig. 7) contains one chain of produced particles, whereas the Pomeron component contains two chains (figs. 4 and 8). However, the single chain in fig. 7 is quite similar to the corresponding one for the Regge-component (fig. 3). Confirmative tests of this idea can be found in refs. 17. This tests, together with the fact that the iterative cascade model [18] explains most of the feature of $e^+e^- \to$ hadron data up to $Q \sim 30$ GeV, suggest that hard hadronic production at present values of Q^2 is dominated by confinement. At higher values of Q^2 the point-like nature of the hard processes may or may not [19] destroy the above similarity.

The main feature of the topological picture is the two-chain nature of the Pomeron component. An interesting test of the existence of these two chains has been proposed in ref. 20. The effect of Bose-Einstein statistics is to enhance the probability that two π^- are very close in rapidity (small $m_{\pi^-\pi^-}$); this is the G.G.L.P. effect [21]. This effect is reduced by the coherence (order) in a multiperipheral chain such as the one in $e^+e^- \to$ hadrons. Thus if the two π^- come from the different chains, the effect should increase. It should even be more important in $\bar{p}p$ annihilations where three different chains of hadrons are produced.

Another test of the overall consistency of the two chain picture has been proposed by the Orsay group [11]. Here the question asked is whether it is possible (taking into account the difference between the CM of each chain and the overall-CM) to reproduce all existing hadronic data assuming that the spectra of each chain is similar to the one in e^+e^-. Let us consider, for definitness, a proton-proton interaction and let x_0 be the fraction of the proton momentum taken by the slow quark. The fast fragment of each colliding hadron will then be $1 - x_0$. (*). It is easy to see that each

---

(*) This condition does not imply that the border lines of the chains are the only constituents of the colliding hadrons to carry energy-momentum. On the contrary, gluons and $q\bar{q}$ pairs are also present (see the last graph in Fig. 8). Such a condition expresses the fact that all the energy of the colliding hadrons is given to the particles produced in the two chains.

chain has then a momentum given by

$$P(x_0) \simeq 1/2 \sqrt{x_0(1-x_0)s}.$$

(We neglect the masses of the constituents at the ends of the chain). The difference between the C.M. of a chain and the overall C.M. is

$$\Delta(x_0) \simeq 1/2 \cdot \ell n(1-x_0/x_0).$$

If $x_0 = 1 - x_0 = 0.5$, the C.M. of a chain coincides with the overall C.M. The hadronic spectrum of a chain (number of particles of type h per unit rapidity) will then be given by

$$N^{\ell}_{qq,q}(y;P) \equiv \frac{dN^h}{dy}(z) = \begin{cases} z \cdot D^{qq\to h}(z) & , z > 0 \\ z \cdot D^{q\to h}(z) & , z < 0 \end{cases} \tag{3.1}$$

Here D^{qq} and D^{q} are standard fragmentation functions of the diquark and quark at the chain ends. Note that $dz/z = dy$ and $z = m^h_T \sinh [y/P_{max}(x_0)]$, where m^h_T is the transverse mass of the produced hadron and $P_{max}(x_0)$ the maximum momentum of hadron h in the chain.

In the general case of a chain having constituents C and C' at its two ends with momentum fraction x and x' respectively, one has instead (see fig. 12)

$$P(x,x') \simeq 1/2 \sqrt{x\, x' s}\ ; \quad \Delta(x,x') = 1/2 \ln\left(\frac{x}{x'}\right) \tag{3.2}$$

$$z = m^h_T \sin h\, [y - \Delta(x-x')/P_{max}(x,x')]$$

$$N_{CC'}[y - \Delta(x,x');\, P(x,x')] = \begin{cases} z\, D^{C\to h}(z) & \text{if} \quad y > \Delta \\ z\, D^{C'\to h}(z) & \text{if} \quad y < \Delta \end{cases} \tag{3.3}$$

In this way, the inclusive spectrum of each chain is defined in terms of quark and diquark fragmentation functions, which can be determined from hard scattering data (see H. Lubatti-lectures). Theoretical information on the $x \to 1$ behaviour of the fragmentation functions can, of course, be obtained from dimensional counting rules [22]. A complete theoretical determination of the fragmentation functions using, among other ingredients, dual graphs in the triple Regge-region can be found in ref. [23]. The spectra for $pp \to hX$ are then obtained just by adding the spectra of the two chains (see fig. 12). This does not assume that the two chains are uncorrelated. Indeed, the rapidity density is a linear quantity and in this case the average of a sum is equal to the sum of the averages no matter whether the added quantities are statistically independent or not. (*) A single experimental number allows then to determine x^0. Using the experimental heights of the inclusive plateau for pp at $\sqrt{s}$ = 20 GeV one obtains $x_0 \simeq 0.05$ [11]. All other features of the pp inclusive spectrum, (as e.g. its width) are then completely specified and turn out to be in agreement with the data (see fig. 13). Moreover, the scaling of the fragmentation functions implies that the energy dependence of the pp spectrum is also determined. One gets in this way a rise of the central plateau with s which is consistent with the observed rise over the ISR energy range. This rise is due to an increasing overlap between the two chains and is computed assuming a constant height for the plateaus of the chains. (**)

Note that essentially the same (small) value of x_0 is obtained in all reactions. Such a small value of x_0 implies that the quark and the diquark in a proton (or the quark and antiquark in a meson) live in a very asymmetric configuration. In the case of a baryon the diquark is very fast (on average) and the loose quark very slow. In the case of a meson both configurations (fast q and slow $\bar{q}$ and vice versa) have to be taken into account in order to reproduce the data. This asymmetry does not imply the existence of long range correlations. Indeed the rapidity difference between two particles at x = 0.95 and x = 0.05 is roughly two units. Such an asymmetry is precisely needed in order to understand the description of the fragmentation regions in eqs. (1.3)-(1.6). Indeed, as far as these equations are

---

(*) I thank A. Krzywicki for an interesting discussion on this point.

(**) Using eq. (3.2) one can check that when s varies over the ISR range, the energy of each chain varies roughly over the Spear range. Therefore, the energy rise of e^+e^- plateau observed at Petra does not affect the calculation of the pp plateau at ISR energies. However, the height of the $\bar{p}p$ central plateau at s = 540 GeV will tell us, within the above model, whether the rise of the e^+e^- plateau is relevant or not in computing the $\bar{p}p$ spectrum. The latter would obviously the case if the rise observed in e^+e^- were due to the opening of a threshold like $c\bar{c}$. We shall come back to this point later on.

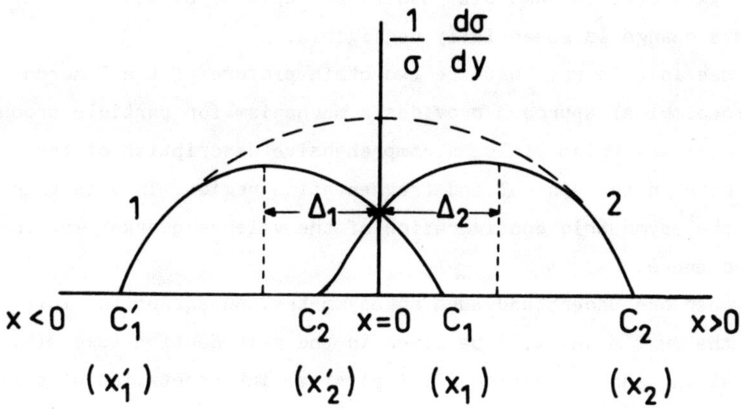

Fig. 12: The rapidity distribution of particles in each chain and their sum (dashed line) C_1, C_2, C'_1 and C'_2 represent the four colored objects at the chains ends and x_1, x_2, x'_1 and x'_2 their (fixed) momentum fractions. Δ is the distance between C.M. of a chain and the overall C.M. (this figure is only meant to illustrate the production mechanism. The actual calculations are presented in fig. 13).

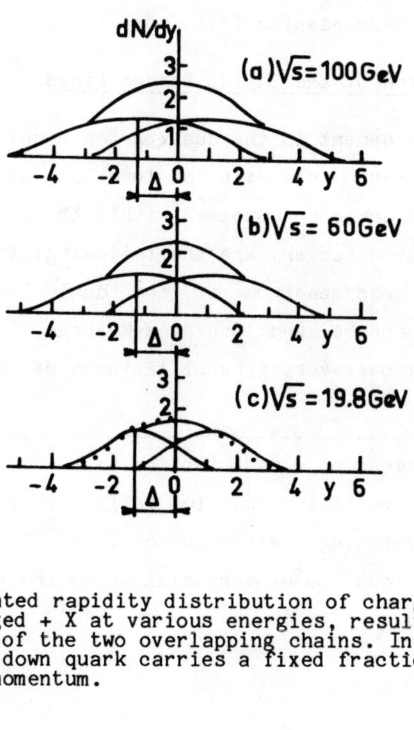

Fig. 13: The calculated rapidity distribution of charged particles in pp $\rightarrow$ charged + X at various energies, resulting from the superposition of the two overlapping chains. In this calculation the slowed down quark carries a fixed fraction (x_0 = 0.05) of the proton momentum.

concerned, the only difference between a quark stopped at $x_0 = 0$ or slowed down at $x_0 = 0.05$ is that D(z) has to be replaced by D(z/0.95). Numerically, such a change is essentially negligible.

We see in this way that the two chain picture of the Pomeron obtained in the topological approach provides a mechanism for particle production in low-p_t processes which yields a comprehensive description of the inclusive spectra both in the central and fragmentation region. This is true provided one has the asymmetric configuration of the valence quarks in the hadron described above.

How can one understand such an asymmetry? An answer to this question within the dual model will be given in the next Section (see also the discussion at the end of section 2). A physical interpretation of this asymmetry in a parton language goes as follows: A valence quark in a hadron has a momentum distribution function which behaves, near x = 0, as $x^{-1/2}$. Thus, the average momentum of a valence quark (border line of one chain) is rather small. By energy-momentum conservation, the average momentum of the border line of the second chain will be rather large, hence an asymmetric configuration. We shall see in the following that replacing the fixed value x_0 of the slow valence quark momentum by a distribution in $x^{-1/2}$ one obtains essentially the same results (*).

4. Momentum-Distribution of the Chains Border Lines

An important development in the dual-parton models occured when it was realized that the momentum distribution functions of the hadron constituents at the chain ends could be obtained within the dual model. The main steps of this important discovery are the following: in 1978, Chiu and Matsuda [24] derived the mass spectrum of the dual loop graph (Pomeron).

Simultaneously, Aurenche and Gonzalez-Mestres [25] obtained the same mass spectrum by using only very general features of the dual-Regge models,

---

(*) In fact, the average x of a distribution in $x^{-1/2}$ is much larger than 0.05. However, the fragmentation function falls with increasing values of x and the result of convoluting a distribution in $x^{-1/2}$ with the fragmentation function turns out to be very similar to the result obtained above with a fixed $x_0 \simeq 0.05$.

namely the dominance of Regge-singularities. More than one year later, the Orsay group [12], Saclay group [13] and Minakata [14] realized almost simultaneously that the above mass spectrum yields, in a rather straightforward way, momentum distribution functions for the border lines of the two chains in the Pomeron, which correspond precisely to the asymmetric configuration of the type described in the previous section - as needed in order to reproduce the experimental data.

In this section I shall sketch the arguments leading to the valence quark momentum distribution functions in the case of a two chain (Pomeron) diagram and indicate how these arguments can be generalized [26] in order to obtain the joint momentum distribution functions of valence and sea-quarks and antiquarks which appear in the multi-chain diagrams corresponding to higher orders in the 1/N expansion (multi-Pomeron cuts). The latter are very important for hadron-nucleus interactions.

a) Two-chain diagrams

Let us consider the two-chain graph for a meson-baryon interaction (fig. 8). Let x_j and $x_{\bar{j}}$ be respectively the x-values of the quark j and the antiquark $\bar{j}$ in a meson. For a proton with indices i, k, l we denote by x_i and $x_{kl} = 1 - x_i$ the x-values of the quark and the diquark, respectively. The corresponding rapidities will be denoted by y_j, $y_{\bar{j}}$, y_i and y_{kl} respectively. Let us consider the three rapidity gaps in fig. 8. Two of them ($y_{kl} < y < y_i$ and $y_j < y < y_{\bar{j}}$) are spanned by a single chain (planar topology) and are thus governed by secondary Regge trajectories of intercepts $\alpha_{i\bar{i}}$ and $\alpha_{j\bar{j}}$, respectively. The third rapidity gap ($y_i < y < y_j$) is spanned by two chains (non-planar topology) and is therefore governed by the Pomeron trajectory.

We are interested here in the behaviour of the valence quark momentum distribution functions near x = 0, i.e. when x_j and x_i in fig. 8 are small. In this situation the three rapidity gaps mentioned above are large (at high energy) and we can write

$$\frac{d\sigma}{dy_i dy_i} \sim \exp\left[\alpha_{i\bar{i}}\,(y_i - y_{kl}) + \alpha_p (y_i - y_i) + \alpha_{j\bar{j}}\,(y_j - y_j)\right] \qquad (4.1)$$

Using the approximate expressions: $y_j - y_{kl} \simeq Y^{lab}_{max}$, $x_j \simeq \exp(y_j - y_{\bar{j}})$, and $-x_i \simeq \exp(-y_i + y_{kl})$ one obtains:

$$\frac{d\sigma}{dx_i dx_j} \sim (-x_i)^{-\alpha_{i\bar{i}}+\alpha_p-1} (x_j)^{-\alpha_{j\bar{j}}+\alpha_p-1} \tag{4.2}$$

In order to obtain the behaviour of the momentum distribution functions near x=1, one has to repeat the above arguments for the case where the antiquark $\bar{j}$ and the diquark (kl) are slow. Using $x_{\bar{j}} = 1-x_j$ and $x_{kl} = 1-x_i$, one finds:

$$\frac{d\sigma}{dx_{kl} dx_{\bar{j}}} \sim [-(1-x_i)]^{-\alpha_{kl\bar{k}\bar{l}}+\alpha_p-1} . (1-x_j)^{-\alpha_{j\bar{j}}+\alpha_p-1} \tag{4.3}$$

Here $\alpha_{kl\bar{k}\bar{l}}$ is the intercept of a $(qq\bar{q}\bar{q})$ exotic trajectory exchanged in the rapidity gap $y_i < y < y_{kl}$ when the diquark (kl) is slow.

Assuming that the momentum distribution function is simply given by the product of the singularities and/or zeros given from eqs. (4.2) and (4.3), one gets:

1) for a valence quark i in a baryon (ikl),

$$\rho_1^q (x_i) \sim x_i^{-\alpha_{i\bar{i}}} (1-x_i)^{-\alpha_{kl\bar{k}\bar{l}}} \sim (x_i)^{-1/2} (1-x_i)^{1.5} \tag{4.4}$$

2) and for a valence quark i or an antiquark $\bar{j}$ in a meson,

$$\rho_1^q (x_i) \sim x_i^{-\alpha_{i\bar{i}}} (1-x_i)^{-\alpha_{j\bar{j}}} \sim (x_i)^{-1/2} (1-x_i)^{-1/2} \tag{4.5}$$

We have used the standard values for the Regge intercepts: $\alpha_p=1$, $\alpha_{i\bar{i}} = \alpha_{j\bar{j}} = 0.5$ (for u and d quarks) and $\alpha_{kl\bar{k}\bar{l}} = -1.5$.

Eqs. (4.4) and (4.5) show that the behaviour near x=0 of a valence quark momentum distribution function is identical to the corresponding one for the non-singlet structure function in deep-inelastic scattering. This may seem natural since the x 0 behaviour is governed in both cases by Regge singularities. However, one could also have expected that the momentum distribution functions of the valence quarks appearing in dual models would rather be governed by the hadronic wave function (and thus have a peak at $x \simeq 1/2$ for a meson, and $x \simeq 1/3$ for a baryon). Obviously, this is not the case. Concerning the behaviour near x=1, it can be seen from eqs. (4.4) and (4.5) that it is different from the corresponding one for the non-singlet structure functions: The power n in the $(1-x)^n$ factor appearing in the structure function is larger than the one in eqs. (4.4) and (4.5) (*). This was to be expected since it is easier for a hadron fragment to be fast when it is 'dressed' than when it is 'bare' (pointlike quark). As a consequence, the $(1-x)^n$ behaviour of the non-singlet structure function is of little relevance for low-p_t physics.

b) Multi-chain diagrams

It is obvious from figs. 5 and 10 that in order to compute multi-chain dual diagram one has to know the joint momentum distribution functions. For instance, the computation of the contributions of the diagrams in fig. 5 requires the knowledge of the joint probability $\rho(x_1,x_2,x_3,x_4)$ of finding in a proton (meson) a valence quark at x_1, a sea quark at x_2, a sea antiquark at x_3 and the diquark (antiquark) at $x_4 = 1-x_1-x_2-x_3$.

We now assume, as we did above for a single quark, that the joint momentum distribution functions are just given by the product of their relevant singularities and zeros. It can be shown [26] that, for valence quarks and diquarks, these singularities and zeros are the same as the corresponding one for a single quark momentum distribution, eqs. (4.4) and (4.5). Also the $x \simeq 0$ behaviour of a sea quark is, for all practical purposes $\sim x^{-1}$ [26]. We can thus write that the joint probability of finding, in a proton, a valence quark x_1, (n-1) sea quark-antiquark pairs at x_2, x_3, ...x_{2n-1} and the diquark at x_{2n} is given by

$$\rho_n(x_1,x_2,..,x_{2n})=C_n f_v(x_1) f_s(x_2).. \, f_s(x_{2n-1}) \, f_{qq}(x_{2n}) \cdot \delta(1- \sum_{i=i}^{2n} x_i) \quad (4.6)$$

Likewise, for a meson

(*) This difference in the power n might be due to the Q^2-dependence resulting from the Altarelli-Parisi evolution equation and/or the higher twist effects.

$$\rho_n(x_1,x_2,\ldots,x_{2n}) = C_n f_v(x_1) f_s(x_2) \ldots f_s(x_{2n-1}) f_v(x_{2n}) \cdot \delta(1 - \sum_{i=i}^{2n} x_i) \tag{4.7}$$

The factors C_n are determined in such a way that, for each value of n, ρ_n is normalized to unity. The $x \simeq 0$ behaviour of valence quarks, sea-quarks and diquarks is given by the functions $f_v(x)$, $f_s(x)$ and $f_{qq}(x)$ respectively, with

$$\begin{aligned} f_v(x) &= (x^2 + \mu^2/P^2)^{-1/4}, \\ f_{qq}(x) &= x^{1.5} \\ f_s(x) &= (x^2 + \mu^2/P^2)^{-1/2} \end{aligned} \tag{4.8}$$

Here the μ is a transverse mass and P the C.M. hadron momentum. The term μ/P is needed in f_s in order to avoid the singularity in x^{-1}. In the numerical calculations we take $\mu = 0.3$ GeV.

c) <u>Comparison of the momentum distributions function used in the various dual fragmentation models.</u>

The momentum distribution functions introduced above for the two-chain diagram are the ones used by the Orsay [12] and Saclay group [13] for pions and protons. Minekata [14] uses the same $x \simeq 0$ behaviour for the momentum distribution function of a valence quark in a hadron as obtained above, namely $x^{-0.5}$. He then assumes that the complete function is given by

$$\rho_1^{q/\pi}(x) \sim \int_0^1 dy\ (xy)^{-1/2}\ \delta(1-x-y) \sim x^{-1/2}(1-x)^{-1/2} \tag{4.9}$$

for a pion projectile, and by

$$\rho_1^{q/p}(x) \sim \int_0^1 dy \int_0^1 dz\ (xyz)^{-1/2}\ \delta(1-x-y-z) \sim x^{-1/2}(1-x)^0 \tag{4.10}$$

for a proton projectile. Obviously, in the case of a pion, this function is identical to the one obtained above. However, for a baryon projectile, the $x \simeq 1$ behaviour used by Minekata is not consistent with the dual model result. (No phenomenological analysis has been performed so far using Minekata's prescription in the proton fragmentation region).

Note also that there is some ambiguity in the determination of the momentum distribution function in the case of a kaon. Indeed, in this case $\alpha_{s\bar{s}} < \alpha_{u\bar{u}}$ (see eq. 4.5). Therefore, the symmetry between x = 0 and x = 1 is lost and the problem arises of determining the relative normalization at the two end points. This can only be done by introducing some extra assumptions. A specific prescription has been given by Hassouni and Napoly [27]. Minekata [14] neglects the difference between $\alpha_{s\bar{s}}$ and $\alpha_{u\bar{u}}$ and uses eq. (4.9) for all flavors.

It has been suggested by both the Orsay and the Saclay group that, eqs. (4.4) and (4.5) could be slightly changed in order to express them as a sum of structure functions. For instance, in the case of a meson projectile, one could use instead of eq. (4.4).

$$\rho^q(x) \sim \frac{1}{\sqrt{x}} \cdot (1-x)^n + \frac{1}{\sqrt{1-x}} \cdot x^n \qquad (4.11)$$

Here the power n is the one governing the x = 1 behaviour of the pion structure function. The two terms in this formula correspond to the structure functions of a quark (with momentum fraction x) and an antiquark (with momentum fraction 1-x). Numerically this formula gives essentially the same results as the factorized forms in eq. (4.4). However, it is not clear that eq. (4.11) is more in the spirit of Q.C.D. Indeed, as stressed above, eqs. (4.4) and (4.5) might be the low $-p_T$ limits of the structure functions measured in hard processes.

5. Formulation of the Model and Numerical Results

The exact formulation of the model presented in this section is the one introduced by the Orsay group [11] [12]. The other two formulations [13] [14] are approximate in that they neglect the difference between the C.M. of a chain and the overall C.M. - and are thus not suitable to study the central region (see section 3). In the fragmentation regions (where such an approximation is harmless) the three versions of the model are essentially equivalent (except for the differences between Minekata and the other two approaches in the baryon fragmentation region, discussed in section 4).

In section 3 we have computed the rapidity distribution of the hadrons produced in one chain, under the assumption that the four colored objects at the chain ends have a fixed value of x (see eq. (3.2) and fig. 12). In the general case, these x-values have the distributions given in eqs. (4.4) and (4.5). Thus the rapidity distribution of particles of type h produced say, in chain 1, is given by

$$N_1^h(y) = \int_0^1 dx_1 \int_0^1 dx_1' \; \rho^{C_1}(x_1)\, \rho^{C_1'}(x_1')\, N_{C_1C_1'}[y - \Delta(x_1,x_1') \; ; \; P(x_1,x_1')] \quad (5.1)$$

Here $N_{C_1C'_1}$ is the corresponding distribution for fixed values of x (see eq. (3.3)) and $\rho^C(x)$ and $\rho^{C'}(x)$ are the momentum distribution functions of the two border lines of chain 1. For the main contribution (Pomeron), where two chains are present, the functions $\rho^C(x)$ are given by eqs. (4.4) and (4.5). As explained in sect. 3, the total rapidity distribution is just the sum of the rapidity distributions of each chain. The model is then completely specified, without any free parameters. Let us discuss the numerical results. We consider first the central region. For pp interactions the charged particle rapidity distribution is well reproduced (height, shape, and energy dependence). (see figs. 14 and 15). The results are rather similar to those obtained in section 3 with a fixed value of $x_0 = 0.05$ for the momentum fraction carried by the slow quark. For $p\bar{p}$ and πp interactions, the charged particle distributions are similar to the one for pp interactions and in agreement with the data. Note however, that one has (fig. 14)

$$\left.\frac{dN}{dy}\right|_{y^*=0}^{\bar{p}p} \qquad \left.\frac{dN}{dy}\right|_{y^*=0}^{\pi p} \qquad \left.\frac{dN}{dy}\right|_{y^*=0}^{pp} \qquad (5.2)$$

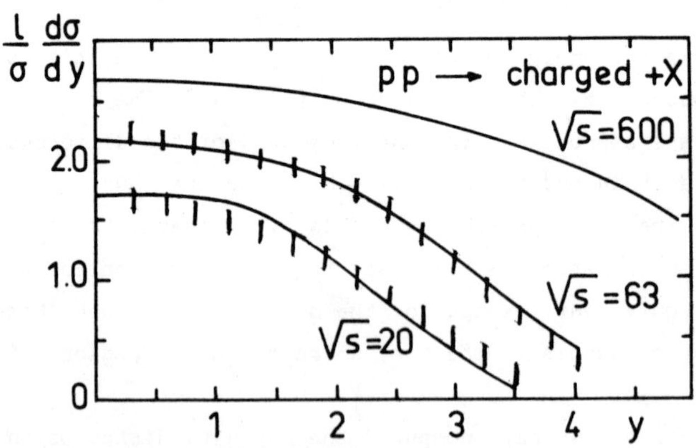

Fig. 14: Same as fig. 13, in the general case where the slowed down quark has a momentum distribution given by eq. (4.4). The dashed area represents the experimental data from ref. 34.

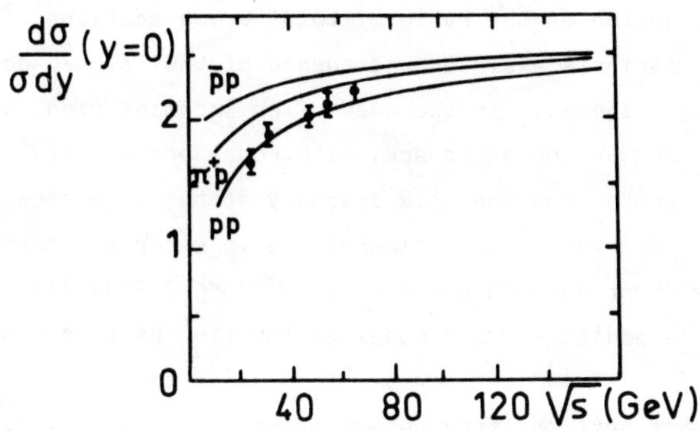

Fig. 15: The height of the central plateau as a function of energy

This is due to the fact that the most probable configuration in pp interactions is that of two overlapping chains of equal length (fig. 12). On the contrary, for $\bar{p}p$ interactions, one can see from fig. 8 that the most probable configuration consists of two chains centered at $y^* = 0$, one of which is very long and the other very short. In the case of a πp interaction both types of configurations are present.

On of the main features of the model is the presence of slowed-down quarks. Thus it is important to look for some signatures of these quarks. The relations in (5.2) are an indirect consequence of their presence. A somewhat more direct signature is provided by the ratios R of π^+ over π^- at $y^* = 0$. A simple quark counting shows that the ratio of the number of slowed-down valence quarks of type u (or $\bar{d}$) divided by the corresponding number of d (or $\bar{u}$) quarks is 4,2,1 and 2/3 for π^+p, pp, a $\bar{p}p$ and π^-p interactions respectively. Thus one has at $y^* = 0$

$$R^{\pi^+p} > R^{pp} > (R^{\pi^-p})^{-1} > R^{\bar{p}p} = 1 \tag{5.3}$$

These ratios are energy dependent and approach one (roughly as $s^{-1/2}$) when $s \to \infty$. At $\sqrt{s}$ = 20 GeV their values, as calculated in the model, are: 1.14, 1.1, 1.05, and 1, respectively. A precise measurement of these ratios would provide an important test of the model.

Before leaving the central region, let me indicate that the model also provides an explanation of the ratio of total cross sections $\sigma^{\pi p}/\sigma^{pp} \simeq 2/3$. The value of this ratio is again a consequence of the 'slowed-down valence quark' phenomenon. Indeed, in the case of an incoming pion, one can slow down either the quark or the antiquark, with equal probability. In the case of an incoming proton one can slow down any of the 3 valence quarks with equal probability. However, the configurations in which a diquark is slowed down are strongly suppressed. Hence a ratio $\sigma^{\pi p}/\sigma^{pp} \simeq 2/3$. This is the only alternative for the additive quark model explanation of this ratio that I am aware of.

Let us consider next the fragmentation region. It is obvious from energy-momentum conservation, that, for large enough x, only one chain can contribute. In the case of $\pi^+ p \to \pi^+ X$ and $\pi^+ p \to \pi^- X$ the main inputs are the $D_{u\to\pi^+}$ (favored) and $D_{u\to\pi^-}$ (unfavored) fragmentation functions, respectively. The results [12] are shown in figs. 16. The input function $D_{u\to\pi^+}(x)$ is also shown. The difference between input and output is very small. We see in this way that eqs. (1.3) and (1.4) are approximately satisfied in the model. Similar results have been obtained by the Saclay groups (fig. 17) and Minakata (fig. 18). The reaction $K^+ \to \pi^\pm$ has been studied by Minakata [14] and Al Hassouni and Napoly [27]. The results are shown in figs. 19 and 20, respectively. I remind you that the two models are different in the case of a kaon projectile (see section 4c).

For the proton fragmentation region, it is obvious from the results mentioned in the Introduction (eqs. 1.5 and 1.6), together with the similarity between the input fragmentation function and the output spectrum demonstrated in fig. 16, that the model will give a correct description of the data - at least at a semi-quantitative level. Numerical results, using as an input the diquark fragmentation functions presented by Lubatti, are not yet available.

So far we have considered single particle inclusive spectra only. In a recent paper Ng, Lassila and Sukhatme [28] studied the reaction $pp \to ph^\pm X$. The analysis follows the model described in the previous sections, the only difference being that the chain where the measured fast proton is produced now has an antiquark (instead of a diquark) at its projectile end. This antiquark is the one left-over after production of the fast proton (see figs. 21). The momentum of this antiquark is $x_{\bar{q}} = 1 - x_1 - x_p$, where x_1 and x_p are the momentum fractions of the slowed down valence quark of the incoming

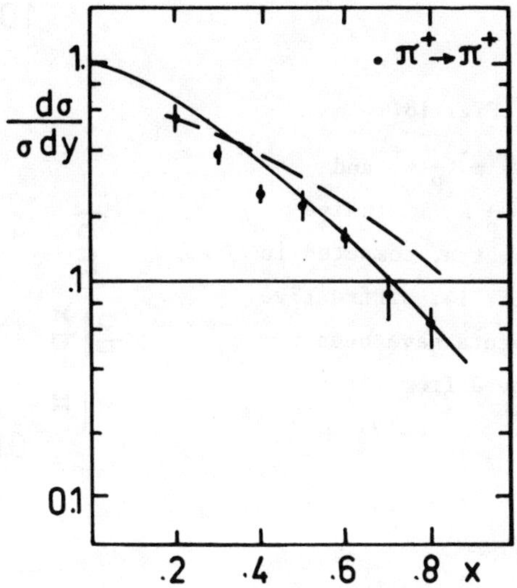

Fig. 16.

The $\pi^+ \underset{p}{\to} \pi^+$ inclusive spectrum computed in ref. 12. The dashed lines show the input $x\,D^{u\to\pi^+}(x)$. Diffractive events have been removed from the data.

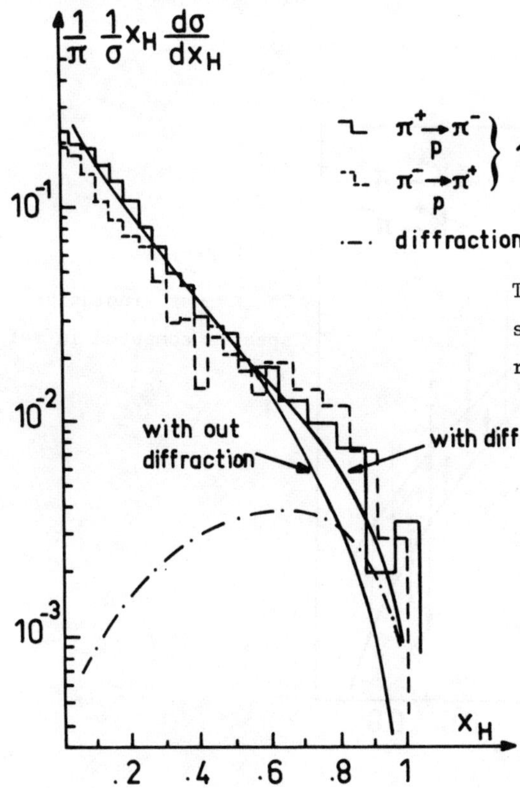

Fig. 17.

The $\pi^\pm \underset{p}{\to} \pi^\mp$ inclusive spectra computed in ref. 13

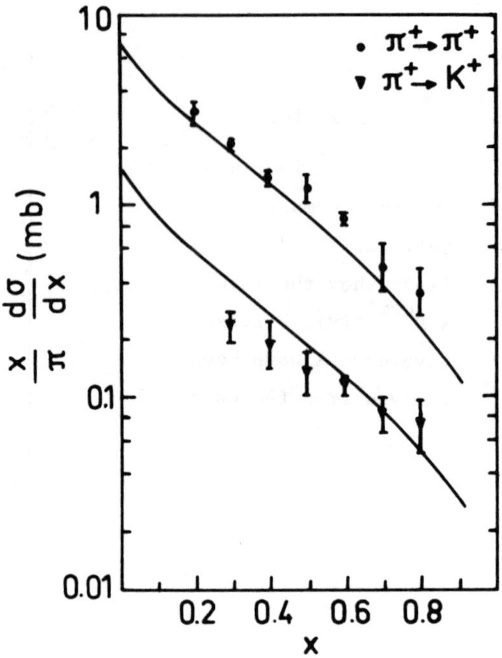

Fig. 18.

The $\pi^+ \underset{p}{\to} \pi^+$ and $\pi^+ \underset{p}{\to} K^+$ inclusive spectra, computed in ref. 14. Diffractive events have been removed from the $^+$ $^+$ data.

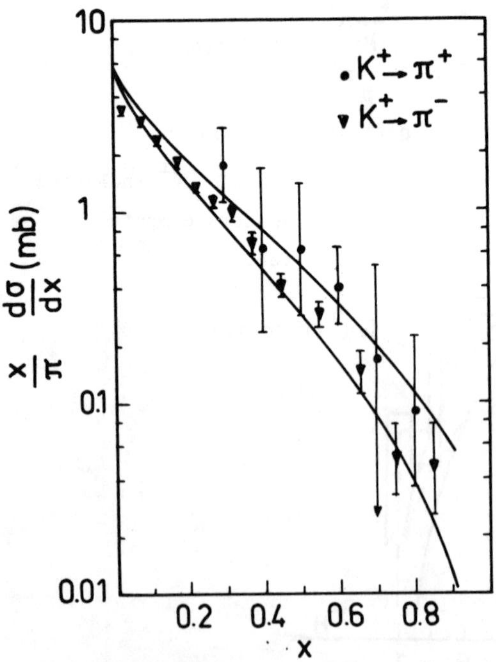

Fig. 19.

The $K^+ \underset{p}{\to} \pi^\pm$ inclusive spectra computed in ref. 14.

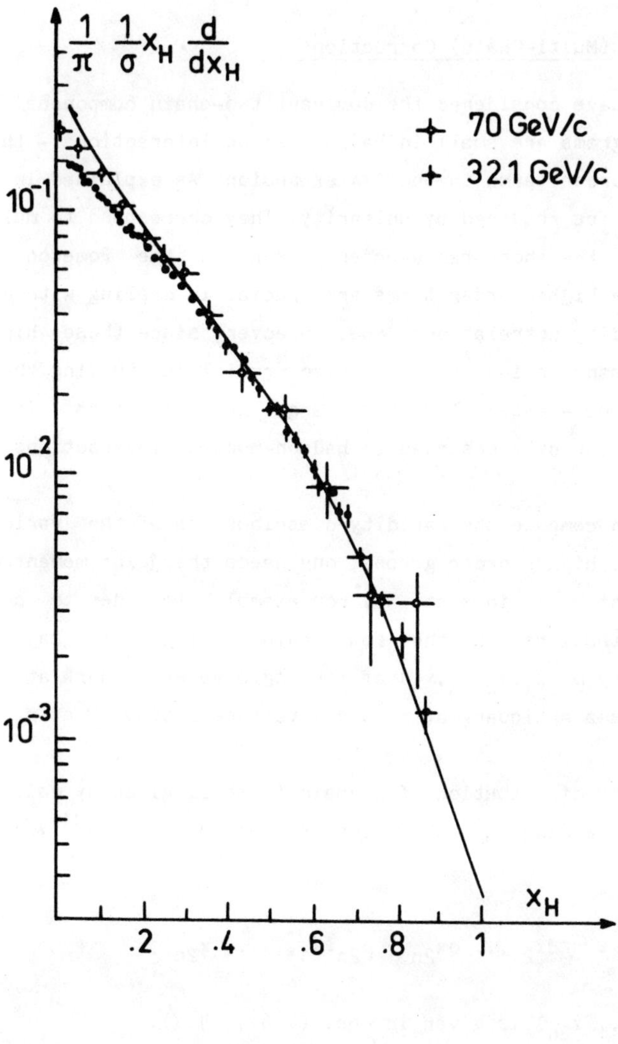

Fig. 20. The $K^+ \underset{p}{\to} \pi^-$ inclusive spectrum, computed in ref. (27).

proton and that of the produced proton respectively. The results are in excellent agreement with the data. They are displayed in figs. 21.

6. Higher-order (Multi-Chain) Corrections

So far we have considered the dominant two-chain component (Pomeron). Multi-chain diagrams are small in hadron-hadron interactions - they correspond to higher order terms in the 1/N expansion. As explained in sect. 2, such diagrams are required by unitarity. They correspond to multi-Pomeron cuts and violate the short-range-order present in the Pomeron component. Therefore, these higher order terms are crucial in dealing with dispersion, long-range rapidity correlations, etc. Moreover, since these higher order terms contain many chains, they are also crucial in studying the topological cross sections - especially for large multiplicities. In addition, these terms are strongly enhanced in hadron-nucleus interactions (see sect. 7).

In order to compute the rapidity distributions of the various chains present in these higher order graphs, one needs the joint momentum distrributions functions given in sect. 4b. For example, in order to compute the rapidity distributions in the four chains of fig. 5 one has to know the joint probability $\rho\ (x_1,x_2,x_3,x_4)$ of finding a valence quark at x_1 at sea quark at x_2, a sea antiquark at x_3 and a valence antiquark at $x_4 = 1 - x_1 - x_2 - x_3$.

The rapidity distribution of a chain is still given by eq. (5.1), except that, in a diagram with 2n chains, the functions $\rho^{C1}(x_1)$ have to be replaced by

$$\rho^{C1}(x_1) = \int dx_2 \ldots dx_{2n} \cdot \rho_{2n}(x_1, \ldots, x_{2n}), \qquad (6.1)$$

where $\rho_{2n}(x_1, \ldots x_{2n})$ is given in eqs. (4.6), (4.7).

Note that although ρ depends on only one variable (all variables other than the one of the constituent at the end of the chain have been integrated over), it depends now on n (the number of chains in the diagram).

It is now easy to compute the single particle rapidity distributions. The formulae are very similar to the ones given in the next section for a hadron-nucleus interaction. One can also compute two-particles correlations - but this requires the extra assumption of statistical independence of the

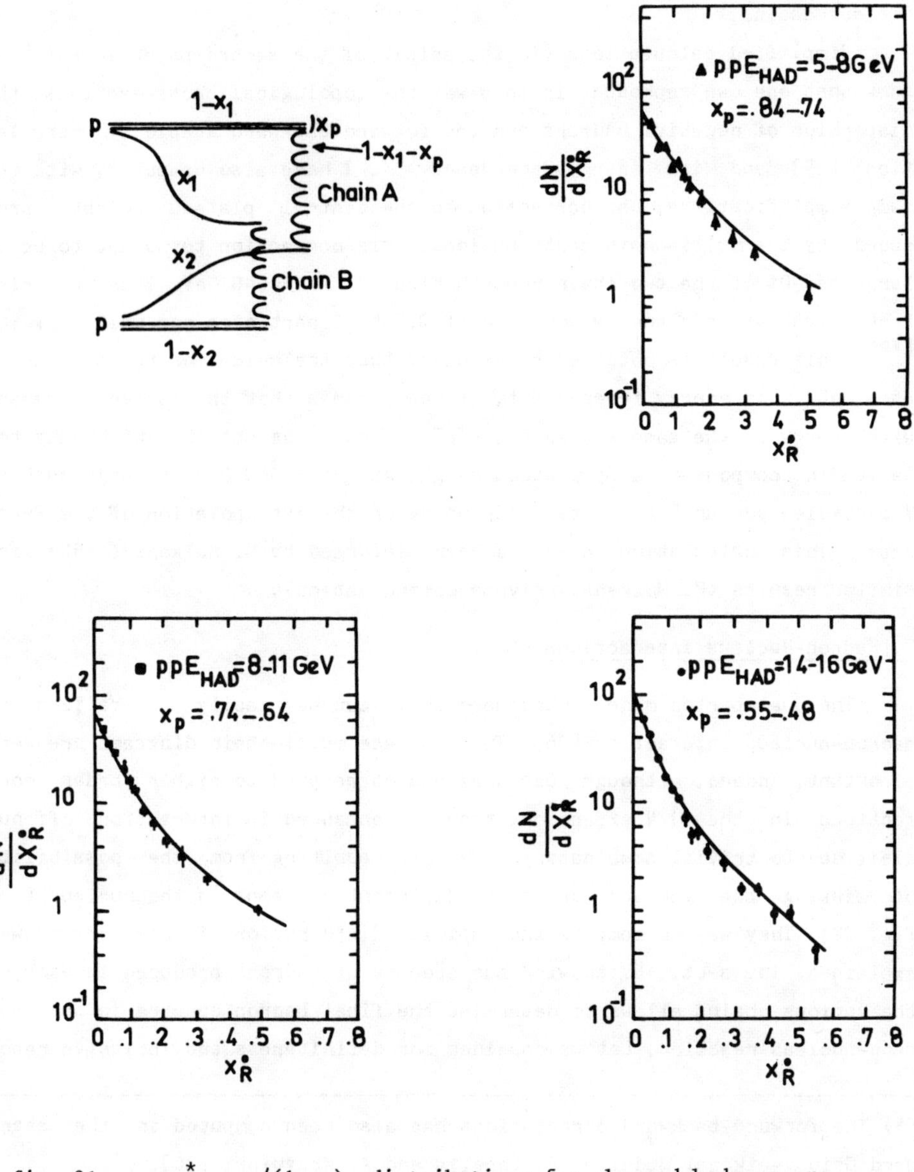

fig. 21. $x_R{}^* = x/(1-x_p)$ distributions for charged hadrons produced in the same hemisphere as the leading proton, for various leading momentum fractions (x_p) and hadron energies ($E_{HAD} = P(1 - x_p)$; $P = \sqrt{s}/2$).

various chains.

Simplified calculations (in the spirit of the second paper in ref. 26) show that one can reproduce in this way the topological cross-sections, the dispersion of negative hadrons and the forward-backward rapidity correlation [29] and its energy-dependence (*). I have also computed, with the same simplifications, the correction to the central plateau height, produced by the multi-chain contributions. This correction turns out to be as large as 30% of the two chain contribution at $\sqrt{s}$ = 540 GeV. Thus, the plateau height at this energy would be of 3.5 to 4 particles per unit rapidity (**). This result is obtained by assuming that the height of the plateau in each chain is energy independent. If one assumes that the latter increases with energy in the same way as the e^+e^- plateau, one obtains, with only the two-chain component, a $\bar{p}p$ plateau height at $\sqrt{s}$ = 540 GeV as large as 5 to 7 particles per unit rapidity - depending on the extrapolation of the Petra data. This calculation has also been performed by H. Mulkens (CERN) with similar results (H. Mulkens, private communication).

7. Hadron-Nucleus Interactions

The dual-parton model introduced above can be easily generalized to hadron-nucleus interaction [26]. In this case multi-chain diagrams are very important. Indeed, although such diagrams correspond to higher order corrections in the 1/N expansion, they are enhanced in interactions off nuclei, due to trivial combinatorial factors resulting from the possibility of linking the various chains to different nucleons in the nucleus (see fig. 22). They way to compute the rapidity distribution of one chain was explained in sect. 6. Knowing the spectra of hadrons produced in each of the various chains allows to determine the final hadron spectra in the hadron-nucleus reaction. Let us consider for definiteness the inclusive reac-

---

(*) The forward-backward correlations has also been computed in the standard Gribov-eikonal model by A. Capella and A. Krzywicki [30].

(**) However, graphs containing three-Pomeron coupling have not been included in this calculation and they might be important a the high energies considered here.

tion p + A → h + X, where A denotes a nuclear target with atomic mass number A. Let σ_n be the cross section for n inelastic collisions with n nucleons of the target. Each inelastic collision corresponds to a Pomeron exchange. Thus a diagram with n inelastic collisions contains 2n chains. The rapidity density of the produced hadron h is given by:

$$\frac{dN^{pA\to hX}}{dy}(y) = \frac{1}{\sigma_{in}^{hA}} \cdot \frac{d\sigma^{pA\to hX}}{dy} = \sum_{n=1}^{A} \frac{\sigma_n}{\sigma_{in}^{hA}} \{ N_1^h(y,n) + N_4^h(y,n) + (n-1)[N_2^h(y,n) + N_3^h(y,n)] \} \quad (7.1)$$

The expressions for the inclusive single particle spectra N_α^h of hadron h, for each type of chains present in hadron-nucleus interaction, are obtained from eqs. (5.1) and (6.1). Eq. (7.1) is obvious if one keeps in mind that, for each value of n, chains 1 and 4 appear only once, whereas chains 2 and 3 appear (n−1) times each.

Neglecting for a moment the n-dependence of N^h (which turns out to be very small throughout the whole central region) one gets from eq. (7.1)

$$\frac{dN^{pA\to hX}}{dy}(y) = N_1^h(y) + N_4^h(y) + (\bar{\nu} - 1)[N_2^h(y) + N_3^h(y)] \quad (7.2)$$

where $\bar{\nu} = A\,\sigma_{in}^{pp}/\sigma_{in}^{pA}$. In order to obtain eq. (7.2), one has to use the definition of $\sigma_{in}^{hA} = \sum_{n=1}^{A} \sigma_n$, together with the well-known A.K.G. cancellation [31]:

$$\sum_{n=1}^{A} n\,\sigma_n = A\,\sigma_{in}^{pp} \quad (7.3)$$

In order to illustrate the qualitative features of the model let us continue to examine the simple situation in which the N_α^h are independent of n. In this case N_1^h and N_4^h are the same as in a proton-proton collision. If one furthermore makes the very crude assumption that $N_1^h(y) = N_3^h(y)$ (in view of the similarity between chains 1 and 3) and neglects chain 2 (which has, at present energies, a low average invariant mass), one obtained from eq. (7.2)

$$R \equiv \left[\int \frac{dN^{hA}}{dy}\,dy \right] / \left[\int \frac{dN^{hp}}{dy}\,dy \right] = \tfrac{1}{2} + \tfrac{1}{2}\,\bar{\nu} \quad (7.4)$$

where the integral extends over the whole central region. Eq. (7.4) is in rather good agreement with present experimental data. The exact calculations presented below, give an even better agreement.

Notice that, when the energy increases, all the chains develop energy independent plateaus of identical height and it is no longer possible to neglect the contributions of chain 2. In this high energy limit, we obtain

$$R = \bar{\nu} \tag{7.5}$$

This result is a general consequence of unitarity in multiple scattering models provided the triple Pomeron coupling is neglected.

Since chains 2 and 3 are essentially the same for all reactions, we can see from eq. (7.2) that the 'excess' of produced particles (as compared to a hadron-nucleon interaction) depends only on the value of $\bar{\nu}$. This so-called $\bar{\nu}$-universality is borne out by the data.

Let us now discuss the projectile fragmentation region (large x). The only important contribution come from chain 4 which contains the fast diquark. Therefore, if N(y;n) were independent of n in the projectile fragmentation region, one would get

$$R(y) = \left(\frac{dN}{dy}\right)^{hA} : \left(\frac{dN}{dy}\right)^{hp} = 1 \tag{7.6}$$

However, it turns out that there is an important n-dependence in this region - which is introduced through the joint momentum distribution functions (eq. 6.1). As a consequence the ration R(y) becomes smaller than 1. If we parametrize it as $R(y) = A^{\alpha(y)}$, $\alpha(y)$ becomes negative in the projectile fragmentation region, and increases, in absolute value, for increasing values of the rapidity (or momentum fraction x) of the produced particle. This effect (called attenuation) is largely due to energy-momentum conservation; the energy taken up by the inelastic collisions within the nucleus is mainly provided by removing energy from the fast constituent at the projectile end of chain 4. As n increases, chain 4 becomes shorter; thus $N_4(y;n)$ decreases. This effect is larger for larger values of y. The 'cross-over' point y_0, where $R(y_0) = 1$, turns out to be independent of both s and A.

The numerical results are presented in figs. 23-32. In the reactions where the fragmentation functions are not known from hard scattering data, we have determined them by a fit to the hydrogen data (or another reference nucleus). The model then give the A-dependence of the data.

In all calculations we have used the σ_n's obtained from the Glauber-Gribov model, in the eikonal approximation, using standard Saxon-Woods nuclear densities with σ^{pp}_{ND} = 30 mb and $\sigma^{p\pi}_{ND}$ = 20 mb. The resulting σ_n's yield a good description of the inelastic pion-nucleus and proton-nucleus cross sections at a 10% accuracy level.

The above model of hadron-nucleus interactions is of a multiple scattering type [32]. However, the introduction of the parton constituents in the framework of D.T.U. makes it much more constrained and enhances its predictive power. In particular it gives a specific partition of the available energy among the various inelastic collisions. This is particularly important in computing the attenuation in the projectile fragmentation region.

Finally, I would like to mention that an interesting model of hadron-nucleus interactions, also based in the two-chain dual parton model of ref. (11 - 12) has been introduced by Chao at al [23]. Although based on the same mechanism of hadron-nucleon interactions as the model described above, the two models are rather different. The model of ref. [33] is based on the following idea. During the time needed for the projectile to travel from one nucleon in the nucleus to the next, chain II, which contains fast particles, has no time to materialize (see fig. 33). One assumes then that this excited state acts as a newly formed projectile, which interacts with another nucleon, as shown in fig. 33. Denoting by $N_I(y)$, $N_{II}(y)$ and $N_{III}(y)$ the rapidity distributions of these three types of chains and assuming that the probability of n inelastic collisions is given by the same σ_n's in the Gribov-Glauber model as above (*), one finds (using eq. 7.1))

(*)The above σ_n's can be obtained from the non-enhanced graphs of the Reggeon calculus in the eikonal approximation. The discontinuity of such graphs corresponding to, say, two inelastic collisions is (at high energies) twice the multiplicity of a single inelastic collision. In the model of ref. [33] the former multiplicity is only 1.5 time the latter (see fig. 33). Thus it is not clear whether one can use the σ_n's of the Gribov-Glauber model. (I thank C. Chiu for many stimulating discussions on this and related topics. A more detailed presentation of this model can be found in Chapter VII).

$$\frac{dN}{dy}^{pA\to hX}(y) = N_I(y) + N_{II}(y) + (\bar{\nu} - 1)\, N_{III}(y) \qquad (7.7)$$

Since $< N_I(y) > \simeq < N_{II}(y) > \simeq N_{III}(y) > \simeq <n>^{pp}/2$, where $<n>^{pp}$ is the pp average charge multiplicity, one finds:

$$< \frac{dN}{dy}(y)>^{pA} = \left(\frac{1}{2} + \frac{\bar{\nu}}{2}\right) < n >^{pp} \qquad (7.8)$$

(In the above equations, the symbol < > denotes the integral over the rapidity region). The results for the rapidity distributions, obtained from eq. (7.7), are shown in fig. 34. It would be interesting to compare the predictions of the models with experimental data in the projectile fragmentation region.

8. Conclusions

We have investigated the role played by the hadrons constituents in multi-hadron production at low-p_T. We have proceeded in the general framework of an S-matrix, and have used, as much as possible, the wisdom contained in dual and Reggeon field theories. These theories suggest a production mechanism consisting in the formation of jets of hadrons, resulting from the fragmentation of various sets of constituents of the colliding hadrons. The joint momentum distribution functions of the sets of constituents responsible for the jets produced in the final state, are completely determined in terms of Regge intercepts. The momentum distribution functions have the same x = 0 behaviour as point like partons (structure functions). This indicates that, in the highly exited states formed in a high-energy collision, the memory of the form factors of 'constituent' quarks (or 'valons') has been lost. The dual-parton fragmentation models have the advantage over the recombination models of being formulated in a very general framework and allowing a unified description of the central and fragmentation regions. Moreover, they can be extended to hadron-nucleus interactions without introducing new physical concepts and/or assumptions. I find it rather remarkable that it is possible, in the very general framework described above, to understand, at a quantitative level, the main features of the available data. However, as I have mentioned in the Introduction, the successes of the recombination model in describing the experimental data in the fragmentation regions are also very impressive, and it is a challenge to the theoreticians in this field to discover the reason(s) why these two models are so successful.

Acknowledgements

I wish to thank Profs. R.T. Van de Walle and S. Ratti for their efficient management and for the stimulating atmosphere they succeeded in creating during the meeting.

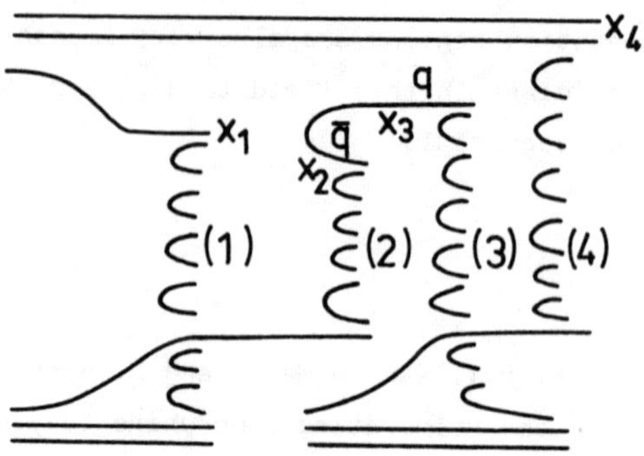

fig. 22. An S-channel discontinuity of a double-scattering diagram in a proton-nucleus interactions

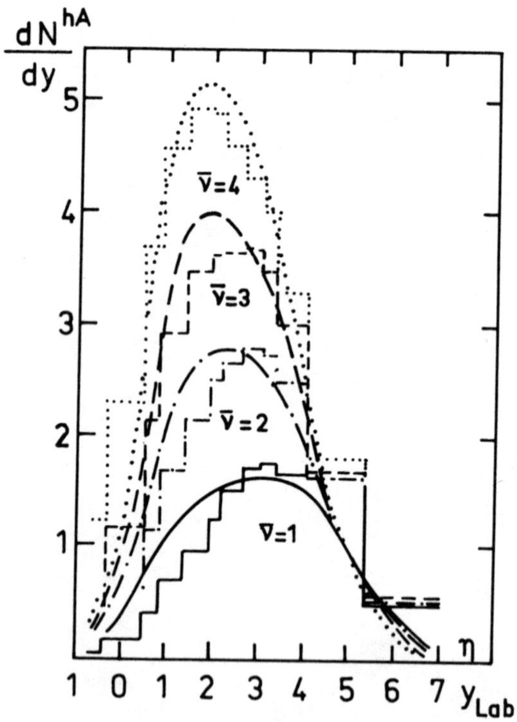

fig. 23. Comparison of our theoretical calculations and the experimental data (35) for the charged differential multiplicities at $\sqrt{s}$ = 20 GeV.

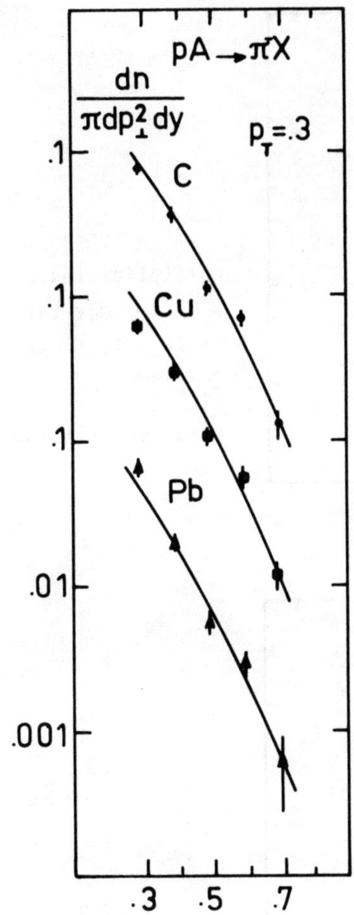

fig. 24

Comparison of our theoretical calculation and the preliminary data (37) for the differential multiplicities at p_L = 100 GeV for π^- production with p_T = 0.3 GeV.

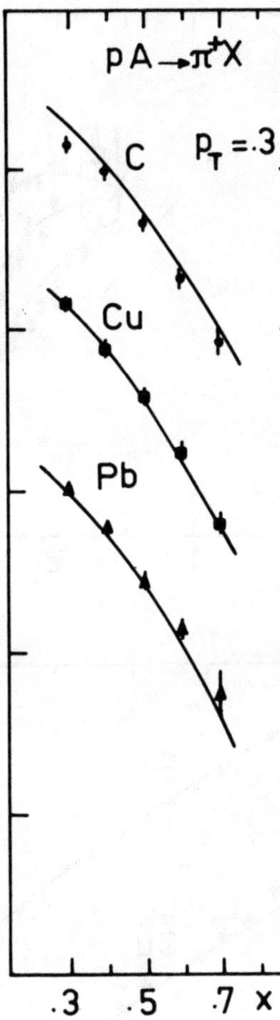

fig. 25

Comparison of our theoretical calculation and the preliminary data (37) for the differential multiplicities at p_L = 100 GeV for π^+ production with p_T = 0.3 GeV.

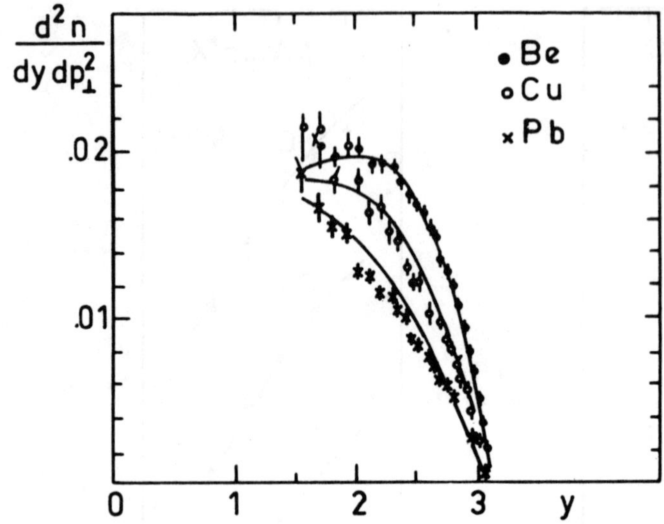

fig. 26

Differential multiplicities at p_L = 300 GeV for $\Lambda°$ production at p_T = 0.
Data from ref.(36)

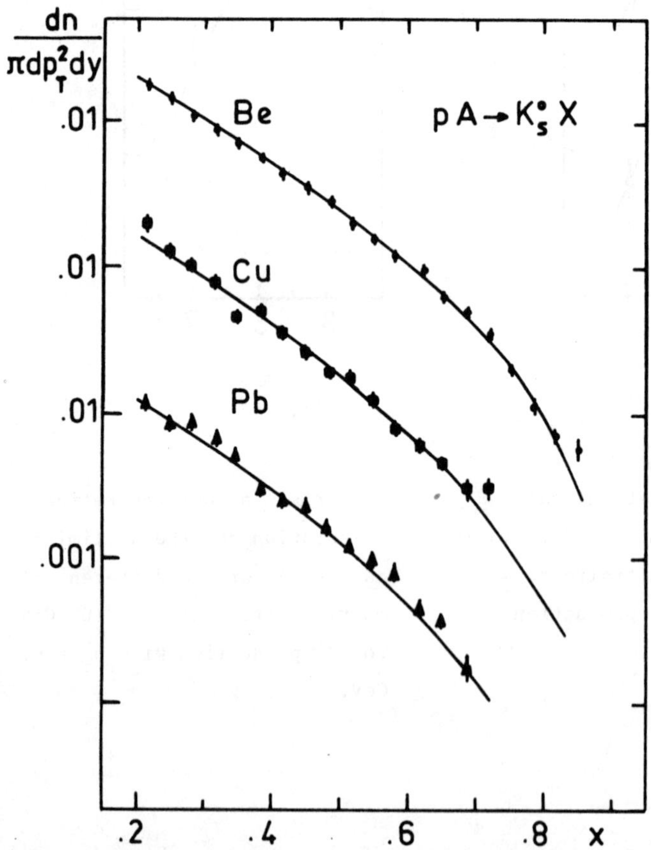

fig. 27

Differential multiplicities at p_L = 300 GeV for K_s° production at p_T = 0.
Data from ref. (36)

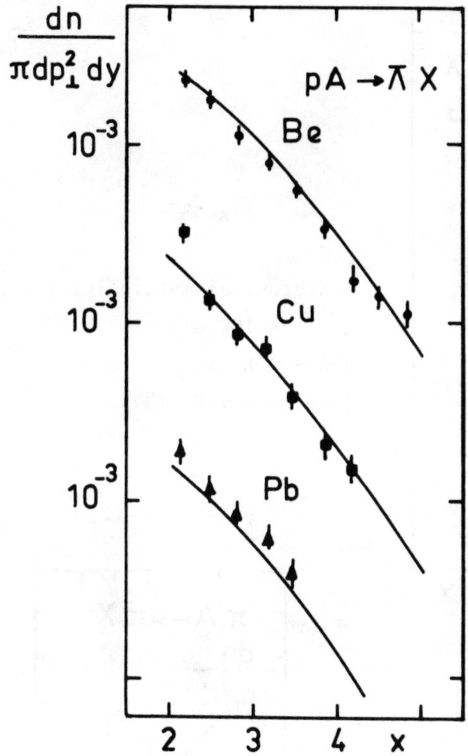

fig. 28

Differential multiplicities at p_L = 300 GeV for $\bar{\Lambda}_o$ production at p_T = 0. Data from ref. (36).

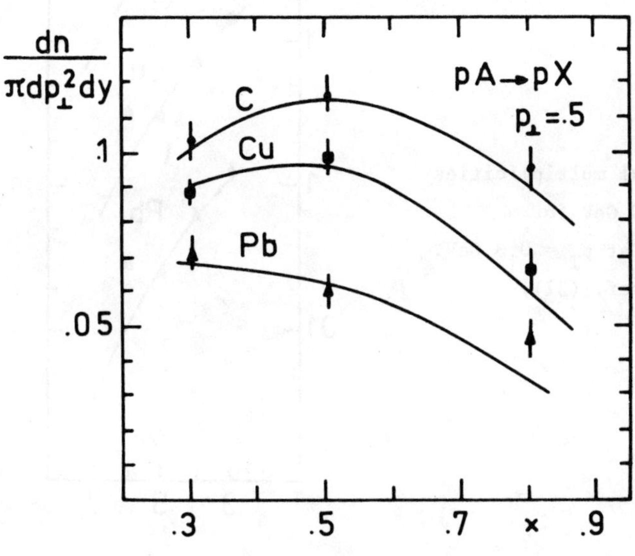

fig. 29

Differential multiplicities at p_L = 100 GeV pA → pX at p_T = 0.5 GeV. Data from ref. (37)

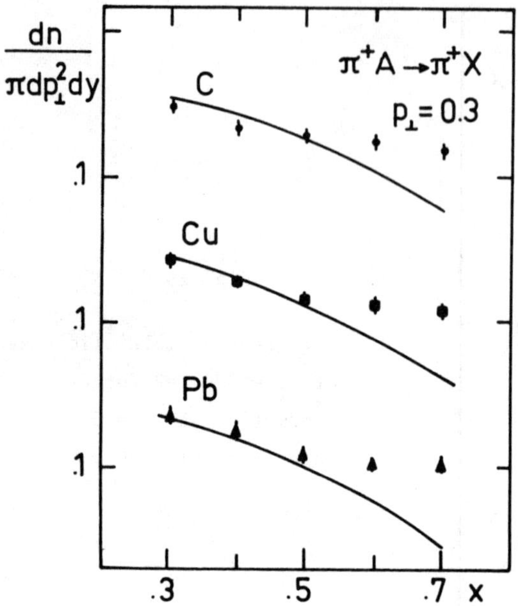

fig. 30

Differential multiplicities at p_L = 100 GeV $\pi^+ A \to \pi^+ X$ at p_T = 0.3 GeV. Data from ref. (37).

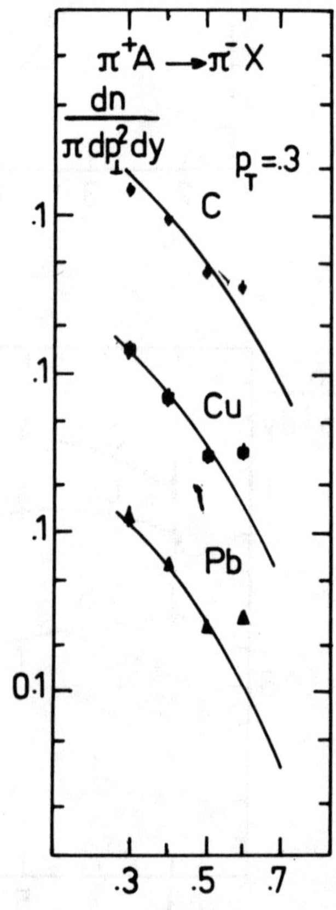

fig. 31. Differential multiplicities at p_L = 100 GeV for $\pi^+ A \to \pi^- X$ at p_T = 0.3 GeV. Data from ref. (37).

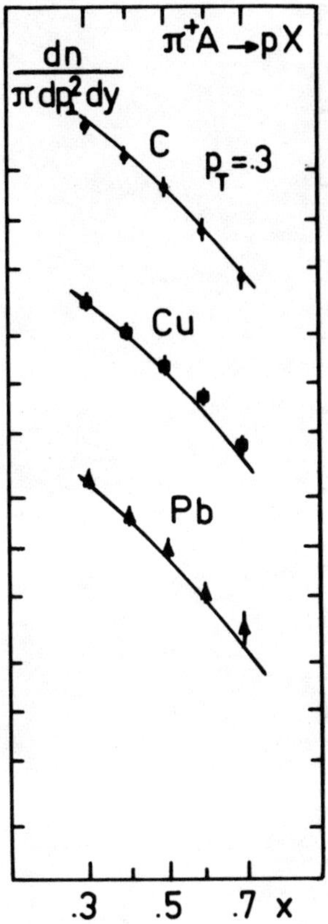

fig. 32

Differential multiplicities at p_L = 100 GeV for $\pi^+A \to pX$ at p_T = 0.3 GeV. Data from ref. (37).

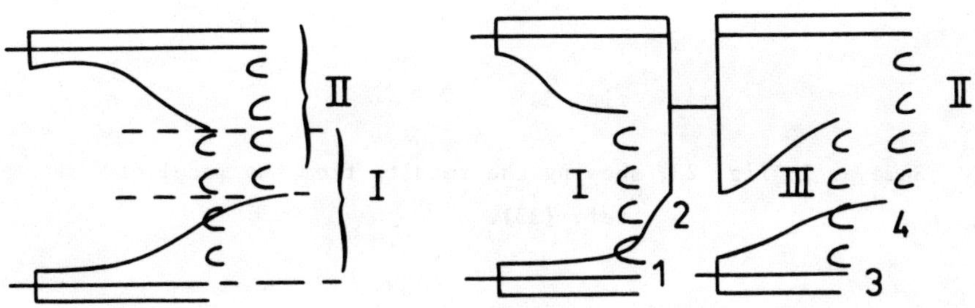

fig. 33. The mechanism of particle production in the model of ref. (33). An example of a graph corresponding to two inelastic collisions within the nucleus is shown.

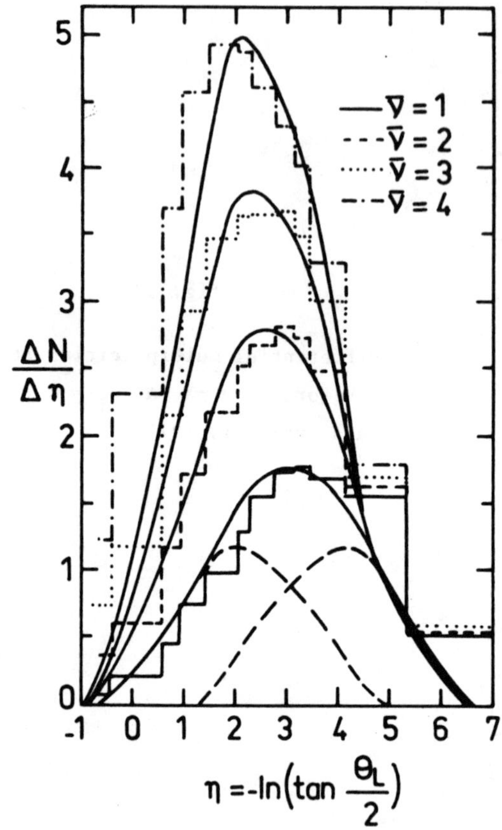

fig. 34

Same as in fig. 23, showing the results from the model of ref. (33).

REFERENCES

[1] See also L. Van Hove. The role of valence quarks in proton fragmentation, CERN preprint TH 2997 (and references therein.)

[2] W. Ochs, Nucl. Phys. B118, 397 (1977)

[3] Brown et al, Phys. Rev. Letters 45, 157 (1980)

[4] For a review see L. Van Hove, Schladming lectures 1979, CERN preprint 2628.

[5] B. Andersson, G. Gustafson and C. Peterson, Phys. Lett. 69B, 221 (1977)

[6] B. Andersson, G. Gustafson, I. Holgersson and O. Månsson, Lund preprint LU TP 80-5

[7] For Non-specialized reviews see G. Veneziano in Proc. XII Rencontre de Moriond 1976, ed. J. Tran Thanh Van and GIF lectures 1977 ; Girardi GIF lectures 1976.

[8] G. Veneziano, Nucl. Phys. B74, 365 (1974) and B117, 519 (1976) ; Chan Hong Mo et al, Nucl. Phys. B86, 470 (1975) and B92, 13 (1975) ; G. Chew and C. Rosenzweig, Nucl. Phys. B104, 290 (1976) and Phys. Rep. 41C, 263 (1978).
X. Artru and G. Menessier, Nucl. Phys. B70, 93 (1974).

[9] G. t'Hooft, Nucl. Phys. B72, 461 (1974).

[10] The concept of diquark is also useful in other domains. See Diquarks, Van de Walle, Erice lectures 1979.

[11] A. Capella, U. Sukhatme, C.I. Tan, J. Tran Thanh Van, Phys. Lett. 81B, 68 (1979).

[12] A. Capella, U. Sukhatme, J. Tran Thanh Van, Particles and Fields 3, 329 (1980).

[13] G. Cohen-Tannoudji, A. El-Hassouni, J. Kalinowski, O. Napoly and R. Peschanski, Phys. Rev. D21, 2689 (1980)

[14] H. Minakata, Phys. Rev. D20, 1656 (1979)

[15] F. Low, Phys. Rev. D12, 163 (1975)

[16] S. Nussinov, Phys. Rev. Lett. 34, 1286 (1975)

[17] J. Dias de Deus and S. Jadach, Act. Phys. Pol. B9, 249 (1978) ;
F. Hayot and S. Jadach, Phys. Rev. D17, 2307 (1978) ;
J. Gasser and U. Sukhatme, Cambridge University report DAMTP 76/3 (1976), unpublished.

[18] A. Krzywicki and B. Petersson, Phys. Rev. D6, 924 (1972)
U. Sukhatme, Phys. Lett. 73B, 478 (1978)
R. Field and R.P. Feynman, Nucl. Phys. B136, 1 (1978).

X. Artru, Orsay preprint LPTHE 78/25

B. Andersson, G. Gustafson, C. Petersson, Lund preprint LU TP 78-5

B. Andersson and G. Gustafson, Lund preprint LU TP 79-2

[19] G. Cohen-Tannoudji and O. Napoly in Proc. XV Moriond Meeting (1980), ed. J. Tran Thanh Van

[20] A. Giovannini and G. Veneziano, CERN preprint TH 2347 (1977) and G. Veneziano, ref. (7)

[21] G. Goldhaber et al. Phys. Rev. 120, 300 (1960)

[22] R. Blankenbecler and S.J. Brodsky, Phys. Rev. D10, 242 (1974); S. Brodsky, G. Farrar , Phys. Rev. Lett. 31, 1153 (1973) ; V. Matveev, R. Muradyan, A. Tavkhelidze, Lett. Nuovo Cim. 7,719(1973)

[23] A. El Hassouni, A. Mantrach, E.G. Oudrhiri-Safiani (to be published in Z. Phys.)

[24] C. Chiu and S. Matsuda, Nucl. Phys. B134, 463 (1978)

[25] P. Aurenche and L. Gonzalez-Mestres, Phys. Rev. D18, 2995 (1978)

[26] A. Capella and J. Tran Thanh Van, Orsay preprint LPTHE 81/2 A simplified version of this model is given in Phys. Lett. 93B, 146 (1980) . See also X. Artru, Orsay preprint 79/8.

[27] A. El Hassouni and O. Napoly, Phys. Rev. D23,193 (1981)

[28] A. Ng, K.E. Lassila and U. Sukhatme, Fermilab Pub 80/90-THY

[29] S. Uhlig et al., Nucl. Phys. B132, 15 (1978)

[30] A. Capella and A. Krzywicki, Phys. Rev. D18, 4120 (1978)

[31] V. A. Abramowski, V.N. Gribov and O.V. Koucheli, Yad. Fiz. 18, 595 (1973) (Sov. J. Nucl. Phys. 18, 308 (1974)).

[32] A. Capella and A. Krzywicki, Phys. Rev. 18D, 3357 (1978) and Phys. Lett. B67, 84 (1977).

[33] W.Q. Chao, C. Chiu, Z. He, D. M. Tow, Phys. Rev. Lett., 44, 518 (1980).

[34] W. Thomé et al., Nucl. Phys. B129, 365 (1977) ; W. Thomé Thesis, III Physikalisches Institut der TH Aachen, Germany.

[35] C. Hallewell et al., Phys. Rev. Lett. 39, 1499 (1977)

[36] K. Heller et al., Phys. Rev. D16, 2737 (1978) and P. Skubic et al. Phys. Rev. D18, 3115 (1978)

[37] A.E. Brenner el al., "A-dependence of Inclusive Hadron Scattering at 100 GeV", Madison, Wisconsin, July 1980 ; See also A. Barton in Proc. XI International Symposium on Multiparticle Dynamics, Bruges 1980, editors E. de Wolf and F. Verbeure.

Chapter VI: THE QUARK-PARTON MODEL - PRODUCTION OF LOW P_T PARTICLES IN NUCLEI

Arnon Dar
Physics Department, Technion-Israel Institute of Technology
Haifa, Israel.

Abstract:

Production of low-p_T particles in nuclei by high energy leptons, hadrons and heavy ions is discussed within the framework of the constituent quark model. Information on the space-time dependence of the hadronization of quarks is derived from the A-dependence and the momentum spectrum of low-p_T particles produced in the projectile and in the target fragmentation regions. Some comments are also made on quarks and high energy nucleus-nucleus collisions.

1. Introduction

Perturbative Quantum Chromo Dynamics (QCD) is very successful in explaining processes where large momentum transfers are involved. However, processes which involve large momentum transfers have very small cross sections and hadron production is completely dominated by the production of hadrons with small p_T. Such soft processes in QCD require a non-perturbative approach which is still lacking and it has not been possible so far to use it to explain the bulk of the experimental data on hadron collisions. A full understanding of low-p_T production may first require the solution of the confinement problem. A breakthrough may come from the study of high energy particle-nucleus collisions or from the study of high energy nucleus-nucleus collisions.[1-7]

Even if the study of high energy particle-nucleus and nucleus-nucleus collisions will not lead to such a breakthrough it will still provide unique and important physical information: The study of leptoproduction of hadrons in nuclei will provide information on the space-time dependence of the hadronization of quarks. The study of multiparticle production in high energy hadron-nucleus collisions will provide information on the structure of hadrons and on the interactions and hadronization of their constituents. Finally, the study of high energy nucleus-nucleus collisions will provide information on new phases of quark matter which may exist in the center of massive neutron stars and also may play an important role in supernova explosions in quasars and in the early stages of the Big-Bang. These points will be demonstrated below in the following order: In Section II I will demonstrate how the study of leptoproduction of hadrons in nuclei can provide information on the space-time dependence of the hadronization of current quarks. Hadroproduction of low-p_T particles in hadron-nucleus collisions and its dependence on the quark substructure of hadrons and nuclei in the central rapidity region and in the projectile fragmentation region will be discussed in Section III. The momentum distribution of quarks in nuclei in a high momentum frame will be discussed in Section IV. In Section V these distributions will be used to explain low-p_T production in the nucleus fragmentation region.

Due to space limitation I will not review the various quark-parton models that have been applied in the literature to production of low-p_T particles in nuclei (some of these models have been discussed by other speakers at this Conference). I will rather show that the constituent quark model description of hadrons and general assumptions about the interactions (additivity) and the hadronization mechanisms (fragmentation and recombination) of quarks are sufficient to understand the observed A-dependence in production of low-p_T particles in nuclei. More details can be found in previous publications by Anisovich, Shabelski and Shekhter[1], by Nikolaev, Ostapchuk and Zoller[2], by Nikolaev and Pokorski[3], by Dar and Takagi[4,5] and by Bialas and Bialas[6] and by Bialas and Czyz[7].

2. Leptoproduction of Hadrons In Nuclei

Information on the space-time dependence of the hadronization of current quarks can be obtained from studying leptoproduction in nuclei.

Consider first leptoproduction from a single nucleon, e.g. the charged current reaction $\nu p \to \mu^- X$. In the quark-parton model it is described by the following diagram

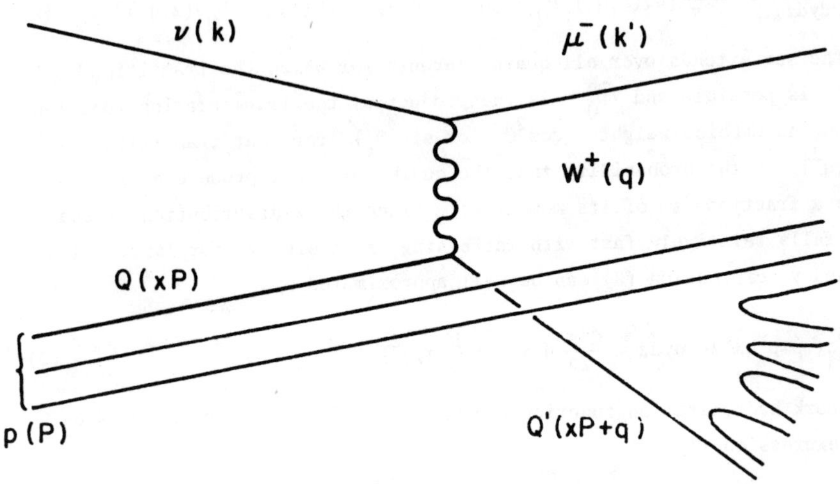

The letters in the parantheses of diagram D1 denote the four momenta of the respective particles. In terms of the Bjorken scaling variables x and y the differential cross section for neutrino-nucleus charge current scattering can be written as

$$\frac{d^2\sigma}{dxdy} = \frac{G^2 s}{2\pi} [\Sigma Q(x,q^2) + (1-y)^2 \Sigma\bar{Q}(x,q^2)] \tag{1}$$

where G is the Fermi constant, s is the total invariant energy squared $s \equiv (k+P)^2$, and Q and $\bar{Q}$ are the quark and anti-quark distribution functions respectively for the various flavours of quarks in the proton. That is, $Q(x,q^2)$ represents the probability of finding a quark with a fraction x of the total momentum P of the proton in a high momentum frame, using a probe of virtual mass q^2.

Due to colour confinement the struck quark and the proton remnant which we call a diquark, fragment into two respective jets with equal and opposite momenta in the W-nucleon c.m. system. If the fragmentation of the quark-diquark system scales, one can write for exclusive hadron production $\nu p \to \mu^- hX$ (in the current fragmentation region) the following expression:

$$\frac{d^3\sigma}{dxdydz} = \frac{G^2 s}{2\pi}\,[\Sigma Q(x,q^2)\,\tilde{D}^h_{Q'}(z,q^2) + (1-y)^2\,\Sigma\bar{Q}(x,q^2)\,\tilde{D}^h_{Q'}(z,q^2)]\ , \qquad (2)$$

where the sum extends over all quark flavours for which the transition $W^+ + Q = Q'$ is possible and $\tilde{D}^h_{Q'}$ is the product of the fragmentation function $D^h_{Q'}$ and the Cabbibo weight ($\cos^2\Theta_c$ or $\sin^2\Theta_c$) for that transition. ($D^h_{Q'}(z,q^2)$ is the probability that the quark Q' will produce a fragment h with a fraction z of its momentum). Since the x-distribution of sea quarks falls relatively fast with increasing x therefore for large values of x and y expression (2) can be well approximated by

$$d^3\sigma(\nu p \to \mu^- hX)/dxdydz \simeq \frac{G^2 s}{2\pi}\, d(x,q^2)\, D^h_u(z,q^2)\ . \qquad (3)$$

The u-quark hadronization function can be obtained by dividing expression (2) by expression (1)

$$D^h_u(z,q^2) \simeq \frac{d^3\sigma(\nu p \to \mu^- hX)/dxdydz}{d^2\sigma(\nu p \to \mu^- X)/dxdy}\ , \qquad (4)$$

The d-quark hadronization function can be obtained in a similar way from $\bar{\nu} p \to \mu^+ hX$:

$$D^h_d(z,q^2) \simeq \frac{d^3\sigma(\bar{\nu} p \to \mu^+ hX)/dxdydz}{d^2\sigma(\bar{\nu} p \to \mu^+ hX)/dxdy}\ . \qquad (5)$$

Diquark hadronization functions can be extracted in a similar way from leptoproduction of hadrons in the target fragmentation region, $z<0$.

Consider now leptoproduction of hadrons in nuclei. According to the quark-parton model a single quark in a nucleon in a nucleus absorbs the four momentum of the intermediate vector boson (γ_V, $W^\pm$ or Z^o) and fragments into final hadrons with a characteristic hadronization time τ . A low-p_T fragment can be produced with a large fraction z of the quark's momentum only if the quark, and afterwards the fragment, escape collisions in the target nucleus. The probability for that depends on their cross-sections and on the distance that they propagate inside the nucleus. Due to time dilation the latter depends also on the energy transfer to the quark. Thus the dependence of the attenuation of leptoproduced hadrons on the mass number (the size) of the target nucleus and on the energy loss of the incident lepton (the energy transfer to the quark) stores information on the cross sections and on the hadronization times of current quarks. In some limiting situations this information can be extracted in a simple way: Consider for instance the case that the energy transfer to the quark is sufficiently high so that due to time dilation the hadronization takes place after the quark emerges at the back of the nucleus. The probability that the deep inelastic scattering of the lepton took place at a point $(\vec{b},z)$ inside the nucleus is proportional to $\rho(\vec{b},z)$, the nuclear density at that point. (The z axis points in the direction of the quark's momentum, $\vec{b}$ is the transverse distance from the center of the nucleus and ρ is normalized such that $\int\rho(\vec{b},z)d^2bdz = A$). The probability of the current quark to escape nuclear collisions when it propagates to the surface of the nucleus is given by $e^{-\sigma^t_{qN}T(b,z)}$ where σ^t_{qN} is the quark-nucleon total cross section and $T(b,z) = \int_z^\infty\rho(b,z')dz'$ is the nuclear thickness that it has to penetrate. The average probability is given by averaging the product of these two probabilities over all points inside the nucleus:

$$R_{\ell A}(z\to 1) = \int d^2b\int dz\ \rho(b,z)e^{-\sigma^t_{qN}(b,z)}\ dz = \sigma^{abs}_{qA}/A\sigma^t_{qN}, \tag{6}$$

where we used the relation

$$\sigma^{abs}_{qA} = \int\ (1-e^{-\sigma^t_{qN}T(b)})\ d^2b \tag{7}$$

for the quark-nucleus total absorption cross section, and the notation

$$T(b) \equiv \int_{-\infty}^{\infty} T(b,z)dz \tag{8}$$

for the nuclear thickness at impact parameter b.

Our prediction is therefore

$$\frac{d^3\sigma(\ell A\to\ell' hX)/dxdydz}{d^2\sigma(\ell A\to\ell' X)/dxdy} = R_{\ell A}\ D^h_q(z,q^2) = \frac{\sigma^{abs}_{qA}}{A\sigma^t_{qN}}\ D^h_q(z,q^2), \tag{9}$$

where D^h_q is given by eqs. (4) and (5).

In Fig. 1 we compare experimental results on $R_{\ell A}$ that were obtained[8] from electroproduction of hadrons in Be, C, Cu and Sn targets, and our theoretical predictions as given by eqs. (6) and (7). In Fig. 1 the full circles and triangles stand for leptoproduction of positive and negative charge hadrons, respectively, by virtual photons with $-q^2<1\ \mathrm{GeV}^2$, while open circles and triangles stand for positive and negative charge hadrons, respectively, that were produced by virtual photons with $-q^2>1\ \mathrm{GeV}^2$. In our theoretical calculations we used eq. (7) with $\sigma^t_{qN} = 13$ mb as suggested by the additive quark model relation $\sigma^t_{qN} = (1/2)\sigma^t_{\pi N} = (1/3)\sigma^t_{pN}$, and the nuclear density functions that were obtained from low energy electron scattering from nuclei.[10]

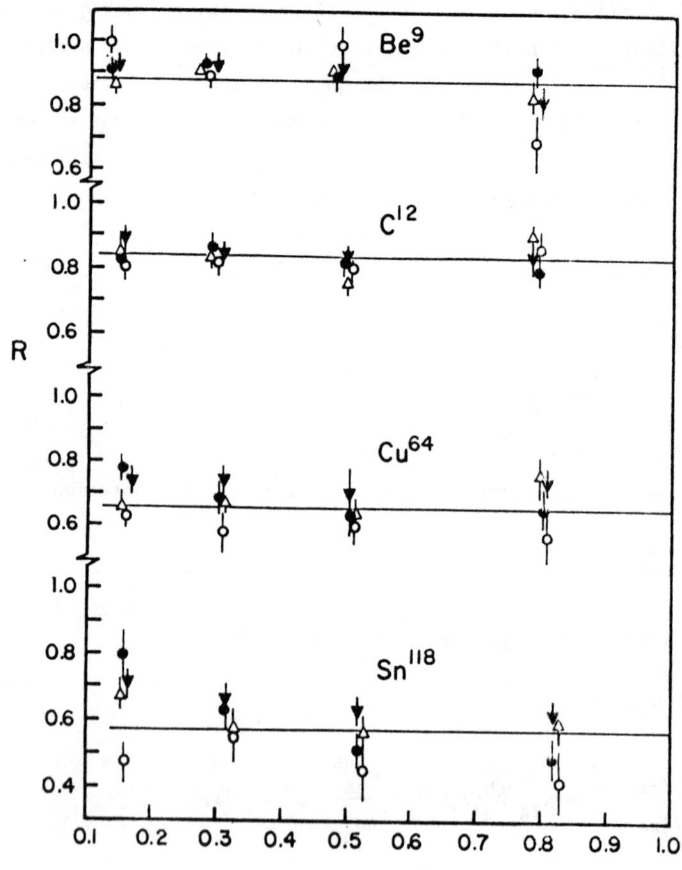

Fig. 1

Fig. 1 demonstrates that good agreement between experiment and theory is obtained for $\sigma^t_{qN} = 13$ mb whereas current quarks are believed to have vanishing cross sections. It is possible that the agreement is because a current quark dresses into a constituent quark (for which $\sigma^t_{qN} = 13$ mb) before it emerges from the parent nucleon and collides with a second nucleon.

If current quarks are not attenuated inside the nucleus before they fragment the attenuation must be due to the attenuation of the fragments. This is possible if the formation time is sufficiently small and the fragmentation occurs inside the nucleus. In particular for very short formation times one obtains

$$R_{\ell A}(z\to 1) = \sigma^{abs}_{hA}/A\sigma^t_{hN}, \tag{10}$$

but when the energy increases, due to time dilation the quarks fragment outside the nucleus and

$$R_{\ell A}(z\to 1) \to 1. \tag{11}$$

According to this picture attenuation is a transient phenomenon which disappears when energy increases. This is demonstrated in Fig. 2 which was taken from the work of Nikolaev[3]. In Fig. 2 $R_{\ell A}$ is represented by an effective power $\alpha(z)$ which is defined by the relation $R_{\ell A}(z) \equiv A^{\alpha(z)}$. The theoretical lines were fitted to the experimental points using $\sigma^t_{qN} = 0$ and a finite formation time.

Measurements of the attenuation of leptoproduced hadrons at higher energies will thus provide valuable information on whether or not current quarks are attenuated in nuclear matter.

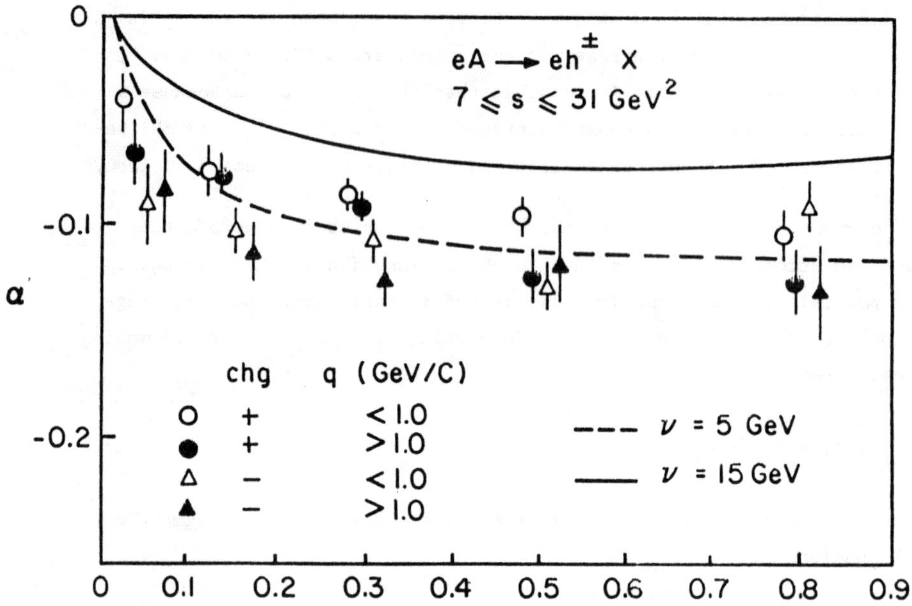

z

Fig. 2

3. Hadroproduction of Particles In Nuclei

The additive quark model provides a simple framework for understanding the A-dependence of multiparticle production from nuclei. The model assumes that hadrons are made of spatially separated constituent quarks. The constituent quarks are made of a valence quark and a virtual cloud of quark-antiquark pairs and gluons. They have the same quantum numbers as those of their valence quark. In inelastic collisions quarks from the projectile and the target collide and fragment. Their fragments populate the central rapidity region. The spectator quarks which did not participate in the collisions and retained their initial momenta fragment and recombine into particles that populate the projectile and target fragmentation retions. Below we shall discuss separately the central region, the projectile fragmentation region and the target fragmentation region and we shall show how the A-dependence of particle production in these regions follows from the quark-parton substructure of particles and nuclei.

A. The Central Rapidity Region

According to the additive quark model hadron-nucleon collisions are dominated by a single collision between a constituent quark from the incident hadron and a constituent quark from the target nucleon, i.e. the number of wounded quarks (quarks that suffered inelastic collisions) in hadron-nucleon collisions is 2. In hadron-nucleus collisions the number of wounded quarks may be larger. Let us denote by $\overline{w}_A$ the average number of wounded quarks in the target nucleus and by $\overline{w}_h$ the average number of wounded quarks in the incident hadron. Then in the central rapidity region the ratio of the average multiplicity in hadron-nucleus collisions and in hadron-nucleon collision is given by

$$R_{hA} \equiv \frac{\langle n\rangle_{hA}}{\langle n\rangle_{hN}} = \frac{\overline{w}_h+\overline{w}_A}{2} , \tag{1}$$

where

$$\overline{w}_A = \frac{3A\sigma_{qh}}{\sigma_{hA}} = \frac{A\sigma_{hN}}{\sigma_{hA}} , \tag{2}$$

and

$$\overline{w}_\pi = \frac{2\sigma_{qA}}{\sigma_{\pi A}} , \quad \overline{w}_N = \frac{3\sigma_{qA}}{\sigma_{NA}} , \ \ldots \tag{3}$$

Note that $\overline{w}_A = \overline{\nu}_A$ where $\overline{\nu}_A$ is the average number of wounded nucleons in A. This is because in hadron-nucleon collisions only one quark in the nucleon is wounded. Eq. (1) holds also for the total multiplicities (central + fragmentation regions) if the number of spectator quarks is proportional to the number of wounded quarks. Indeed in the target the ratio of spectator to wounded quarks, s_A/w_A, is 2:1 independent of A, because each wounded nucleon contains only one wounded quark. The projectile however suffers multiple collisions in the target nucleus and the ratio of s_h, the number of spectator quarks, and w_h, the number of wounded quarks in the projectile depends on A. Consequently eq. (1) holds only approximately for the total multiplicities.

We note also that the quark model prediction $R_{hA} = (\omega_h + \overline{\nu}_A)/2$ is not much different from the prediction[11] $R_{hA} = (1+ \overline{\nu}_A)/2$ which ignores the quark substructure of the projectile and the target. This is demonstrated in Fig. 3 where we compare both predictions with the experimental data of Elias et al.[12] on the average multiplicities produced by π, k and p collisions with nuclei at 50 GeV/c and 100 GeV/c incident momentum.

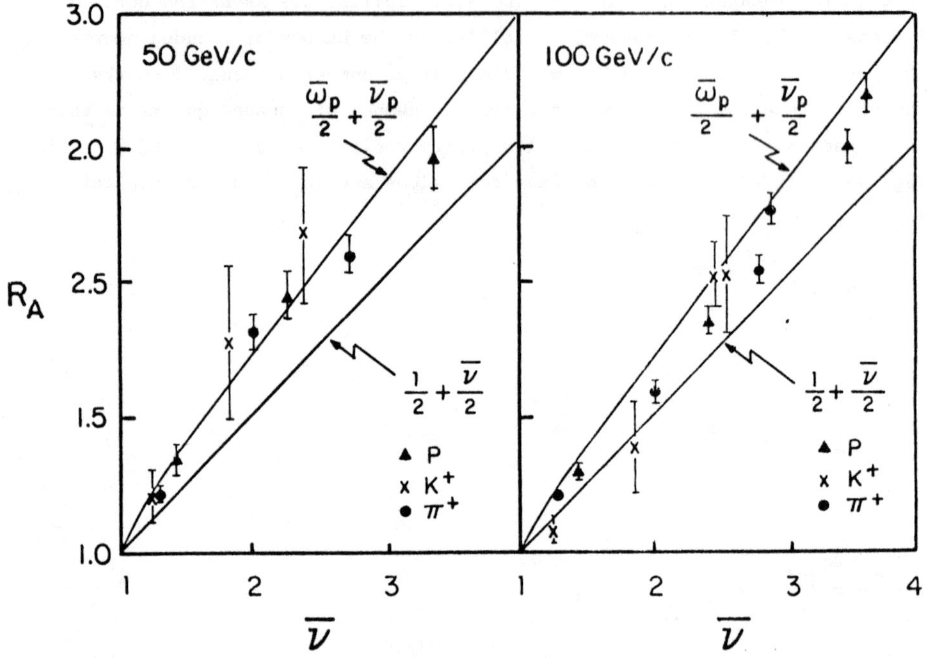

Fig. 3

The difference between the predictions of models that count wounded quarks and models that count wounded nucleons is much larger when one considers heavy ion collisions. The multiplicity ratio for A+B collisions in the central rapidity region is given by[7]

$$R_{AB} \equiv \frac{\langle n \rangle_{AB}}{\langle n \rangle_{pp}} = \frac{3B\sigma_{qA} + 3A\sigma_{qB}}{2\sigma_{AB}} \tag{4}$$

if one counts wounded quarks, while it is given by[13]

$$R_{AB} \equiv \frac{\langle n \rangle_{AB}}{\langle n \rangle_{pp}} = \frac{B\sigma_{NA} + A\sigma_{NB}}{2\sigma_{AB}} \tag{5}$$

if one counts wounded nucleons. The difference between eq. (4) and eq. (5) for light ions is small, e.g. for $\alpha + \alpha$ collisions eq. (5) yields $R_{\alpha\alpha} = 1.78 \pm 0.22$ while eq. (4) yields $R_{\alpha\alpha} = 1.6 \pm 0.20$. Preliminary ISR data on 62 + 62 GeV $\alpha\alpha$ collisions[14] give $R_{\alpha\alpha} = 1.7 \pm 0.1$. The only available experimental data on high energy $(E/A > 10\text{ GeV})$ heavy ion collisions is from studies of cosmic ray collisions. However, these studies suffer from low statistics and uncertainties in the estimates of the energy, mass and charge of the incident particle. Most of the attention in these studies was focused on central collisions. Experimentally, central collisions of a projectile A with a target B when $A < B$ are defined as those collisions where no trace is found in the projectile fragmentation region of nucleons which escaped collision. Theoretically, central collisions are defined as those collisions where the impact parameter is smaller than the difference between the radii of the colliding nuclei. For such collisions $(A < B)$

$$R_{AB} = \frac{3B\sigma_{qA} + 3A\sigma_{qB}}{2\sigma_{pB}} \tag{6}$$

if one counts wounded quarks, while

$$R_{AB} = \frac{B\sigma_{pA} + A\sigma_{pB}}{2\sigma_{pB}} \tag{7}$$

if one counts wounded nucleons. There are some indications in cosmic ray data[15] that eq. (6) describes the multiplicity in the central rapidity region better than eq. (7). For instances in a Ca + Pb collision at 300 GeV/c per nucleon $dn/d\eta$ reached 170 in the central rapidity region which is about 100 times its value in pp collisions at that energy. Eq. (6) predicts that $R_{CaPb} = 93 \pm 10$ while eq. (7) predicts that $R_{CaPb} = 53 \pm 5$.

B. The Projectile Fragmentation Region

A striking similarity exists between the momentum distribution of low p_T mesons, which are produced by fast nucleons and which contain a valence u or d quark, and the momentum distributions of these quarks in fast nucleons as extracted from deep inelastic scattering of leptons and from hadroproduction of lepton pairs off nucleons[16]. This is demonstrated in Figs. 4-7 for $pp \to \pi^+X$, $np \to \pi^+X$, $n\pi \to \pi^+X$ and $pp \to k^+X$ where the incident nucleon and the produced meson both contain a valence u-quark, and for $pp \to \pi^-X$, $np \to \pi^-X$ and $n\pi \to \pi^-X$ where the incident nucleon and the π^- both contain a d-quark.

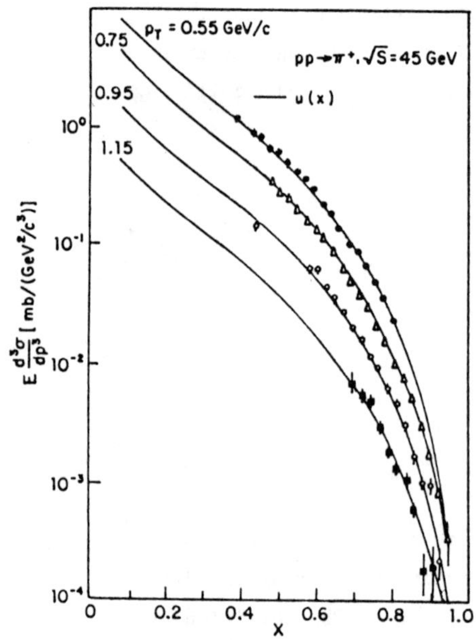

Fig. 4

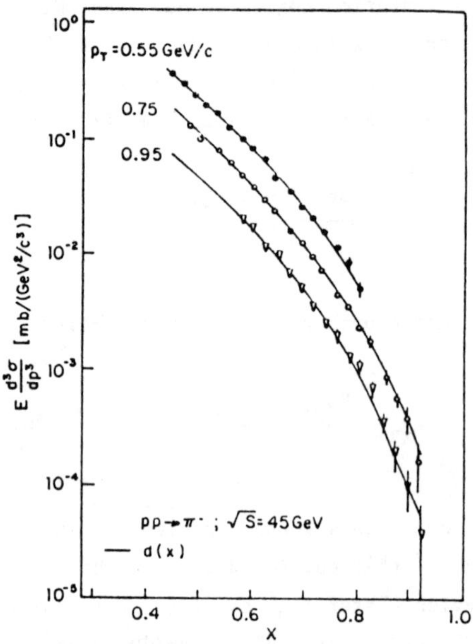

Fig. 5

The experimental points on proton fragmentation[13] in Figs. 4,5,7 are those of Singh et al.[17], while the experimental points on neutron fragmentation are those of Hanlon et al.[18]. The lines represent the x-distribution of u and d quarks in nucleons. They were obtained from the following functional parametrizations:

$$u_p(x) = d_n(x) = \frac{35}{16} \frac{(1-x)^3}{\sqrt{x}} , \tag{1}$$

$$d_p(x) = u_n(x) = \frac{315}{256} \frac{(1-x)^4}{\sqrt{x}} , \tag{2}$$

which fit well[19] the experimental data on deep inelastic lepton scattering and on hadroproduction of lepton pairs off nucleons.

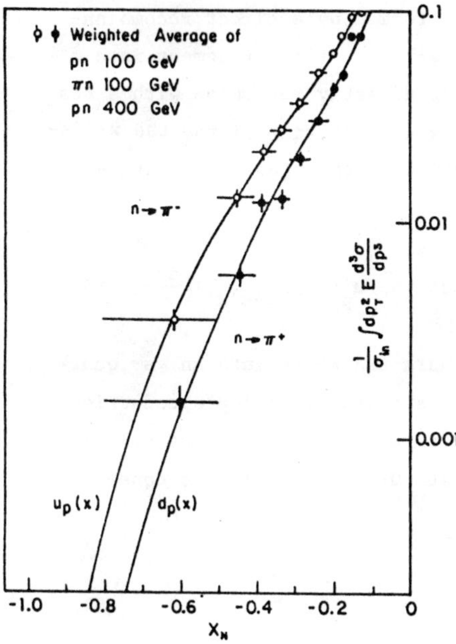

Fig. 6

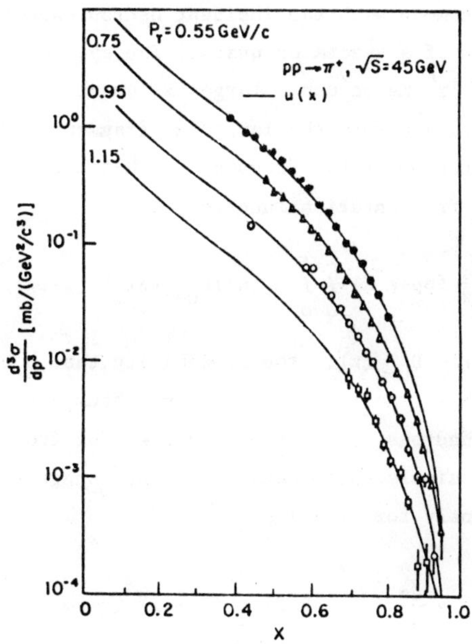

Fig. 7

Figs. 4-7 suggest[16] the following quark-parton picture of production of particles with low p_T and large x: The incident hadron fragments when one of its constituent quarks (valons) collides in the target. Those valence quarks in the incident hadron that are part of valons that escaped collisions in the target, (which we shall call spectator quarks) retain their original fraction x of the projectile momentum. They can first fragment and then recombine or directly recombine with sea quarks $(x \sim 0)$ in the valon and form a low p_T - large x fragment. If the recombination probability is x-independent than the x-distribution of a fragment formed by a direct recombination of a spectator quark is the same as that of the spectator quark in the incident hadron, e.g.

$$x \frac{d\sigma}{dx} (pp \to \pi^+ X) \sim u_p(x). \tag{3}$$

However, a large x-low p_T fragment which does not contain any valence quark in common with the incident hadron cannot be formed by a direct recombination of a spectator quark. The spectator quark must first fragment into a low x meson and a large x quark which can directly recombine with a sea quark and form the large x fragment. The x-distribution of the the x-distribution of the fragment is then a convolution of the spectator quark and its fragmentation function, e.g.

$$x \frac{d\sigma}{dx} (pp \to k^- X) \sim \int_0^1 \int_0^1 u_p(x') D_{u\to s}(zx') \delta(x - zx') dz dx' = \int_x^1 u_p(x') D_{u\to s}(\frac{x}{x'}) \frac{dx'}{x'}, \tag{4}$$

where $D_{u\to s}(z)$, the probability that a u-quark fragments into an s quark with a fraction z of its momentum, can be extracted from leptoproduction of hadrons, e.g. from $\nu p \to \mu^- kX$ or from $e^+e^- \to kX$. Both, experimental data and simple space considerations[4] suggest that $D(z) \sim z(1-z)$ and consequently for $x \to 1$

$$x \frac{d\sigma}{dx} (pp \to k^- X) \sim (1 - x)^5. \tag{5}$$

Fig. 8 shows that relation (4) is in good agreement with the experimental data of Singh et al.[17]

Several authors consider the success of the simple recombination models to be accidental and they propose alternative models for low p_T - large x production which they claim also explain successfully the experimental data (see for instance the papers presented by Capella, by Gunion and by Sukhatme at this Conference). In order to subject all these models to additional and perhaps discriminative tests one can examine their predictions for low p_T - large x production off nuclei. Here in particular we shall examine the predictions of the simple recombination model for the

Fig. 8

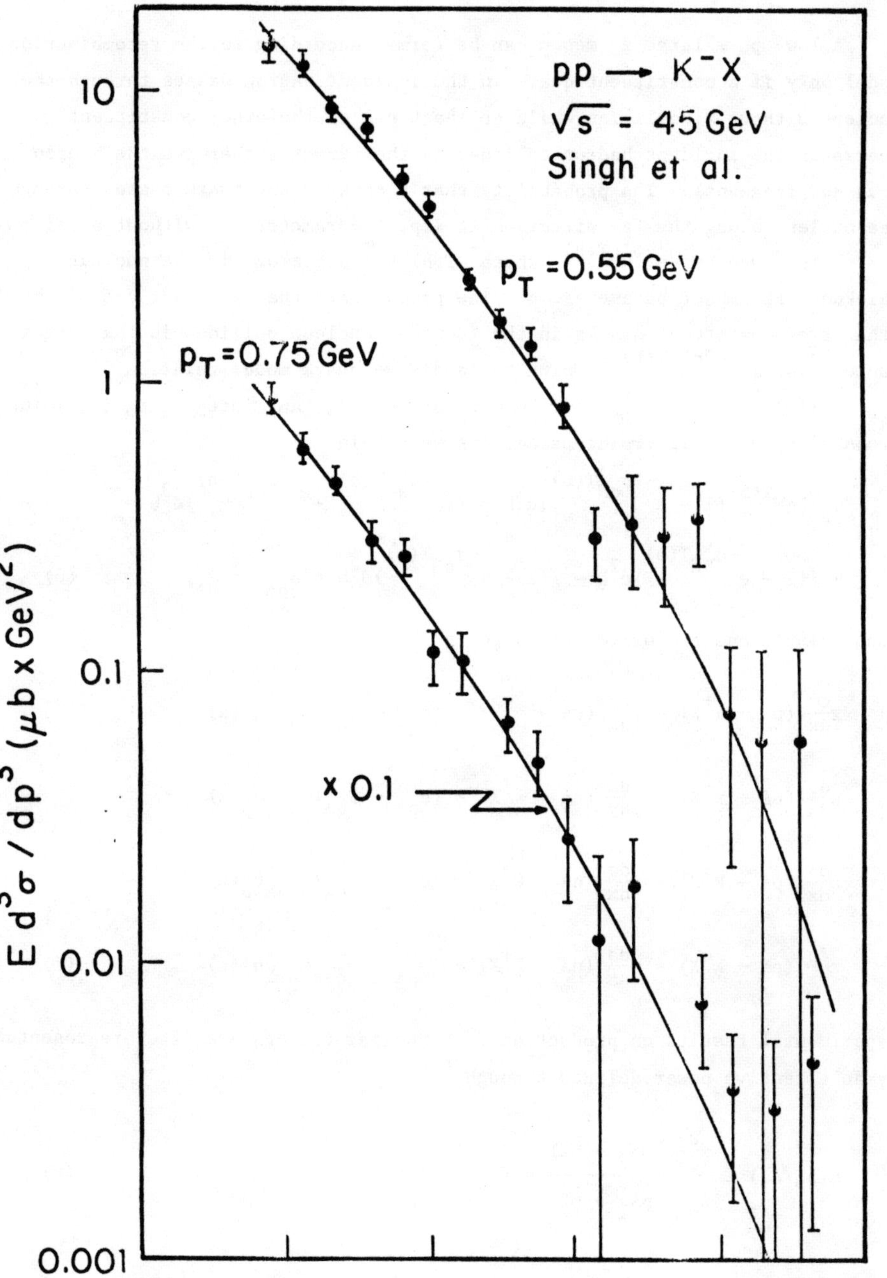

pp → K⁻X
√s = 45 GeV
Singh et al.
p_T = 0.55 GeV
p_T = 0.75 GeV
x 0.1
10
1
0.1
0.01
0.001
0.4
0.5
0.6
0.7
0.8
0.9
X
E d³σ / dp³ (μb x GeV²)

dependence of the momentum spectrum of fast fragments on the mass number A of the tartet nucleus.

A low p_T - large x meson can be formed according to the recombination model only if a constituent quark in the incident hadron passes through the nucleus without a collision while at least one of the other constituent quarks in the incident hadron collides in the target (otherwise the hadron does not fragment). The probability that a constituent quark passes through the nucleus along the z direction at impact parameter b without a collision is given by $e^{-\sigma^t_{qN}T(b)}$ where $T(b) = \int_{-\infty}^{\infty} \rho(b,z)dz$ is the nuclear thickness at impact parameter b. The probability that at least one of the other two constituent quarks in the incident nucleon collides in the target nucleus is $1-e^{-2\sigma^t_{qN}T(b)}$. Using the additive quark model relation[9] $\sigma^t_{qN} = (1/2)\sigma^t_{\pi N} = (1/3)\sigma^t_{NN}$ for q = u or q = d, and integrating the joint probability over all impact parameters we obtain

$$\int e^{-\sigma^t_{qN}T(b)}(1 - e^{-2\sigma^t_{qN}T(b)})d^2b = \int(e^{-\sigma^t_{qN}T(b)} - e^{-\sigma^t_{NN}T(b)})d^2b =$$

$$= \int(1 - e^{-\sigma^t_{NN}T(b)})d^2b - \int(1 - e^{-\sigma^t_{qN}T(b)})d^2b = \sigma_{pA} - \sigma_{qA}. \qquad (6)$$

Consequently one can write for large x

$$x\frac{d\sigma}{dx}(pA \to \pi^+X) = x\frac{d\sigma}{dx}(nA \to \pi^-X) = (\sigma_{pA} - \sigma_{qA})C_{u\to\pi}u_p(x)$$

$$x\frac{d\sigma}{dx}(pA \to \pi^-X) = x\frac{d\sigma}{dx}(nA \to \pi^+X) = (\sigma_{pA} - \sigma_{qA})C_{d\to\pi}d_p(x)$$

$$x\frac{d\sigma}{dx}(pA \to k^+X) = x\frac{d\sigma}{dx}(nA \to k^oX) = (\sigma_{pA} - \sigma_{qA})C_{u\to k}u_p(x)$$

$$x\frac{d\sigma}{dx}(pA \to k^oX) = x\frac{d\sigma}{dx}(nA \to k^+X) = (\sigma_{pA} - \sigma_{qA})C_{d\to k}d_p(x) \qquad (7)$$

Experimental results on production from nuclear targets are often represented by an effective power defined through

$$R(A_2/A_1) \equiv \frac{x\frac{d\sigma}{dx}(pA_2 \to MX)}{x\frac{d\sigma}{dx}(pA_1 \to MX)} = \left(\frac{A_2}{A_1}\right)^{\alpha(x)} \qquad (8)$$

Thus the simple quark fragmentation/recombination model predicts that for non diffractive production of mesons at large x by fast nucleons the A-dependence is given by

$$R(A_2/A_1) \equiv (A_2/A_1)^{\alpha} = (\sigma_{pA_2} - \sigma_{qA_2})/(\sigma_{pA_1} - \sigma_{qA_1}) \tag{9}$$

Note that eq. (9) predicts an effective power α which is x-independent, but which is A-dependent! In Fig. 9 we compare the theoretical prediction for $\alpha(x)$ as given by eq. (9) and experimental results on $\alpha(x)$ that were obtained from measurements of low-p_T $\pi^{\pm}$ and $k^{\pm}$ production by 19 and 24 GeV/c protons[20] and of low-p_T k_s^o production by 300 GeV/c protons[21] from Be and Pb targets. In the theoretical calculations of α we used

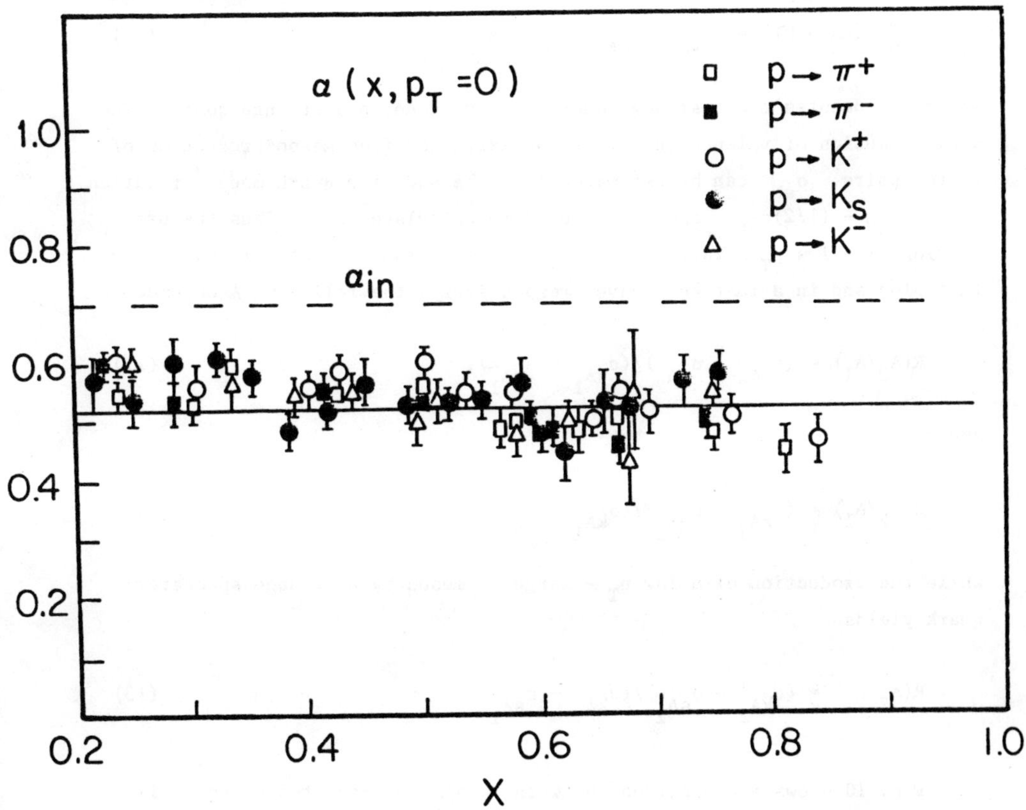

Fig. 9

Agreement between experiment and theory is good although more accurate experimental data is necessary before a more definite conclusion can be drawn.

For non diffractive production of low p_T - large x mesons by fast mesons the quark fragmentation/recombination model yields similar expressions, e.g.

$$x \frac{d\sigma}{dx} (\pi A \to MX) = (\sigma_{\pi A} - \sigma_{qA}) C_{q\to M}\, q_\pi(x)$$

$$x \frac{d\sigma}{dx} (kA \to MX) = (\sigma_{kA} - \sigma_{qA}) C_{q\to M}\, q_k(x)$$

$$x \frac{d\sigma}{dx} (kA \to MX) = (\sigma_{kA} - \sigma_{sA}) C_{s\to M}\, s_k(x) \tag{10}$$

where q denotes a non strange quark and s denotes a strange quark. The x-distribution of valence quarks can be extracted from meson production of lepton pairs. σ_{sN} can be estimated from the additive quark model[9] relation $\sigma_{sN} = \sigma_{kN} - (1/2)\sigma_{\pi N}$ and can be used to calculate σ_{sA}. Thus the production of a low p_T - large x meson by a non strange spectator quark in a fast pion and in a fast kaon have, respectively, the following A-dependence

$$R(A_2/A_1) = (\sigma_{\pi A_2} - \sigma_{qA_2})/(\sigma_{\pi A_1} - \sigma_{qA_1}), \tag{11}$$

and

$$R(A_2/A_1) = (\sigma_{kA_2} - \sigma_{qA_2})/(\sigma_{kA_1} - \sigma_{qA_1}), \tag{12}$$

while the production of a low p_T - large x meson by a strange spectator quark yields

$$R(A_2/A_1) = (\sigma_{kA_2} - \sigma_{sA_2})/(\sigma_{kA_1} - \sigma_{sA_1}) \tag{13}$$

Fig. 10 shows a comparison[5] between $\alpha(x)$ as predicted by eq. (11) and $\alpha(x)$ that was obtained by Brenner et al.[22] from their experimental data on $\pi^+ A \to \pi^\pm X$ and $\pi^+ A \to k^\pm X$, A = C, Al, Cu, Ag and Pb_s at p_{LAB} = 100 GeV/c.

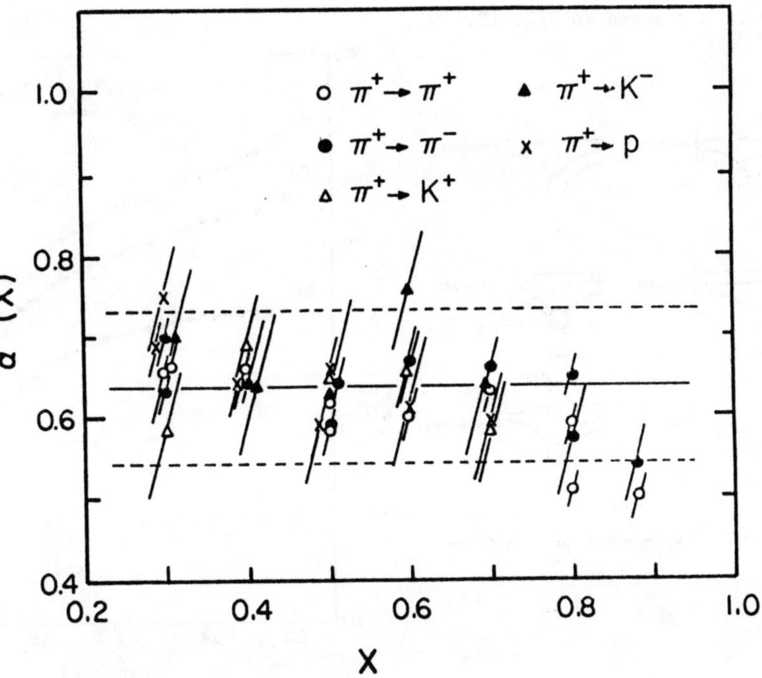

Fig. 10

Non diffractive production of baryons that contain two valence quarks in common with the incident nucleon, such as $pp \to \Lambda x$, can proceed via a recombination of a single quark or a diquark in p as indicated in Fig. 11. The resulting A-dependence is

$$x\frac{d\sigma}{dx}(pA \to \Lambda X) = (\sigma_{pA} - \sigma_{\pi A})C_{D\to\Lambda}D_p(x) + (\sigma_{pA} - \sigma_{qA})C_{q\to\Lambda}(u_p(x) + d_p(x)) \quad (14)$$

where $C_{D\to\Lambda}$ is the probability that a ud-diquark recombines into Λ, and $C_{q\to\Lambda}$ is the probability that a u or a d quark recombines into Λ. $D_p(x)$ is the x-distribution of the ud-diquark in p. Simple phase-space considerations suggest that $D_p(x) \sim x(1-x)$. We used this form of $D_p(x)$ and the forms of $u_p(x)$ and $d_p(x)$ as given by eqs. (3) to fit eq. (14). to the experimental data of Heller et al. on Λ production by 300 GeV/c protons from Be, Cu and Pb targets[21]. (The only free parameters are the two constants $C_{D\to\Lambda}$ and $C_{q\to\Lambda}$). The comparison between theory and experiment is presented in Fig. 12.

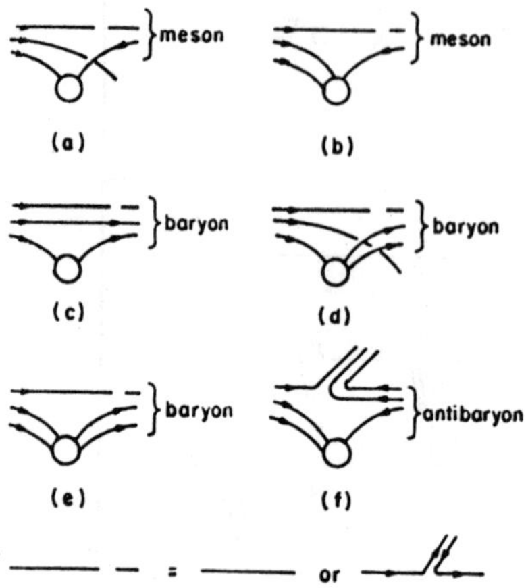

Fig. 11

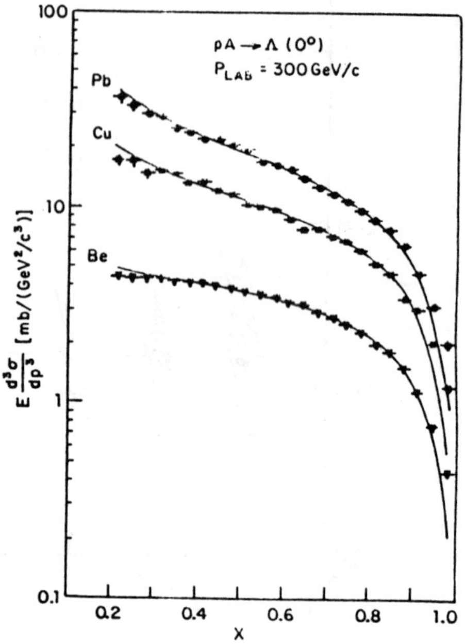

Fig. 12

Fig. 12 demonstrates that the x-dependence and the A-dependence of Λ production in pA collisions are both well reproduced by eq. (14). Note that $u_p(x)/D_p(x) \sim (1-x)^2/x\sqrt{x}$, consequently for $x \to 1$

$$x\frac{d\sigma}{dx}\,(pA \to \Lambda X) \simeq (\sigma_{pA} - \sigma_{\pi A})C_{D\to\Lambda}\; D_p(x) \tag{15}$$

and

$$R(A_2/A_1) = (\sigma_{pA_2} - \sigma_{\pi A_2})/(\sigma_{pA_1} - \sigma_{\pi A_1}) \tag{16}$$

In Fig. 13 we compare $\alpha(x)$ that was calculated from eq. (14) and experimental results[21] on low-p_T Λ production by 300 GeV/c protons and on low-p_T n production by 400 GeV/c protons, from Be and Pb targets. Agreement between theory and experiment is excellent.

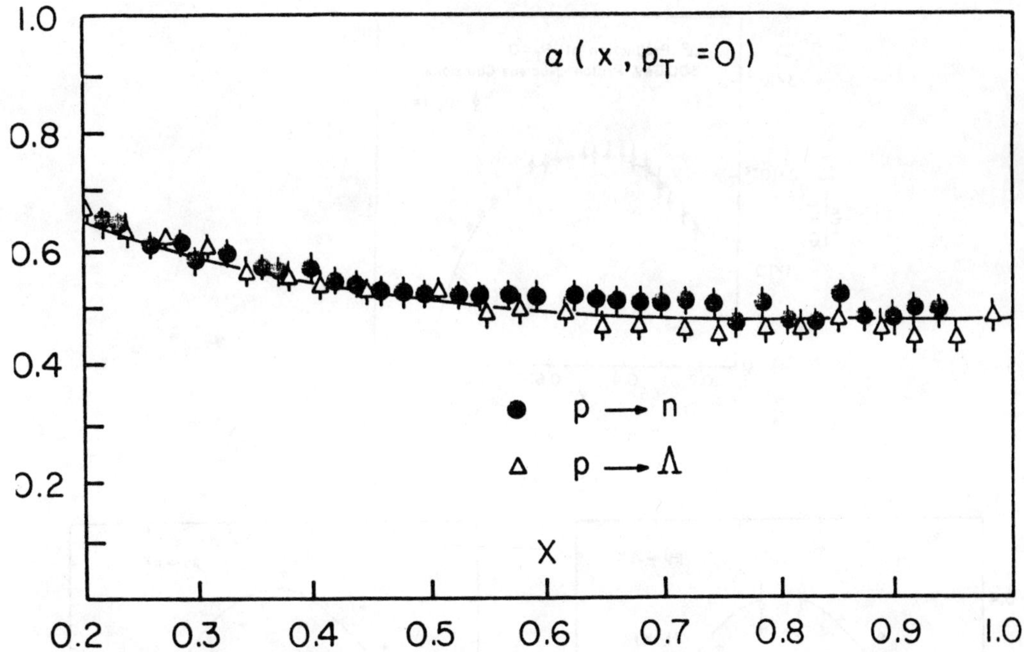

Fig. 13

Bialas and Bialas[6] have used eq. (14) and the experimental data on low-p_T Λ production[21] to estimate $D_p(x)$ from the data. Fig. 14 shows their results. I have added to their figure a plot of the function $Cx(1-x)$ which indeed fits nicely their points.

Deviations from the simple predictions of the recombination model are expected. Around $x=1$ they can come from exclusive channels. At smaller x values and at non zero p_T values they can come from formation and decay of resonances and/or from the recombination of wounded quarks, i.e. of quarks that suffered collisions in the target. It is possible to estimate all these effects but their inclusion requires additional assumptions and the model looses its simplicity and its predictive power.

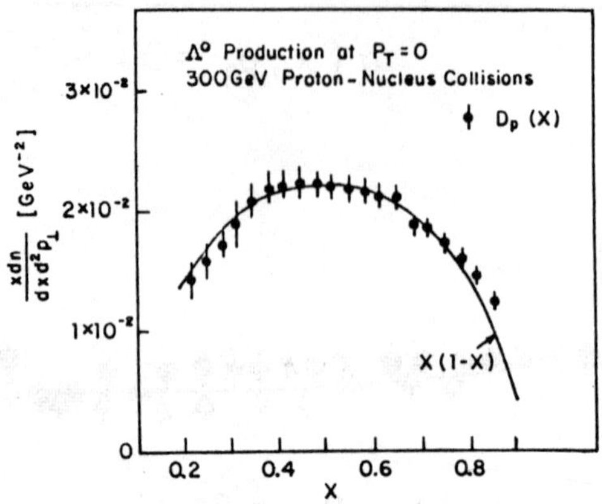

Fig. 14

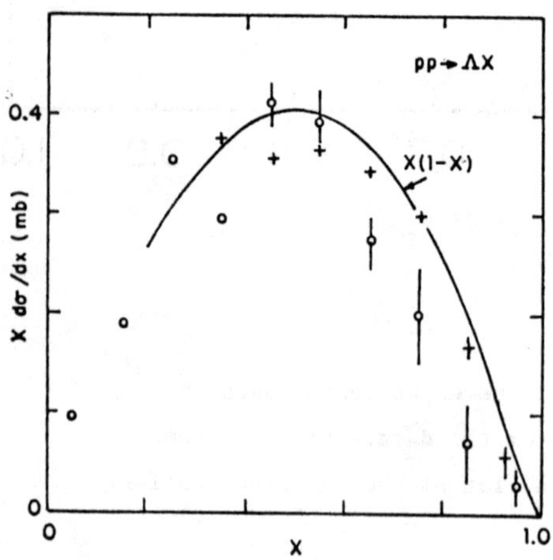

Fig. 15a

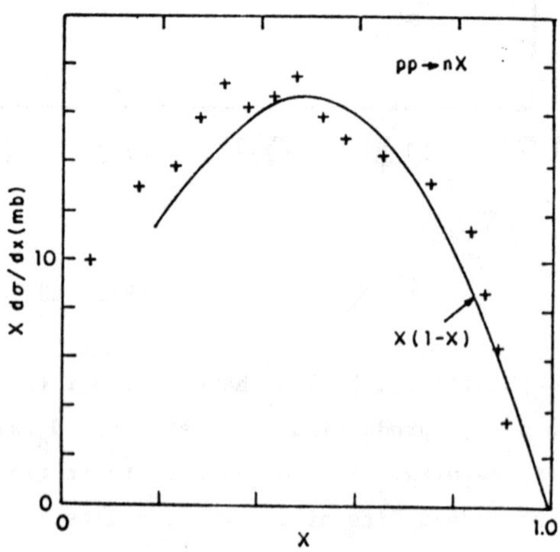

Fig. 15b

Figs. 15a, 15b compare the "diquark structure function" $x(1-x)$ and the x distribution of low-p_T Λ and n produced in the high energy reactions $pp \to \Lambda X$[23] and $pp \to nX$[24].

4. The Quark-Parton Structure of Nuclei

Low-p_T production in the projectile fragmentation region was explained in the previous chapter in terms of the momentum distribution of quarks and gluons in the projectile when it is viewed in a high momentum frame. To extend the quark fragmentation/recombination model to low-p_T production from nuclear targets in the target fragmentation region one needs to know the momentum distribution of quarks and gluons in nuclei in a high momentum frame. These distributions can be obtained, of course, from deep inelastic scattering of leptons in nuclei. However, present measurements cover only the range of small values of x_A (x_A is the fraction of the total momentum of the nucleus which is carried by the quark) i.e. $x_A << 1/A$. (Nuclear structure functions apparently fall very fastly with x_A and one needs very high statistics experiments to explore the region $x_A > 1/A$).

Theoretically, one would like to start from a theory of quarks and their interactions, such as QCD and compute the properties of their bound states, i.e., mesons, nucleons and nuclei. Obviously, it may be some time before this program can be carried out. Meanwhile, instead of waiting with folded arms one can try to see whether QCD inspired guesses of the nuclear structure functions can explain low-p_T production in the nuclear fragmentation region. Such an attempt which so far has been quite successful is described below.

In QCD coloured quarks interact strongly via exchange of coloured gluons. These interactions are believed to bind and confine three coloured quarks to give colour singlet nucleons. Being colour singlets the nucleons cannot interact strongly via an exchange of single coloured gluon. Thus binding forces between nucleons in nuclei are small compared with binding forces within nucleons (the average binding energy per nucleon in a nucleus is about 8 MeV while the nucleon rest mass is about 1 GeV). Therefore it is useful to consider a nucleus in its rest frame as a bound state of point like nucleons. However, when a nucleus moves with a relativistic velocity its nucleons look as if they are composed of valence quarks and of a neutral sea of gluons and quark-anti-quark pairs. The momentum distribution of quarks and gluons in a fast nucleus may be estimated by convoluting the quark and gluon distributions in nucleons, that were extracted from deep inelastic leptoproduction from isolated nucleons, with the momentum distribution of the nucleons inside a fast nucleus.

The momentum distribution of nucleons in a nucleus that moves with a relativistic velocity cannot be extracted from its non-relativistic wave function. In fact, a knowledge of this distribution requires the solution of a complicated relativistic quantum mechanical many body problem. Solution of such relativistic bound state problems have not been found yet. Instead attempts were made to determine these distributions by extracting the leading behaviour from the lowest order diagrams in preturbation theory[26]. The following "counting rules" were obtained: If one denotes by x_{NA} the fraction of the total momentum of a nucleus A that is carried by a single nucleon N inside a fast nucleus, then for large values of x_{NA} its momentum distribution behaves like

$$N_A(x_{NA}) \sim (1 - x_{NA})^{2T(A-1)-1}, \qquad (1)$$

where T depends upon the basic nucleon-nucleon interaction; T = 1 for a renormalizable interaction between the nucleons, such as a vector exchange with point interactions, T = 3 for a nucleon-nucleon interaction which is mediated by the exchange of vector mesons with a monopole form factor at each vertex, etc. Although initially the counting rules with T = 3 were shown to correctly describe the fragmentation of certain relativistic nuclei, it was soon realized that they cannot describe the fragmentations of nuclei heavier than carbon. The powers observed in backward production of fragments from heavy nuclear targets by high energy incident protons are much smaller that those predicted with either T = 1 or T = 3.

However, the counting rules were obtained under the assumption that the nucleus is fully broken up (into A nucleons for T = 1, and into 3A valence quarks for T = 3) during the collision. Such an assumption may be reasonable for high energy central (small impact parameter) nucleus-nucleus collisions, or for high energy particle collisions with very light nuclei, but it is unreasonable and in conflict with experimental observations in collisions between high energy particles and medium and heavy nuclei. The multiplicity distribution of charged nuclear fragments in such collisions was studied extensively by emulsion techniques. It was found to drop rapidly with increasing number of nuclear fragments, i.e. a nucleus very rarely dissociates completely. Usually only the part of the nucleus that interacts directly with the projectile dissociates completely. The Blankenbecler-Schmidt counting rules will be applied to this part.

We note that when a nucleus is viewed from a high momentum frame it is Lorentz contracted. All its nucleons that lie within an imaginary tube with a nucleon cross section σ, that is drawn along the momentum axis, are contracted into a small volume with the same cross section. Due to the Lorentz contraction all nucleons in the contracted tube will overlap and communicate easily via exchange of massless gluons, while their communication in the transverse direction will not be affected. Lorentz contraction will simultaneously increase the longitudinal momentum components of the nucleons due to the uncertainty principle and reduce the strength of their interactions due to asymptotic freedom. Thus the nucleons in a Lorentz contracted tube will behave like a free Fermi gas. The total momentum of the tube will therefore be distributed among its constituent nucleons according to longitudinal phase space rules. The quark structure functions in nuclei will then be obtained by convoluting their structure functions in nucleons with the momentum distribution of nucleons in nuclei in a high momentum frame.

Let us denote by p_q, p_N and p_t, respectively, the momenta of a quark, of a nucleon and of a tube in a nucleus in a high momentum frame and let us also define the momentum fractions

$$x_{qN} \equiv p_q/p_N, \quad x_{Nt} \equiv p_N/p_t, \quad x_{qt} \equiv p_q/p_t = x_{qN}x_{Nt}.$$

We shall first assume that the structure functions of quarks in nucleons and of nucleons in nuclei obey Bjorken scaling. If quarks are confined to nucleons in a tube, then their momentum distribution in a tube is given by

$$q_t(x) = \int_0^1 \int_0^1 q_N(x_{qN})N_t(x_{Nt})\delta(x - x_{qN}x_{Nt})dx_{qN}dx_{Nt}, \tag{2}$$

i.e.

$$q_t(x) = \int_x^1 q_N(x')N_t(\frac{x}{x'})\frac{dx'}{x'}, \tag{3}$$

where $x \equiv x_{qt}$ and $x' \equiv x_{qN}$, $q_N(x')$ is the momentum distribution of quarks of type q in a nucleon and $N_t(x_{Nt})$ is the momentum distribution of nucleons in a tube. They satisfy the normalization conditions

$$\int_0^1 q_N(x)dx = i_q(N), \tag{4}$$

$$\int_0^1 N_t(x)dx = i_N(t), \tag{5}$$

where $i_q(N)$ is the number of quarks of type q in a nucleon and $i_N(t) \equiv i$ is the number of nucleons in the tube. Since we assume that the nucleons in a tube carry almost all its momentum (because it is coloured, a single gluon cannot be exchanged between colourless nucleons) we have also the approximate sum rule

$$\int_0^1 xN_t(x)dx = 1. \tag{6}$$

Simple phase space arguments suggest

$$N_t(x) \simeq i(2i-1)(2i-2)x(1-x)^{2i-3}. \tag{7}$$

This choice of $N_t(x)$ automatically satisfies the normalization condition (5) and the momentum sum rule (6). It also follows from the dimensional counting rules of Blankenbeckler and Brodsky[27], combined with conditions (5) and (6), or if the interaction between the nucleons in a tube is described by a field theory with a renormalizable interaction such as $\lambda\phi^4$, vector exchange, etc.[26].

We now substitute eqs. (B6), (B7) and (7) into (3) and obtain the structure functions of quarks in a tube of i <u>protons</u>

$$u_i(x) = \frac{35}{16}\frac{3}{2i+1}\frac{(1-x)^{2i+1}}{\sqrt{x}} F(2.5,\ 2i-2;\ 2i+2;\ 1-x), \tag{8}$$

$$d_i(x) = \frac{315}{256}\frac{12}{(2i+1)(2i+2)}\frac{(1-x)^{2i+2}}{\sqrt{x}} F(3.5,\ 2i-2,\ 2i+3;\ 1-x). \tag{9}$$

Similarly, since by isospin invariance $d_n = u_p$ and $u_n = d_p$, where n stands for neutron, for a tube that contains $(Z/A)i$ protons and $((A-Z)/A)i$ neutrons we obtain

$$u_i(x) = (Z/A)u_i(x) + ((A - Z)/A)d_i(x), \tag{10}$$

$$d_i(x) = (Z/A)d_i(x) + ((A - Z)/A)u_i(x). \tag{11}$$

$F(a,b,c;z)$ in eqs. (8) and (9) is the confluent hypergeometric function, represented by the integral

$$F(a,b,c;z) = \frac{1}{B(b,c-b)} \int_0^1 t^{b-1}(1-t)^{c-b-1}(1-tz)^{-a}dt. \tag{12}$$

where

$$B(u,v) = \frac{\Gamma(u)\Gamma(v)}{\Gamma(u+v)}. \tag{13}$$

Note that $F(a,b,c;0) = 1$ (provided that $\mathrm{Re}(c-a-b) \gtrsim 0$) and thus for $x \to 1$

$$u_i(x) \sim (1-x)^{2i+1}. \tag{14}$$

The power $2i+1$ can be obtained directly from the dimensional counting rules if one assumes that the elementary spectators accompanying a valence quark in a nucleon are the two other valence quarks in the hit nucleon and all the other nucleons in the tube (i.e. $n_s = 2 + (i-1)$ and thus

$2n_s - 1 = 2i + 1$).

The probability to find i out of A nucleons inside a tube of cross section σ at impact aparameter b is given by

$$P(i,A;b) = \binom{A}{i}\left(\frac{\sigma T(b)}{A}\right)^i\left(1 - \frac{\sigma T(b)}{A}\right)^{A-i}, \tag{15}$$

and the probability to find a tube of i nucleons at any impact parameter is

$$P(i,A) = \int P(i,A;b)d^2b \Big/ \sum_{i=1}^{A} \int P(i,A;b)d^2b \tag{16}$$

where

$$\int \sum_{i=1}^{A} P(i,A;b)d^2b = \int\left[1 - \left(1 - \frac{\sigma T(b)}{A}\right)^A\right]d^2b = \sigma_{pA}^{in} \tag{17}$$

The quark distribution in nuclei can now be written as

$$q_A(x_A) = \frac{\sigma_{pA}}{\sigma_{pp}} \sum_{i=1}^{A} P(i,A)q_i(x_i), \tag{18}$$

where $\sigma_{pA}/\sigma_{pp} = A/\bar{\nu}$ is the average number of "tubes" in A, $x_A \equiv p_q/p_A$ is the fraction of the total momentum of the nucleus carried by the quark and $x_i \equiv p_q/p_t = ix_A$.

The quark distributions (8), (9) in nuclei are considerably different from the corresponding distributions in isolated nucleons. As a consequence production processes which depend on quark distributions in the target exhibit a non-trivial A-dependence. To demonstrate that we have plotted in Fig. 16 the ratio $F_2^{Pb}(x)/F_2^C(x)$. We have used the relation

$$F_2^A(x) = \Sigma\, x\, e_q^2\, q_A(x), \tag{19}$$

where e_q is the charge of the quark q, and $q_A(x)$ is given by eq. (18). Note that for $x_N > 0.5$ the ratio is much different than 207/12 as expected from a naive quark model of the nucleus (a model where the nucleus consists of A stationary nucleons). Moreover, eq. (18) yields a considerable probability for a quark in a nucleus to carry momentum much larger than the average momentum per nucleon in the nucleus. Such momenta lead to subthreshold production of particles and to cumulative production of particles, i.e. to production of masses and momenta which cannot be produced in collisions between isolated nucleons. In section V we shall make explicit use of the quark distributions in nuclei as given by eq. (18) to predict production of low-p_T particles in the fragmentation region of the target nucleus.

Fig. 21

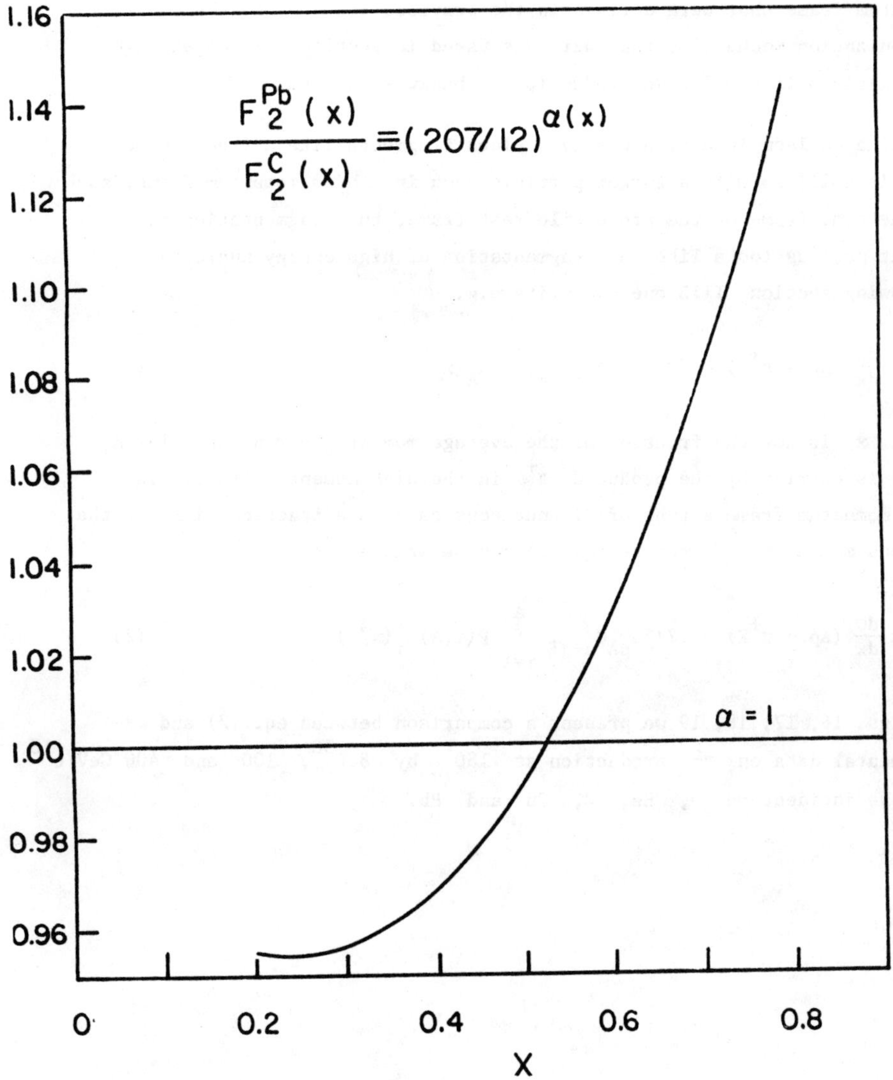

5. The Target Fragmentation Region

The most striking feature of particle production in the fragmentation region of the target nucleus is the production of particles at backward angles with momenta far exceeding those possible in collisions with a free stationary nucleon. This phenomenon was first observed by Baldin and co-workers[28] and later by many other groups[29]. Here we shall show that the internal (collective) motion of quarks within nuclei can explain cumulative production, i.e. the momentum distributions of quarks in nuclei in a high momentum frame that were derived in the previous section and the quark hadronization mechanisms that were discussed in section III can explain remarkably well cumulative production at backward angles.

If a nucleon in a high energy nucleus fragments like a free nucleon when it collides with a target particle then in a high momentum frame, such as the c.m. frame or the projectile rest frame, the fragmentation of the target nucleus looks like the fragmentation of high energy nucleons. Following section IIIB one can write e.g.

$$x \frac{d\sigma}{dx}(Ap \to \pi^+X) = (\sigma_{pp} - \sigma_{qp})C_{u\to\pi^+}\, u_A(x) \tag{1}$$

where x is now the fraction of the average momentum per nucleon in A, which is carried by the produced π^+ in the high momentum frame. In a high momentum frame a tube of i nucleons carries a fraction i/A of the nucleus momentum. Therefore eq. (1) can be written as

$$x \frac{d\sigma}{dx}(Ap \to \pi^+X) = (2/3\sigma_{pA}C_{u\to\pi^+} \sum_{i=1}^{A} P(i,A)u_i(x/i) \tag{2}$$

In Figs. 16, 17, 18, 19 we present a comparison between eq. (2) and experimental data on $\pi^\pm$ production at 180° by 8.6[30], 100 and 400 GeV/c[18] protons incident on D, He, C, Cu and Pb.

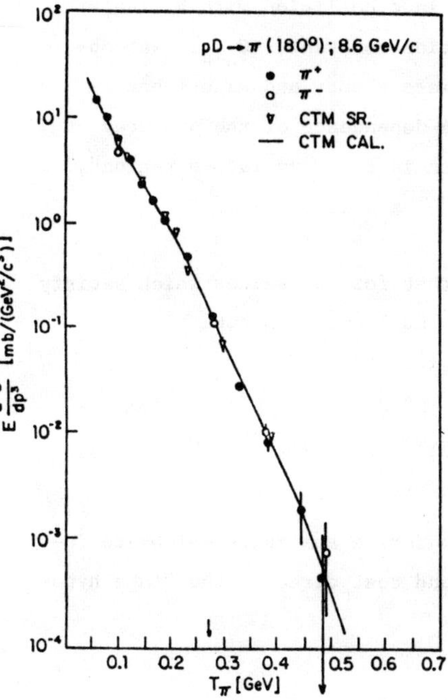

Fig. 16

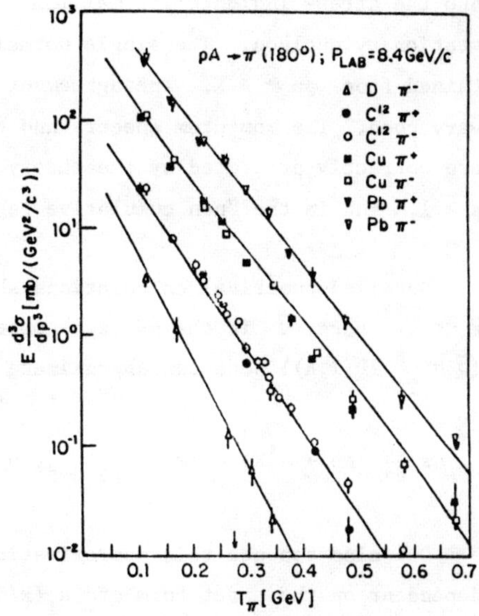

Fig. 17

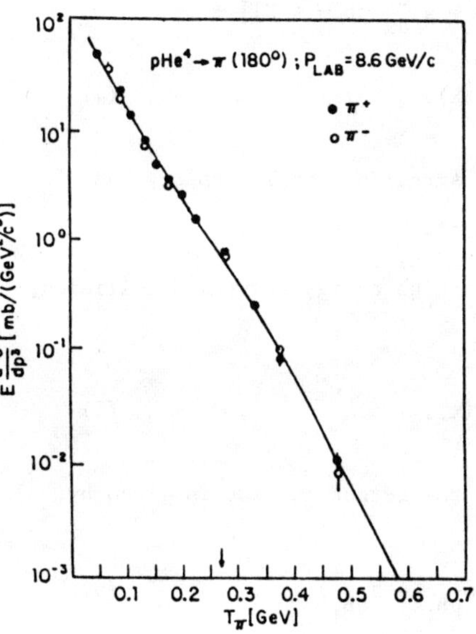

Fig. 18

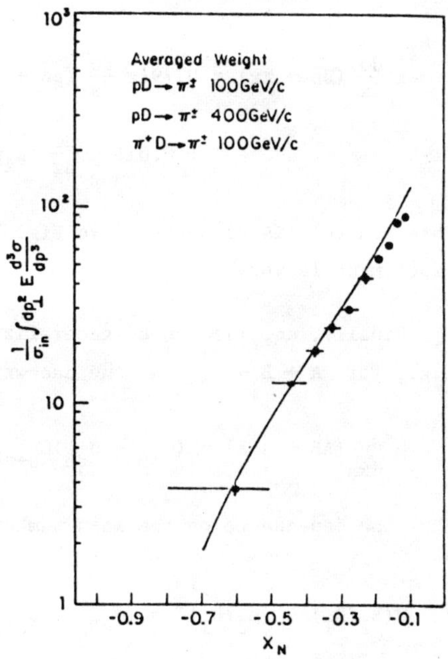

Fig. 19

In Figs. 16–18 $T_\pi = E_\pi - M_\pi$ is the kinetic energy of the produced π, and the arrows indicate its maximum value in a collision with a single stationary nucleon. The single normalization constant, $C_{u\to\pi}$, was obtained from $pp \to \pi^+X$. The agreement between theory and experiment is very good. The momentum spectra and the A-dependence of the produced π are correctly predicted by the theory, both in the "cumulative region", $x > 1$, and in the "non cumulative region", $x \leqslant 1$.

Detailed numerical calculations show that for x values which satisfy $x \ll \bar{\nu}$ where $\bar{\nu}$ is the average number of nucleons in a tube ($\bar{\nu} = \sum_{i=1}^{A} iP(i,A)$) one can approximate eq. (2) by

$$x \frac{d\sigma}{dx} (Ap \to \pi^+X) = (2/3)\sigma_{pA} C_{u\to\pi^+} u_{\bar{\nu}}(x/\bar{\nu}). \tag{3}$$

One also can use this approximation to derive sum rules which are independent on the exact form of $u_i(x/i)$ and test directly the "tube hypothesis", e.g.,

$$x \frac{d\sigma}{dx} (Dp \to \pi x) \simeq 0.79[x \frac{d\sigma}{dx} (pp \to \pi^+X) + x \frac{d\sigma}{dx} pp \to \pi^-X)] + \\ + 0.016\, x \frac{d\sigma}{dx} (A_\ell p \to \pi X). \tag{4}$$

This sum rule is also tested in Fig. 16. Agreement between theory and experiment is very good.

Finally, eq. (1) can be generalized to high energy heavy ion collision, e.g., for $A + B \to \pi^+ + X$ one can write

$$x \frac{d\sigma}{dx} (AB \to \pi^+X) = (\sigma_{pB} - \sigma_{qB}) C_{u\to\pi^+} u_A(x). \tag{5}$$

Thus the dependence on the mass number of the target nucleus is given by

$$R(B_2/B_1) \equiv (B_2/B_1)^\alpha = (\sigma_{pB_2} - \sigma_{qB_2})/(\sigma_{pB_1} - \sigma_{qB_1}). \tag{6}$$

In Fig. 20 eq. (6) is compared with experimental results[31] on $\pi^{\pm}$ production at 2.5^{o} by D, α and C ions at 1.05 GeV and 2.1 GeV incident kinetic energy per nucleon. Agreement between experiment and theory is very good in spite of the relatively low energies.

From our discussion here it seems that the x-distributions of low-p_T π^+ and π^- produced in the target fragmentation region are proportional to the structure functions of u and d quarks, respectively, in the target nucleus. A direct measurement of the nuclear structure functions in the cumulative region $(x > 1)$ by deep inelastic scattering of leptons in nuclei or by hadroproduction of lepton pairs in nuclei is highly desirable to verify this.

TABLE A

A	σ_{qA}(mb)	$\sigma_{\pi A}$(mb)	σ_{pA}(mb)
p	13	26	39
He^4	47		120
Li^6	69	123	174
Be^9	99	152	210
C^{12}	128	181	248
Al^{27}	254	342	445
Cu^{64}	510	647	796
Sn^{119}	826	1045	1259
Pb^{208}	1253	1548	1812
U^{238}	1389	1740	2052

Summary of total absorption nuclear cross sections that were used in this paper. The quark cross sections were calculated from eq. (17), section IV, using the nuclear density functions of reference 10. $\sigma_{\pi A}$ and σ_{pA} are from reference 32.

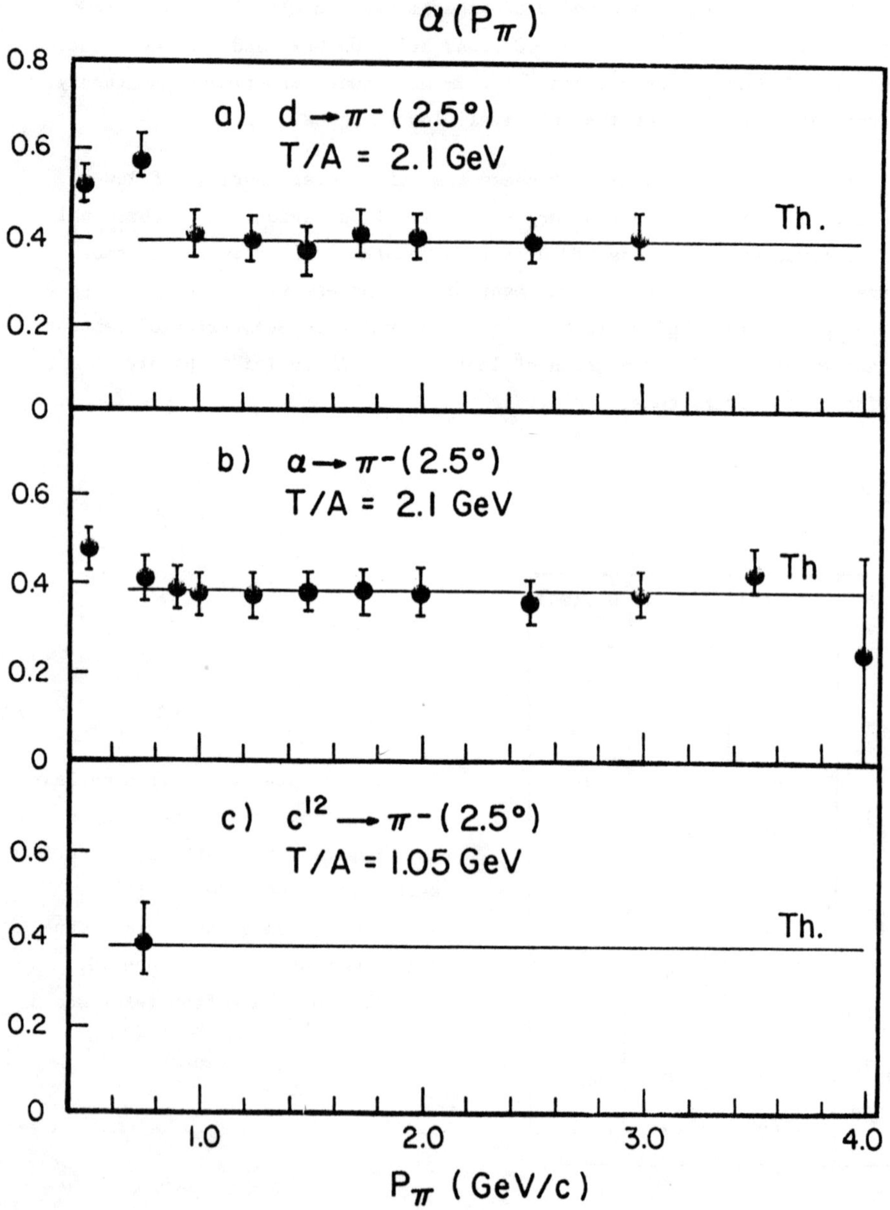
α(P$_\pi$)
a) d → π⁻ (2.5°)
T/A = 2.1 GeV
Th.
b) α → π⁻ (2.5°)
T/A = 2.1 GeV
Th
c) C^{12} → π⁻ (2.5°)
T/A = 1.05 GeV
Th.
0.8
0.6
0.4
0.2
0
1.0
2.0
3.0
4.0
P$_\pi$ (GeV/c)

Fig. 20

References and Footnotes

1. V.V. Anisovich, Yu.M. Shabelski and V.M. Shekhter, Nucl. Phys. B133, 477 (1978).

2. N.N. Nikolaev, A. Ya. Ostapchuk and V.R. Zoller, Preprint CERN-TH, 2541 (1978).

3. N.N. Nikolaev and S. Pokorski, Phys. Lett. 80B, 290 (1979). N.N. Nikolaev, Particles and Fields, 4C, 291 (1980).

4. A. Dar and F. Takagi, Phys. Rev. Lett. 44, 768 (1980). See also G. Berlad, A. Dar and G. Eilam, Phys. Rev. D22, 1547 (1980).

5. F. Takagi, Tohoku University Preprint TU/80/215.

6. A. Bialas and E. Bialas, Phys. Rev. D20, 2645 (1979). A. Bialas, Acta Physica Polonica B11, 475 (1980).

7. A. Bialas and W. Czyż, Acta Phys. Pol. B10, 83 (1979); Cracow Preprint INP 1109/PH and TPJU-7/80.

8. L.S. Osborne et al., Phys. Rev. Lett. 40, 1624 (1978).

9. E.M. Levin and L.L. Frankfurt, JETP Letters 2, 108 (1965); H.J. Lipkin and F. Sheck, Phys. Rev. Lett. 16, 71 (1966).

10. S. Barshay et al., Phys. Rev. C11, 360 (1975).

11. A. Dar and J. Vary, Phys. Rev. D6, 2412 (1972).

12. J.E. Elias et al., Phys. Rev. D22, 13 (1979).

13. See for instance A. Dar, I. Otterlund and E. Stenlund, Phys. Rev. D20, 2344 (1979), and references therein.

14. M.C. Albrow and M. Jacob, CERN Report EP, 5924; TH, 2412 (1981).

15. See for instance A. Dar, Phys. Lett. 87B, 136 (1979) and references therein.

16. See for instance L. Van Hove, Acta Phys. Austriaca, Suppl. 21, 621 (1979) and references therein.

17. J. Singh et al., Nucl. Phys. B 140, 189 (1978).

18. J. Hanlon et al., (Stony Brook-Carnegie Mellon-Fermilab-Technion Collaboration), to be published.

19. For the sake of simplicity we neglect violations of scaling.

20. J.V. Allaby et al., CERN Report 70-12 (1970).
T. Eichten et al., Nucl. Phys. B44, 333 (1972).

21. K. Heller et al., Phys. Rev. D16, 2737 (1977).

22. A.E. Brenner et al., Fermilab Preprint Conf-80/47-Exp. (1980).

23. H. Kichimi et al., Phys. Rev. D20, 37 (1979).
S. Erhan et al., Phys. Lett. 85B, 447 (1979).

24. V. Blobel et al., Nucl. Phys. B135, 379 (1978).

25. G. Berlad, A. Dar and G. Eilam, Phys. Rev. D22, 1547 (1980).

26. I.A. Schmidt and R. Blankenbecler, Phys. Rev. D15, 3321 (1977).

27. R. Blankenbecler and S.J. Brodsky, Phys. Rev. D10, 2973 (1974);
J. Gunion, Phys. Rev. D10, 242 (1974).

28. A.M. Baldin et al., Sov. Journ. Nucl. Phys. 18, 41 (1974).

29. For a complete list of references see for instance G. Berlad et al., Phys. Lett. 93B, 86 (1980).

30. A.M. Baldin et al., Sov. Jour. Nucl. Phys. 20, 629 (1975).
A.M. Baldin, Particles and Nuclei 8, 429 (1977) and references therein.

31. J. Papp et al., Phys. Rev. Lett. 34, 601 (1975); J. Papp, Ph.D. Thesis, Report LBL-3633 (1975).

32. S.P. Denisov et al., Nucl. Phys. B61, 62 (1973).

Chapter VII: QCD-CONFINEMENT, DTU-MODEL AND HADRON-NUCLEUS COLLISIONS

Charles B. CHIU
Center for Particle Theory, Department of Physics,
University of Texas at Austin, U.S.A.

Abstract: Complementary discussions on the QCD flux tube model and the DTU model in connection with our previous work are given. We also show that our recent hadron-nucleus collision model has two important suppression mechanisms for particle production. Within the projectile cascade approximation, our model leads to the prediction of an approximate $\bar{\nu}$-universality.

During this conference, especially in the lectures of Capella and Sukhatme, you have already seen that there have been considerable attempts[1] to understand the small p_T-processes from the notions of color confinement and the dual-topological-unitarization (DTU) formalism. I would like to make several complementary comments on this approach especially in connection with our previous work. Topics I wish to comment on are: the color flux tube model, the DTU model and our extension of these ideas to hadron-nucleus collisions.

1. The Flux Tube Model

The folklore of string picture for quark confinement[2,3] has been with us for several years. One imagines that quarks are confined through QCD-gluon-flux tube. The breaking of this tube occurs when $q\bar{q}$ pair is created within the tube. This provides a very natural mechanism for quark confinement. Within this picture, it is not possible to free the end-quark by transferring more and more energies to the tube. All what the increase of energy does is to create more and more tubelets--the hadrons between the two ends. The flux tube picture naturally leads to a fixed-p_T hadronic jet. Here the term fixed-p_T refers to the fact, that the p_T-distribution does not vary as a function of the rapidity variable y. However, this distribution is expected to vary as a function of the Feynman variable x. I shall return to this point later on.

In high energy collisions, we have seen the footprints of QCD flux tube all over. It constitutes the bulk of the final state particles in hadronic processes. It is plausible that those jets produced in hard processes, such as large-p_T hadronic processes and e^+e^- collisions, have two components: the perturbative-QCD component or the angular component and the small p_T-component of present concern. The very similarity between e^+e^- jets and small p_T-hadronic jets presented in this conference suggests that, at least up to the presently available energies, the flux-tube jet is still an important component in hard processes.

Let me now turn to the color tube model considered by Casher Neuberger and Nussinov (CNN),[3] and review some general features of the model. Similar features are also present in the string models considered by the Lund group and by others.[4] Consider a color flux tube with $q\bar{q}$ at two ends having a tube stretched out between them, say along the positive z axis. Treat this system as a parallel plate capacitor with plate charges $\pm g$ and field strength ε. Then the probability for the pair production per unit time per unit volume is given by[5]

$$\rho = \frac{P}{TV} = \frac{g^2\varepsilon^2}{\pi^2} \sum_{n=1}^{\infty} \frac{1}{n^2} \exp\left[\frac{-n\pi m^2}{g\varepsilon}\right] , \tag{1}$$

where m is the constituent quark mass. For $m = m_{u,d} \approx 0.35$ GeV, the sum converges rapidly. For our discussions below, it is sufficient to write

$$\rho \propto \exp\left[\frac{-\pi m^2}{g\varepsilon}\right] . \tag{2}$$

The factors $m^2/g\varepsilon$ in the exponent can be motivated by treating the pair creation process as a quantum mechanical tunneling problem.[6] Consider a virtual pair originally trapped in a barreir at the origin. To pair produce, this virtual pair must acquire at least their rest mass energy. For definiteness we focus on, say the quark. The barrier height at $z = 0$ is mc^2. As the quark is "dragged" along by the field ε, it acquires an energy $zg\varepsilon$. So at z the effective barrier seen by this quark is reduced to $V = mc^2 - zg\varepsilon$. For the production of a pair with zero kinetic energy, the exponential factor in the WKB amplitude is given by

$$\int_0^{z_0} \sqrt{2mV} = \int_0^{z_0} \sqrt{2m(mc^2 - zg\varepsilon)} = \frac{2\sqrt{2}}{3} \cdot \frac{m^2}{g\varepsilon} , \quad z_0 = \frac{mc^2}{g\varepsilon} . \tag{3}$$

One thus obtains the desired factor $m^2/g\varepsilon$. Within the WKB method, CNN have in fact shown that, with relativistic kinematics when all the appropriate kinematical and statistical factors have been taken into account, Eq. (1) can be reproduced exactly.

Furthermore, for the present tunneling problem, the transverse momentum commutes with the hamiltonian. Consequently the mass can be replaced by the transverse mass ($m_T^2 = m^2 + k_T^2$). In turn the momentum spectrum in the rest frame of the pair is given by

$$\frac{dn}{dk_T^2} \propto \exp\left(-\frac{\pi}{g\varepsilon} m_T^2\right) \propto \exp\left(-\frac{\pi}{g\varepsilon} k_T^2\right) \tag{4}$$

The momentum distribution of Eq. (4) can also be formally obtained through manipulating a divergent sequence of Feynman diagrams.[7] Making use of the technique of Borel sum, one finds that the pair production amplitude has the following general form:

$$A \propto \exp\left[-\frac{\text{const}}{g\varepsilon}(E'^2 + 2ik_z E')\right], \text{ with } E'^2 = m_T^2 + k_z^2 \ . \tag{5}$$

This leads to

$$E' \frac{dn}{dk_T^2 dk_z} = \frac{dn}{dy' dk_T^2} \propto |A|^2 \propto \exp\left[-\frac{\pi}{g\varepsilon}(k_T^2 + k_z^2)\right] \ . \tag{6}$$

Notice in the pair rest frame, the cutoff in the quark momentum is both in the transverse direction as well as in the longitudinal direction.

One assumes that as a pair is produced, the flux tube is broken. The successive pair creations lead to further production of the tubelets or the hadrons. Convoluting over the transverse momenta of the quark and the antiquark at the ends of each tubelet, one arrives at the momentum distribution of the hadrons, which is given by:

$$\frac{dn}{dy dp_T^2} \propto \exp\left(-\frac{a}{2} p_T^2\right) \ , \text{ with } a = \frac{\pi}{g\varepsilon} \tag{7}$$

Let me now come back to my earlier remark on the terminology of the "fixed" p_T-distribution. For a given y-bin, the model gives

$$\left.\frac{d\sigma}{dp_T^2}\right|_y \equiv \Delta y \frac{d\sigma}{dy dp_T^2} \sim \exp\left(-\frac{a}{2} p_T^2\right) \tag{8}$$

However, for a given x-bin, the model will give instead

$$\left.\frac{d\sigma}{dp_T^2}\right|_x \equiv \int_{y(x,p_T^2)}^{y(x+\Delta x,p_T^2)} dy \frac{d\sigma}{dy dp_T^2} = \Delta x \frac{\Delta y}{\Delta x} \cdot \frac{d\sigma}{dy dp_T^2} \sim [m^2 + p_T^2 + (xE)^2]^{-\frac{1}{2}} e^{-\frac{ap_T^2}{2}} \ , \tag{9}$$

where E is the maximum possible hadronic cm momentum. This gives rise to an x-dependent p_T-distribution. The plot of the mean value $\langle p_T^2 \rangle_x$ vs x is not a constant. It will show a "seagull effect." So once again, when we talk about the fixed p_T-jet, we refer to properties of p_T-distribution for a fixed y-bin, not for a fixed x-bin.

If one identifies the flux tube as a dual string, the string constant is $g\varepsilon = \pi^2\alpha'$, where $\alpha' \approx 1\ \mathrm{GeV}^{-2}$. The predicted parameter-free p_T-dependence is in qualitative agreement with the e^+e^- data. I refer the reader to ref. 8 for further discussion on this point.

Within the color flux tube model, the pion production time defined in the rest frame of the pair can be estimated in the following way. CNN found that $P \approx 10^{-3}\ \mathrm{GeV}^4$. So $1/\tau_0 = PV$, where the available volume is approximately given by $\pi r^2 \tau_0$. Assume $r \approx 1/m_\pi$, we get[9] $1/\tau_0 = 0.4$ GeV. The space-time description of the tube fragmentation is now straightforward. In any rest frame within the tube, one sees the following universal picture. The two ends of the tube are moving away with essentially the speed of light. The pair with a rapidity y will be created at the dilated time $t = \tau_0 \cosh y$. This leads to the "slow-to-fast" pattern for pair productions. This frame independent symmetry is a very general property in high energy multiparticle productions. For instance a similar picture is also in the hydrodynamical model,[10] and also in string models considered by the Lund group and by others.[4] This pattern implies for instance, in the cm system, one would see the inside-out production pattern, while in the rest frame of the right-moving valence quark, one sees the right-to-left fragmentation pattern, and so on. We shall return to some of these string features below.

2. The DTU Model

For the small-p_T processes, there have been various phenomenological models available, e.g. fragmentation models and recombination models, which have been quite successful in describing the momentum spectrum of hadrons in various processes. Where does DTU formalism stand among these models?

I would like to echo the point of view expressed by Capella et al. and Cohen-Tannoudji et al.,[1] that DTU is not just yet another model, piecemealed to describe high energy production phenomena. DTU is a viable theoretical model which provides a plausible description for hadron dynamics. The DTU program[11,12] you recall starts from the dual resonance model. This dual model reproduces resonance pole features at low energies and Regge behavior at high energies. This model also satisfies finite energy sum rule duality constraints and the crossing asymmetry. These are important features of hadron data which we have learned over the past decades. These features, let me reiterate, have been elegantly incorporated in the dual model. DTU program goes one step beyond the dual model, in that it implements unitarity through the dual topological expansion.[11] I find it remarkable that DTU with its global sophistication, still can provide us with detailed insights on high energy production phenomena. We now proceed to see how the flux tube discussed in section 1 may be incorporated into the DTU framework.

a. Flux tube production in DTU

Let me begin with the spectrum of mesonic system. We all know that mesons are observed in SU_3 singlets and octets. This suggests that mesons are made out of $3\bar{3}$, the quark and antiquark, or the nonexotic states. The nonexoticity feature is not only for the stable mesons, it is also for mesonic resonances extending to large masses.

Now when two mesons collide at high energies, how should we describe the excited intermediate states? DTU gives the following description. It predicts that the intermediate excited states are either one pair of $q\bar{q}$ excited resonance states, or two pairs of $q\bar{q}$ excited resonance states. Other forms of intermediate excitations are suppressed according to the topological expansion criterion. The $q\bar{q}$ and the two pairs of $q\bar{q}$ excitation states are intermediate states of the planar and the cylinder duality diagrams. They are shown in Figs. 1a and 1b respectively.

We recall the dynamical proposal of Lee and Veneziano.[12] (Their argument is simple to understand but it is only qualitative. For quantitative arguments, the reader should consult ref. 11 for details.) The planar contribution gives rise to a cross section which behaves like $s^{\alpha-1}$. This cross section is built up by the planar multiperipheral (MP) diagrams (e.g. see Fig. 2). In particular:

$$\sigma_{pl.} \sim s^{2\alpha_{in}-2}\sum_{\ell}(g^2 \ln s)^{\ell}/\ell! = s^{2\alpha_{in}-2+g^2} \sim s^{\alpha-1} \quad . \tag{10}$$

The bootstrap dynamics is introduced through identifying the input and the output trajectories, i.e. $\alpha_{in} = \alpha$. This then leads to the relation:

$$g^2 = 1 - \alpha \, . \tag{11}$$

The multiparticle productions for the nonplanar diagram or the cylinder diagram (e.g. Fig. 1b), as we shall explain below, correspond to those MP diagrams obtained by making all possible cross-insertions on the propagators (see e.g. Fig. 3). For a final state with $\ell + 1$ particles, there are altogether $2^{\ell} - 1$ diagrams with different cross-insertion patterns. Corresponding total cross section is

$$\sigma_{nonpl.} \sim s^{2\alpha_{in}-2}\sum_{\ell}\frac{(g^2 \ln s)^{\ell}}{\ell!}\,2^{\ell} \sim s^{2\alpha-2+2g^2} = s^{\alpha_P - 1} \, . \tag{12}$$

This implies that at $t = 0$, the Pomeron trajectory zero intercept:

$$\alpha_P = 2\alpha - 1 + 2g^2 = 1 \ . \tag{13}$$

To sum up, the dynamical picture introduced here leads to the understanding of why the total cross section at high energies is approximately constant. Furthermore, an inspection on the untwisted nonplanar diagram illustrated in Fig. 3c also reveals that the nonplanar diagram corresponds to two-chain fragmentations, in contrast to one-chain fragmentation for the planar diagram (see Fig. 2b).

In the dual resonance model language, the excited $q\bar{q}$ system labeled by m in Fig. 1a has the spectrum of an excited string. So the bootstrap ansatz essentially amounts to identifying this intermediate excited system with multiparticle final states shown in Fig. 2. This suggests the following sequence of events. During the meson-meson collision, there is first the creation of a $q\bar{q}$ string or a $q\bar{q}$ flux tube. Subsequently this tube breaks into color-singlet tubelets.

Next the cylinder diagram. The correspondence between the intermediate $q\bar{q}\bar{q}q$ lines of Fig. 1b to multiparticle final states of nonplanar MP diagrams (e.g. Fig. 3) has been argued in the context of mass distributions.[13] In particular, one may identify the $q\bar{q}$ labeled by m_1 and the $\bar{q}q$ by m_2 in Fig. 1b with the corresponding untwisted two-chains shown in Fig. 3c. One has found that the mass distribution functions $\rho(m_1,m_2)$ for the two cases are practically identical. So analogous to the planar case, the cylinder case has the sequence: first the production of the two $q\bar{q}$ pairs (or two color tubes), then they evolve into two-chains of multiparticles.

b. A parton picture for DTU Pomeron

Technically speaking the DTU cylinder diagram does not necessarily imply that immediately after the collision the corresponding two tubes produced must be two spatially well separated entities. For instance following qualitative picture is also within the spirit of DTU. Consider a meson-meson collision process. In their cm system, the valence partons (quark and antiquark) of the projectile and those of the target pass by each other. Each valence parton has a long tail of vacuum polarization extending down to the central wee region. In this region, the vacuum polarizations of the right-movers and those of the left-movers overlap each other (Fig. 4a). Parton annihilations in the wee region eventually lead to the inside-out pair creations (Fig. 4b,c).

For definiteness let us identify each pair-production configuration with a nonplanar MP diagram. At asymptotic energies, DTU tells us for a fixed number of final particles, the cross section contributed by a planar diagram is identical to that contributed by a nonplanar diagram which has a specific pattern of crosses for its propagators. Furthermore, when we consider one specific pattern, on the average the probability of pair production per unit y-interval for the cylinder case is identical to that for the planar case. However, the planar case has only one-chain fragmentation. This implies that the longitudinal momentum of the newly created quark should always be aligned along the field direction, say the positive z direction. On the other hand, for the cylinder case, owing to the simultaneous presence of the $q\bar{q}$ field along the positive z direction and that along the negative z-direction, there is a large fluctuation. We imagine that both the positive z and the negative z alignments are allowed here. This is illustrated in Fig. 4b. The very presence of the two possible orientations here corresponds to, in the dual model language, the choice of the crossed and the planar propagators. So for the cylinder case, there will be $2^{\ell} - 1$ patterns associated with $\ell + 1$ particles produced. This would then lead to effectively a two-chain fragmentation spectrum (e.g. see Fig. 4c).

c. Comments on the phenomenological applications

I shall refer the phenomenological application of the DTU model to ref. 1, except mentioning two points, which will facilitate our discussions in the following section. Consider pp collision. The above consideration for a $q\bar{q}$ system, or a $3\bar{3}$ system for a meson can be naturally extended to a quark (3) and a diquark ($\bar{3}$) system for a baryon. This is partly motivated by the fact that DTU predicts[1,11] that the diquark in the projectile should carry most of the projectile momentum, while that in the target proton should be essentially at rest in the target rest frame.[1] Furthermore, DTU predicts that the inner boundaries of the flux tubes are given by:[13,14]

$$\frac{dn}{dy_1 dy_2} \sim \exp[-(\alpha_p - \alpha)(Y - y_2)]\cdot\exp[-(\alpha_p - \alpha)y_1] \tag{14}$$

This leads to the two-displaced chain (2DC) spectra for the pp case (see Fig. 5a,c). A similar argument for the $p\bar{p}$ case, leads to a short-long-chain (SLC) spectra (see Fig. 5b,d).

In Fig. 5, the shaded area near the two ends of the kinematic boundaries is the triple Regge region. A triple Regge diagram which corresponds to the inclusive diagram for the Pomeron contribution is illustrated in Fig. 6. In this domain the fragmentation function is predicted to behave like $(1-x)^n$. For the case illustrated $n = 2\alpha - 1$, where for $a \to c\ X$, $t = (p_c - p_a)^2$. The very t-dependent behavior has been confirmed: in $\pi^{\pm}p \to \pi^{0}X$ data.[15a] The power behavior extracted from the data comapred quite favorably with that expected from the rho trajectory. This situation is illustrated in Fig. 7. Notice there is a strong t-dependence. Analysis on the t-dependence of the power n at lower x values will be very useful. This is especially in view of the fact that the power n extracted from the various counting rules is a t independent quantity. Next we extend DTU to hadron-nucleus collisions.

3. A Hadron-Nucleus Collision Model[16-18]

Consider the passage of a high energy proton through a nucleus. Immediately after the first collision, according to the DTU model two pairs of color singlet tubes are formed. We shall refer to the tube tied to the leading diquark as the excited projectile system (EPS) and the one tied to the diquark of the target nucleon as the excited target system (ETS). We consider the ansatz that: the excited system undergoes collisions with downstream nucleons in the target in a similar manner as that for a nucleon-nucleon collision. In this context there are two important mechanisms which naturally suppresses the particle productions.

(1) Suppression of the intermediate EPS fragmentations: As the excited projectile system undergoes a collision, the initial EPS is to be replaced by a new EPS in the final state which may participate the next collision. Since in the target rest frame, the EPS fragmentations occur only after some dilated time has elapsed, being energetic the EPS tube is expected to fragment, at least for the most part, outside of the nucleus. Consequently only one EPS fragmentation will be materialized. Our intuitive picture differs from that of ref. 19 in that at high energies they have in effect $\bar{\nu}$ EPS's, where $\bar{\nu}$ is the number of inelastic collisions. So this suppression mechanism here is absent in their model.

(2) Saturation of production for a target nucleon: Once the valence partons of a target nucleon are tied to flux tubes, they are no longer available for initiating further multiparticle productions. This then prevents additional hadron production from an already "wounded" target nucleon. In other words, each target nucleon can only be wounded once.

Denote K to be the average number of nucleons which the projectile or the projectile system encounters as it traverses through the nucleus. The hadron-nucleus inelastic cross section is approximately given by

$$\sigma_{hA} = C_k \, A^{2/3} \, [1 - (1 - P)^k] \, , \tag{15}$$

where P is the probability that the projectile hadron undergoes an inelastic collision. This expression is in good agreement with the data.[20,21]

To define the quantity $\bar{\nu}$, we need to make use of the projectile cascade approximation, i.e. consider only those downstream interactions with the EPS, but ignore those with ETS's. Based on the ansatz given above, we also assume that the cross sections $\sigma_{EPS,N}$ and σ_{NN} are similar. Within these two assumptions, the average number of collisions is then given by the usual expression:

$$\bar{\nu} = A\sigma_{hN}/\sigma_{hA} \quad . \tag{16}$$

For comparison with the data below, we also recall the definition:

$$R = \frac{\langle n\rangle_{hA}}{\langle n\rangle_{hN}} \quad . \tag{17}$$

We first compare our model with proton-nucleus data. The situation for the ratio R is shown in Fig. 8. The solid line corresponds to taking the leading projectile cascade approximation and the dotted line takes into account the further interactions of the ETS's with the unwounded downstream target nucleons. We see that the correction is in the right direction, although at present energies, this is not large. The comparison between the differential multiplicity dn/dy within the EPS cascade approximation is illustrated in Fig. 9. Here for the $\bar{\nu} = 1$ case, the 2DC spectrum of Fig. 5a is adjusted to fit the data. The subsequent $\bar{\nu}$-dependent y-distributions are parameter-free predictions of the model. We see that the agreement with the data is reasonable.

Next consider the $\bar{p}$A case. In the same projectile cascade approximation, we observe that the differential multiplicity distribution for $\bar{p}$-nucleus should be similar to that for p-nucleus at the same $\bar{\nu}$. The situation is illustrated in Fig. 10. Within the plateau approximation the very similarity may be deduced by inspection of Fig. 10c,d.[17,18].

Furthermore, our conclusion for the $\bar{\nu}$-universality for p- and $\bar{p}$-projectiles can also be extended to meson projectile, since the meson-initiated spectrum is the average of the pp-type (2DC) and the $\bar{p}$p-type (SLC) spectra. To be more precise the universality here states that for $\bar{\nu}_{pA} = \bar{\nu}_{\bar{p}A'} = \bar{\nu}_{MA''}$, in the plateau approximation,

$$\left.\frac{dn}{dy}\right|_{PA} \approx \left.\frac{dn}{dy}\right|_{\bar{p}A'} \approx \left.\frac{dn}{dy}\right|_{MA''} \quad , \tag{18}$$

and the ratio

$$R \approx \frac{\bar{\nu} + 1}{2} + \frac{(\bar{\nu} - 1)}{2(Y - y_1)}\Delta \quad , \tag{19}$$

where Δ is the rapidity interval at the low rapidity end of EPS, within which the flux tube may fragment during its travel between the two successive collisions. In our analysis for Figs. 8, 9, we have set $\Delta = 0$. For further details of our model, in particular the comparison of our model with those of others and the extension to nucleus-nucleus collisions, I refer the reader to ref. 18.

I would like to thank A. Capella, G. Gustafson, R. Hwa, W. Kittel, H. Mietinen, U. Sukhatme and D. Tow for invaluable discussions. My appreciations to Professor R. Van de Walle for a special provision which enabled me to participate in this conference. This work was supported in part by the U.S. Department of Energy.

References

1. See this proceedings: Lectures by Capella and by Sukhatme and references quoted therein. See also G. Cohen-Tannoudji et al., Phys. Rev. D21, 2699 (1980) and earlier references quoted therein.
2. See for example: A. Casher, J. Kogut and L. Susskind, Phys. Rev. D10, 732 (1974).
3. A. Casher, H. Neuberger and S. Nussinov, Phys. Rev. D20, 179 (1979).
4. B. Andersson, G. Gustafson and C. Peterson, Z. Phys. C3, 223 (1980). X. Artru, Orsay preprint, LPTHE 79/8, and references quoted therein.
5. J. Schwinger, Phys. Rev. 82, 664 (1951).
6. E. Brezin and C. Itzykson, Phys. Rev. D2, 1191 (1970).
7. C. B. Chiu and S. Nussinov, Phys. Rev. D20, 945 (1979).
8. C. B. Chiu, Proc. of the Moriond Conference, Les Arcs, 1979.
9. C. B. Chiu, Proceedings of the 20th Int. Conf. on High Energy Physics, Madison, 1980, part 2, p. 80.
10. E. V. Shuryak, Phys. Rev. 61, 71 (1980); see also C. B. Chiu, E.C.G. Sudarshan and K. H. Wang, Phys. Rev. D12, 902 (1975).
11. G. Veneziano, Nucl. Phys. B74, 365 (1974); Phys. Lett. 52B, 220 (1974); H. M. Chan et al., Nucl. Phys. B86, 479 (1975); G. F. Chew and C. Rosenzweig, Phys. Rep. 41C, 263 (1978).
12. H. Lee, Phys. Rev. Lett. 30, 719 (1973); G. Veneziano, Phys. Lett. 43B, 413 (1973).
13. C. B. Chiu and S. Matsuda, Nucl. Phys. B134, 463 (1978).
14. P. Aurenche and L. Gonzalez-Mestres, Phys. Rev. D18, 2995 (1978).
15. Experimental data illustrated in the figures.
    a. Figure 6: Caltech-LBL collaboration, Nucl. Phys. B145, 45 (1978).
    b. Figs. 8 and 9: W. Busza et al., Proc. of the 18th Int. Conf. on

High Energy Phys., Tbilisi, 1976.

16. W. Q. Chao, C. B. Chiu, Z. He and D. M. Tow, Phys. Rev. Lett. 44, 518 (1980).
17. C. B. Chiu and D. M. Tow Phys. Lett. 97B, 443 (1980).
18. C. B. Chiu, Z. He and D. M. Tow, CPT preprint, 1981.
19. A. Capélla and J. Tran Thanh Van, Phys. Lett. 93B, 146 (1980).
20. G. Bialkowski, C. B. Chiu and D. M. Tow, Phys. Rev. D17, 862 (1978).
21. See also M. Hossain and D. M. Tow, Phys. Rev. D21, 1842 (1980).

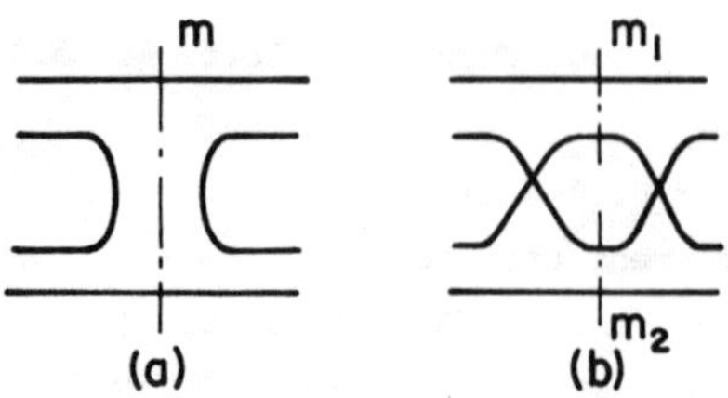

Fig. 1. Quark-duality diagrams.
a. planar excitation: m
b. cylinder excitations: m_1, m_2.

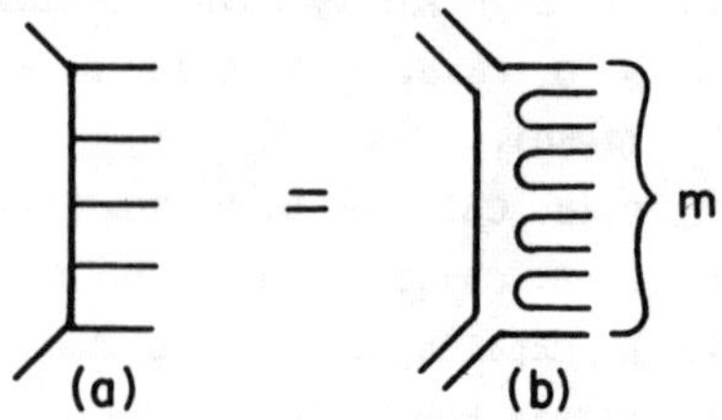

Fig. 2. A typical planar MP diagram.
In a, hadrons are solid lines.
In b, hadrons are $q\bar{q}$ ribbons.

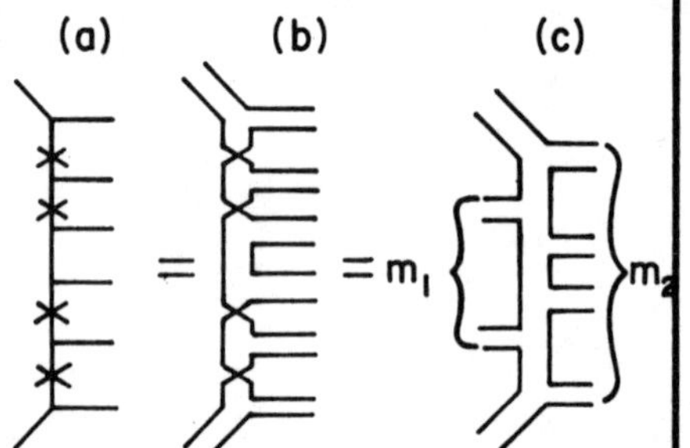

Fig. 3. Nonplanar MP diagram.
a,b. Illust. of cross-insertions.
c. Equivalent untwisted diagram.

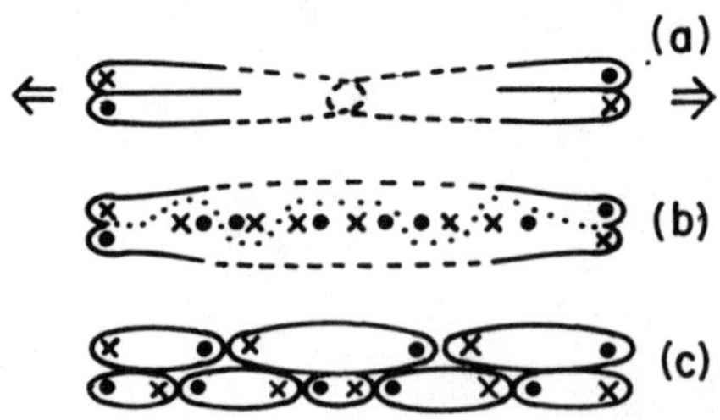

Fig. 4. Sequence of parton pictures.
a. Overlapping vacuum polarization.
b,c. Creation of $q\bar{q}$ pairs & hadrons.
Quarks are dots and antiquarks crosses.

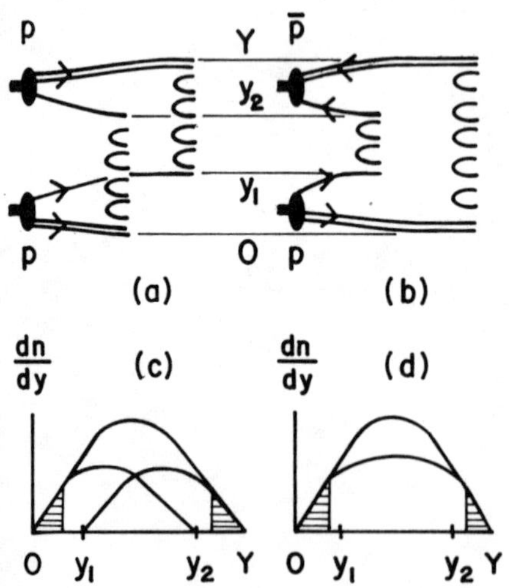

Fig 5. Spectra for pp and $\bar{p}$p.
a,c. Two-displ. chain, 2DC-spectra.
b,d. Short-long chain, SLC-spectra.

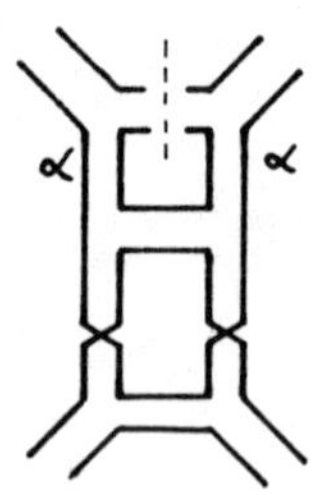

Fig. 6

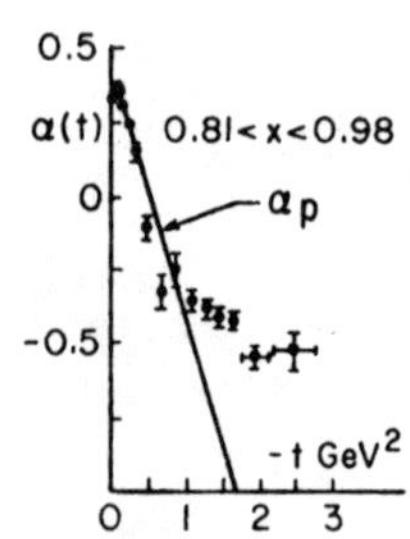

Fig. 7

Fig. 6. A triple Regge (TR) diagram.
Fig. 7. Confirmation of TR behavior in $\pi^{\pm}p \to \pi^0 X$ (from ref. 15a).

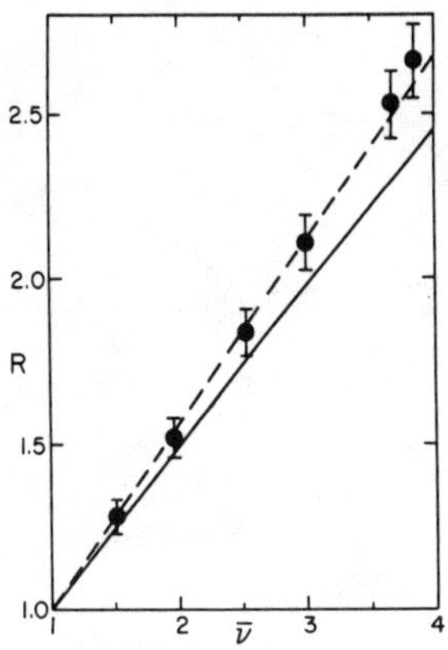

Fig. 8. Ratio R, pA data at 200 GeV (ref. 15b).
Solid line: EPS cascade approx.
Dotted line: ETS cascades included.

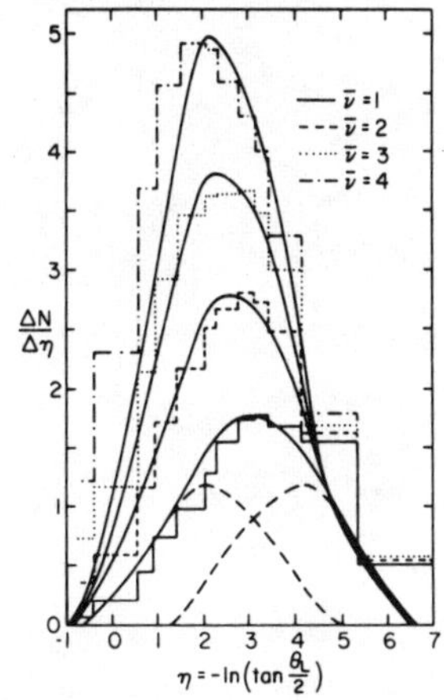

Fig. 9. $\bar{\nu}$- and y-dependence of dn/dy for $\bar{\nu} \geq 2$ compared to the pA data of ref. 15b, at 200 GeV.

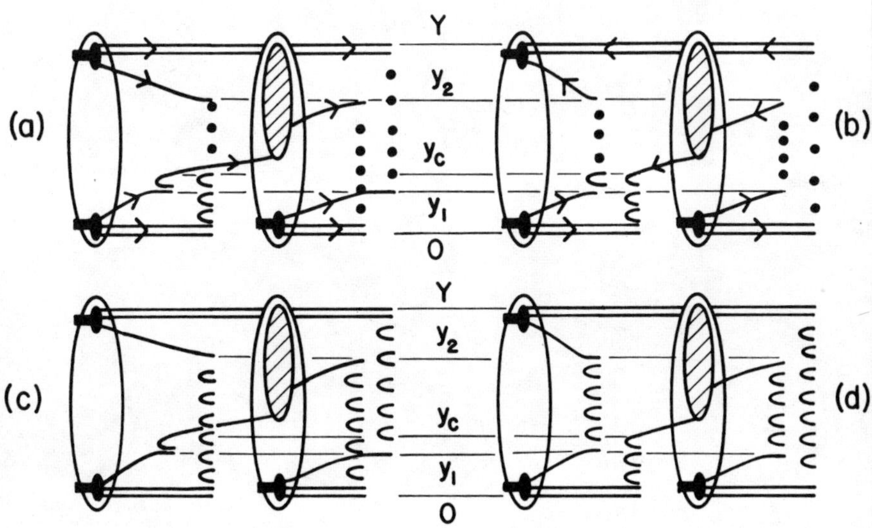

Fig. 10. Hadron-nucleus collisions as a function of time for $\bar{\nu}$ = 2.
a and c are for the pA case; b and d are for the $\bar{p}$A case.
a,b illustrate fragmentations immediately after second collision.
c,d illustrate final fragmentations. Notice that within plateau approximation final spectra of c and those of d are the same.

Chapter VIII: QCD AND LOW-p_T PHYSICS

J.F. GUNION
University of California
Davis, California

Abstract:

A review of the use of perturbative quantum chromodynamics to understand the major aspects of low-p_T physics is presented. Emphasis is on total cross sections, central region multiplicity (in particular, relative to e^+e^- multiplicity) and the fragmentation region. Some comparisons with other approaches are made, especially where QCD suggests that these approaches may be incomplete.

1. Introduction:

In these lectures we will see that perturbative calculations within quantum chromodynamics provide useful guidelines in understanding low-p_T physics. On occasion semi-quantitative and, even, rigorous calculations are possible; but in general non-perturbative effects of confinement could alter our conclusions. Nonetheless we will discover that the perturbative QCD approach to low-p_T phenomena provides a remarkably unified and coherent picture of the gross observables of hadronic interactions: total cross sections, central region multiplicity, and fragmentation region distributions. We will focus on each of these subjects in turn, emphasizing the physics insight to which the QCD calculations give rise. Considerable progress has been made in all areas since last year[1].

2. Total Cross Section - The Pomeron

The underlying mechanism responsible for the collision of two hadrons has been a subject of debate for many years. Within the framework of QCD it is simplest and most natural to consider the exchange of the elementary quanta of the theory, quarks or gluons. The lowest order (in α_s) mechanism for a hadronic collision is the exchange between the two colliding hadrons of a single colored gluon, as first suggested by Low and Nussinov,[2,3] (Fig. 1). The amplitude for production of the color jet final state is first order in α_s; the total cross section (summed over final states) is proportional to the imaginary part of the two-gluon exchange contribution to the forward amplitude. Thus the "Pomeron" of Regge theory corresponds to two-gluon exchange; since gluons carry no quantum numbers other than color it is clear that this two-gluon-exchange Pomeron has all the required short range order properties with regard to flavor.

Of course in a perturbative calculation of the forward amplitude it is necessary to consider all diagrams[3] which contribute in the same order in α_s. For π-π scattering there are actually 16 diagrams, four of which appear in Fig. 2. Since the gluon couples with opposite signs to the antiquark and quark of a color singlet state (think of a chargeless e^+e^- positronium state and a photon) the diagrams in which such couplings interfere have a negative sign with respect to ("diagonal") diagrams in which the two gluons are emitted and absorbed on the same quark line. In addition, the interference diagrams require that the momentum ($\ell_\perp$ in Fig. 2) carried by the exchanged gluons be routed through the hadron wave function; this gives rise to a q-$\bar{q}$ form factor $f_{q\bar{q}}(\ell_\perp^2)$ not present for the diagonal diagrams. The total cross section is[3]

$$\sigma_T^{\pi\pi} = \frac{8}{9} \int \frac{d^2\ell_\perp}{\ell_\perp^4} \alpha_s^2 \, (2) \, [1-f^{I}_{q\bar{q}}(\ell_\perp^2)] \, (2) \, [1-f^{II}_{q\bar{q}}(\ell_\perp^2)] \ . \tag{1}$$

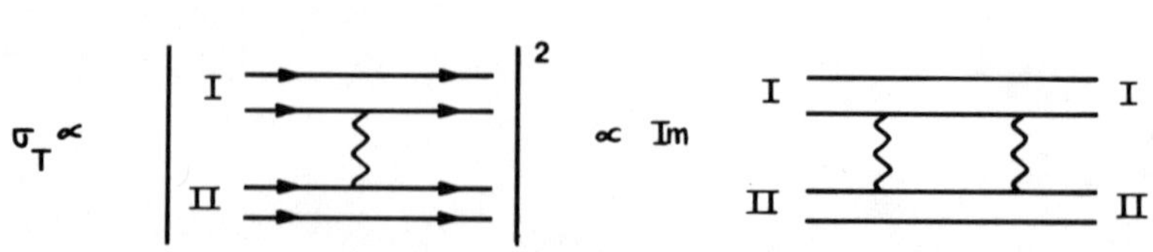

Figure 1 The Low-Nussinov gluon exchange model for the "Pomeron" contribution to the total cross section.

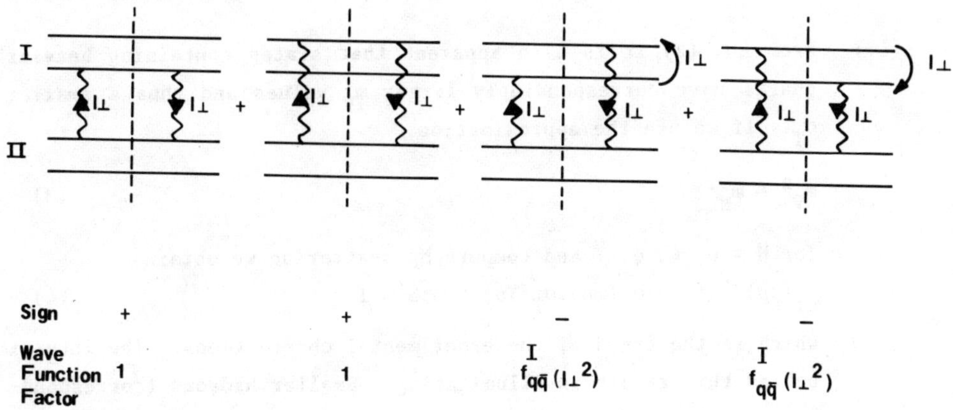

Figure 2 Four of the 16 diagrams contributing to the imaginary part of the forward π_I-π_{II} scattering amplitude. Relative signs and wave function factors are indicated.

This result has a number of remarkable features.

(i) The cross section is constant, independent of the collision energy, s; this is due to the vector nature of the exchanged gluons.

(ii) There is a ["diagonal" - "interference"] structure for each of the colliding hadrons which vanishes as $\ell_\perp^2 \to 0$, thus regularizing the infrared divergence, $1/\ell_\perp^4$, coming from the gluon propagators.

(iii) For each meson there is a factor of (2) which counts the number of constituents. For protons the (2) is replaced by a (3) and the q-$\bar{q}$ form factor, $f_{q\bar{q}}(\ell_\perp^2)$, is replaced by a q-q form factor, $f_{qq}(\ell_\perp^2)$. Simple diagrams for these two form factors yield

$$f_{qq}(\ell_\perp{}^2) \approx f_{q\bar{q}}(\ell_\perp{}^2) \approx F^M(4\ell_\perp{}^2) \sim \frac{1}{1+\frac{4\ell_\perp{}^2}{m_V{}^2}} \tag{2}$$

when the quarks and antiquark all have the same mass. F^M is the electromagnetic form factor of the q-$\bar{q}$ meson and $m_V{}^2$ is the vector dominance mass appropriate to a bound state formed from quarks of a given type and mass. Thus one obtains quark counting for total cross sections on a specifically graphical basis.

(iv) From Eq. (2) it is also apparent that states containing heavier quarks have correspondingly larger m_V values and thus a smaller σ_T. If we use the approximation

$$m_V^2 = m_M^2 \tag{3}$$

for $M = \rho, \phi, \psi, \Upsilon$ and compute Mp scattering we obtain

$$\sigma_T(\rho p):\sigma_T(\phi p):\sigma_T(\psi p):\sigma_T(\Upsilon p) = 6:5:3:1 \tag{4}$$

which is the trend of the experimental observations. The intuition behind this result is illuminating. Smaller hadrons (corresponding to bigger m_V^2 values) couple more weakly to the exchanged colored gluon. The cross section actually vanishes (roughly) proportional to $\langle r^2\rangle_I$ when $\langle r^2\rangle_I \to 0$. In comparison a geometrical type cross section is finite and proportional to $\langle r^2\rangle_{II}$ when $\langle r^2\rangle_I \to 0$.

(v) Finally the normalization of the cross section is reasonable. If α_s is taken as a constant and $m_V^2 = m_\rho^2$ is used, for πp scattering one obtains $\sigma_T^{\pi p} \simeq 25$ mb for $\alpha_s \sim 2$. The α_s value is of the size expected for low transverse momentum phenomena; it is not required to be overwhelmingly large nor is it in the range $\alpha_s < .5$ associated with truly short distance phenomena.

Clearly the most unique feature of a strong interaction scattering mechanism which explicitly requires color exchange is that the amplitude must vanish for a color singlet hadron when the hadron's color radius and hence its color-multipole moments approach zero. This leads to a number of unexpected phenomena. A particularly interesting example is provided by the study of diffraction excitation of a hadron by a nuclear target.[4] Consider a pion incident on a nucleus. We <u>require</u> that it traverses the nucleus without transferrring net color to <u>any</u> of the nucleons it interacts with. Thus the π must exchange (in lowest order) two gluons with each struck nucleon, Fig. 3. The struck nucleons will not be color excited and will not radiate colored quanta. The struck nucleons may be excited to higher levels, of course, but the experimental trigger we envision requires a large separation in rapidity between the target region and the projectile region; the latter contains the diffractively excited "pion" of mass $\mathcal{M}$. We write the excited pion state in terms of the initial pion state as

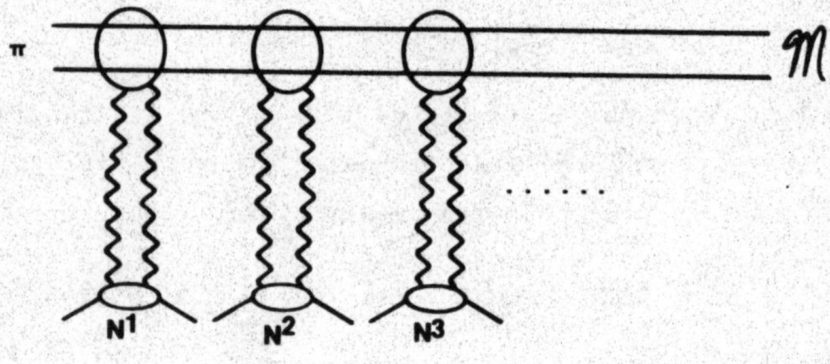

Figure 3. A picture of the transmission of a pion state through a nucleus. It must interact via 2-gluon exchange with each nucleon in order to transfer no net color to each one.

$$\psi'(b,r_\perp,x) = S_A(b,r_\perp)\ \psi_\pi(r_\perp,x) \tag{5}$$

where S_A is the nuclear collision S-matrix appropriate to the above trigger, $r_\perp$ is the transverse separation of the q and $\bar{q}$ in the pion state and $x=(k_q^o+k_q^3)/(p_\pi^o+p_\pi^3)$ (k_q=quark momentum, p_π=π momentum - $p_\pi^{1,2}$=0). For small $r_\perp$, $S_A \to 1$; since this portion of the π is colorless and has small size it does not couple to gluons. At large $r_\perp$, $S_A \to 0$. We estimate, for large A,

$$S_A(b,r_\perp) \sim \exp\left[-\frac{r_\perp^2\ A^{1/3}}{20}\ \text{GeV}^2\right]. \tag{6}$$

It is the variation of S with $r_\perp$ which leads to diffractive excitation.[5] (An analogue would be a plane wave incident on a pinhole; the wave spreads out on the far side of the aperture.) The enhancement of small $r_\perp$ causes an enhancement of large $k_\perp$, the q-$\bar{q}$ relative transverse momentum, and large $\mathcal{M}^2$. The enhancement is much greater for a QCD Pomeron model than in other approaches. Using (5) and (6) we can estimate the cross section for producing a q-$\bar{q}$ orthogonal to the original pion provided we know $\psi_\pi(0,x)$. This is calculated rigorously in QCD from the $\pi\to\mu\nu$ decay rate[6] as

$$\psi(0,x) = \sqrt{12\pi}\ x(1-x)f_\pi,\quad f_\pi \sim .1\ \text{GeV}\ . \tag{7}$$

The cross section for diffractive production of inelastic states is equal to the elastic scattering cross section of the projectile on the target, multiplied by the probability that a state of size $r_\perp \gtrsim 0$ is present and is transmitted ($S_A \neq 0$). Combining (5), (6), and (7), squaring, integrating over b and x, and converting to $k_\perp$, the relative transverse momentum between the q and $\bar{q}$ of the π diffractive state, we obtain

$$\frac{d\sigma_A}{d^2k_\perp} \ (\pi \to \text{orthogonal diff. state}) \sim 50 \text{ mb GeV}^{-2} \exp[-10\ k_\perp^2/A^{1/3}\ \text{GeV}^2] \tag{8}$$

and

$$\sigma_A \ (\pi \to \text{orthogonal diff. state}) \underset{\sim}{\sim} 16 \text{ mb } A^{1/3}. \tag{9}$$

The expected $\langle k_\perp \rangle$ and cross section size are both far bigger than would be typical of other approaches with slower $r_\perp$ variation (see Ref. 4).

Finally note that if charm states of the proton are intrinsically small one obtains[4]

$$\sigma_A \ (\text{diff. charm production}) = \sigma_A^{el} \ P_{charm}, \tag{10}$$

where P_{charm} is the probability that the projectile contains a small intrinsic charm state.[7] A value $P_{charm} \underset{\sim}{\sim} 2\%$[8] yields charm production cross sections of the same general size as those seen at Fermilab.[9]

Thus we see that a color based mechanism for hadronic collisions agrees with basic total cross section phenomenology, but has many unusual and striking features. In the following sections we will presume that the Low-Nussinov (abbreviated L-N hereafter) gluon exchange diagram is the dominant interaction and derive consequences for the central multiplicity and fragmentation functions of a typical hadronic collision. Again QCD will play a special role, yielding many novel and unexpected effects.

3. Central Region Multiplicity

As already outlined, the colored gluon exchange mechanism for hadronic collisions leads to a primordial final state consisting of rapidly separating color octet jets. This situation is similar to that encountered in e^+e^- annihilation where the primordial final state contains separating quark and antiquark color triplet states. In perturbative quantum chromodynamics the evolution of the color jets in the annihilation case is described using the "jet calculus" formalism.[10] In this approach final state particle production is initiated by perturbatively produced radiation which begins immediately after the creation of the q and $\bar{q}$ jets. Eventually sufficient numbers of perturbatively produced quanta are present that on-shell hadrons can be produced from overlapping color quanta -- the perturbatively produced radiation products materialize as hadrons through non-perturbative means. Nonetheless, this framework suggests that it is the underlying perturbatively produced color quanta which largely determine the distribution and multiplicity of final state particles. In the L-N gluon exchange model of a hadronic collision a similar sequence of events occurs. The

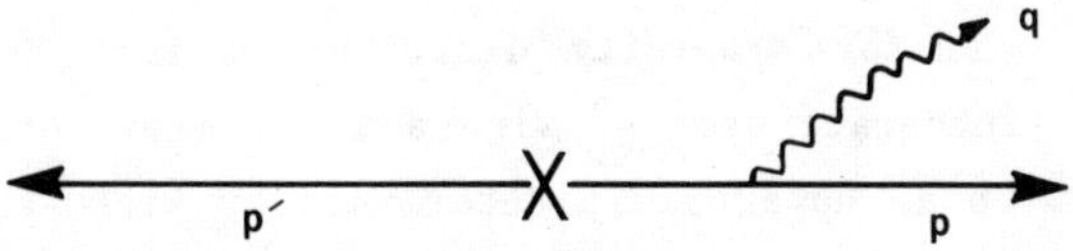

Figure 4. Gluon emission in an e^+e^- collision.

octet color separation leads initially to color radiation/bremsstrahlung followed by non-perturbative hadron formation. The relationship between the final states of e^+e^- annihilation and of a hadron collision will depend upon the comparative distribution of radiation products in the two cases.[11]

In this section I wish to outline results[12] of this comparison at the simplest level -- a calculation of the gluon radiation multiplicity distribution from one gluon emission diagrams. The comparison focuses on the central plateau region. In e^+e^- annihilation there is only one contributing diagram (in an appropriately chosen gauge for the emitted gluon), Fig. 4. Defining light cone momenta ($k^+ = k^0+k^3$, $k^- = k^0-k^3$)

$$q = (q^+, q^-, \vec{q}_\perp) = \left(y\sqrt{s},\ \frac{q_\perp^2}{y\sqrt{s}},\ q_\perp,\ 0\right)$$

$$p' = \left(0, \sqrt{s} - \frac{q_\perp^2}{y(1-y)\sqrt{s}},\ 0,\ 0\right) \tag{11}$$

-- y is the fractional "+" momentum carried by the gluon relative to the primordial source jet with "+" momentum $\sqrt{s}$ -- one obtains

$$\frac{dn}{d^2q_\perp dy} \equiv \frac{\dfrac{d\sigma^{1\text{-gluon}}}{d^2q_\perp dy}}{\sigma^{0\text{-gluon}}} = \frac{C_F\alpha_s}{\pi^2}\,\frac{1}{y\,q_\perp^2} \tag{12}$$

where $C_F\,\alpha_s$ is the effective color "charge" of the radiating quark -- $C_F=4/3$. The $\frac{dy}{y}\,\frac{d^2q_\perp}{q_\perp^2}$ type distribution is the characteristic radiation bremsstrahlung distribution of a vector gauge theory. In terms of the rapidity of the emitted gluon,

$$\eta = \frac{1}{2}\,\ell n\,\frac{q^+}{q^-} = \ell n\,(y\sqrt{s}/q_\perp), \tag{13}$$

we have

$$\frac{dn}{d^2q_\perp d\eta} = \frac{C_F\alpha_s}{\pi^2}\,\frac{1}{q_\perp^2} \tag{14}$$

corresponding to a uniform rapidity distribution for the emitted gluon radiation. If we integrate over $q_\perp^2$ with a lower limit Λ^2, determined by confinement, and with an upper limit determined by kinematics, $q_\perp^2 \sim s$, we obtain

$$\frac{dn}{d\eta} = \frac{C_F\alpha_s}{\pi} \ln(s/\Lambda^2). \tag{15}$$

If we also integrate over the allowed rapidity range, $-\ln(\sqrt{s}/m) \leq \eta \leq \ln(\sqrt{s}/m)$ where m is a typical hadron mass, one finds

$$\langle n\rangle = \frac{C_F\alpha_s}{\pi} \ln^2 s + C_1 \ln s + C_2; \tag{16}$$

C_1 and C_2 are model dependent constants. The result of Eq. (15) corresponds to a rising rapidity plateau and (16) predicts a rapidly rising total gluon multiplicity. We have, of course, ignored possible variation with q^2 of α_s. The generalization to higher orders in α_s leads to both a moving coupling constant and higher order final states associated with "double-log" behavior. These calculations[13] suggest an even more rapid rise of multiplicity

$$\langle n\rangle \sim a+b \exp\,[c(\ln(s/\Lambda^2))^{\frac{1}{2}}]. \tag{17}$$

So far only the calculations analogous to (15) and (16) have been performed in the hadron case[12] and thus we focus on these low order results.

The connection between these gluon multiplicity calculations and the observed hadron multiplicity is tenuous. It is conventional[11] to presume that

$$dn\ (\text{hadron}) \approx dn\ (\text{gluon})\ . \tag{18}$$

Thus forms (16) and (17) have been used to fit the observed e^+e^- charged particle multiplicities. Assuming that $\frac{1}{2}$ the produced hadrons are neutral (as in a simple gluon → quark pair → hadron model) one obtains from the $\ln^2 s$ term of (16)

$$\alpha_s \approx 1 \tag{19}$$

and the observed plateau height is fitted by Eq. (15) provided.

$$\Lambda \sim .2\ \text{GeV}. \tag{20}$$

Assumption (18) is almost certainly too extreme, however, and one should be prepared, at current energies, for non-perturbative effects to contribute not only to the constant term of <n> but also through kinematics and phase space expansion to the increase with s of <n>. This has been illustrated by the Monte Carlo calculations of the DESY collaborations[14] in which the Feynman-Field jet model[15] produces much of the observed rise in <n> purely through the increasing phase space for more massive resonances. Nonethe-

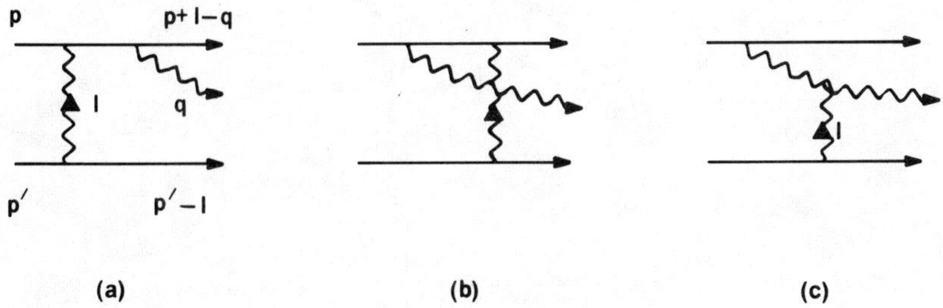

Figure 5. Gluon radiation diagrams induced by a L-N gluon exchange collision between two quarks: a) post-emission; b) and c) pre-emission.

less differences in the distribution of underlying perturbatively produced gluon radiation in the e^+e^- annihilation and hadron collision cases should ultimately produce corresponding differences in final state hadron multiplicities.

Thus we turn to the hadron collision calculation.[12] We begin with the radiation resulting from a quark-quark collision via one gluon exchange. We employ a center of mass frame for the colliding quarks, see Fig. 5,

$$
\begin{aligned}
p &= (\sqrt{s},0,0,0) \\
p' &= (0,\sqrt{s},0,0) \\
q &= \left(y\sqrt{s},\ \frac{q_\perp^{\,2}}{y\sqrt{s}},\ q_\perp,\ 0\right) \qquad (21) \\
\ell &= (\ell^+,\ell^-,\vec{\ell}_\perp)
\end{aligned}
$$

where we have neglected masses. The conditions $(p'-\ell)^2 \simeq 0$ and $(p+\ell-q)^2 \simeq 0$ force ℓ_+ and ℓ_- to be of order $(1/\sqrt{s})$ --

$$
\begin{aligned}
\ell_+ &\simeq -\ell_\perp^{\,2}/\sqrt{s} \\
\ell_- &\simeq (q_\perp^2/y + \ell_\perp^{\,2} - 2\ell_\perp\cdot q_\perp)/(1-y)\sqrt{s}.
\end{aligned} \qquad (22)
$$

A gauge choice for the emitted gluon q is made such that only the three illustrated diagrams, (a)-(c), contribute to the central region. One obtains

$$\frac{dn}{dy\,d^2q_\perp} = \frac{\dfrac{d\sigma^{1\text{-gluon}}}{dy\ d^2q_\perp}}{\sigma^{0\text{-gluon}}} = \frac{\alpha_s}{\pi^2}\frac{1}{y}\int \frac{d^2\ell_\perp}{\ell_\perp^4}\left\{\frac{\vec{q}_\perp - y\vec{\ell}_\perp}{\left[\frac{(y\vec{\ell}_\perp-\vec{q}_\perp)^2}{(1-y)}\right]} C_{(a)} - \frac{\vec{q}_\perp}{q_\perp^2} C_{(b)} \right. \tag{23}$$

$$\left. + \frac{\vec{\ell}_\perp - \vec{q}_\perp}{\left[\frac{(q_\perp-\ell_\perp)^2}{(1-y)}\right]} C_{(c)}\right\}^2 \Bigg/ \int \frac{d^2\ell_\perp}{\ell_\perp^4}$$

Note that each diagram enters with a different color factor, $C_{(a)-(c)}$. Diagram (a) could be viewed as "final state emission" of the gluon; in diagrams (b) and (c) the gluon appearing in the final state is created virtually in the initial state and subsequently either it or the secondary quark scatters with the target quark, p'. These processes occur at the amplitude level, however, and interfere with one another in the multiplicity calculation! The $\int d^2\ell_\perp/\ell_\perp^4$ divergence associated with the exchanged gluon is only a temporary artifact of the quark level calculation -- imagine a fixed $\ell_\perp$ for purposes of illustration.

To understand the implications of (23) consider first the QED-like, abelian theory situation with $C_{(c)} = 0$ and $C_{(a)} = C_{(b)}$. In the central region, corresponding to $\eta = \ell n(y\sqrt{s}/q_\perp) \gtrsim 0$ or $y \gtrsim q_\perp/\sqrt{s}$, (a) and (b) clearly cancel; in fact $dn/d\eta$ has a finite limit as $s \to \infty$ only in the two "fragmentation" regions where $|\eta| \gtrsim \ell n\,(\frac{\sqrt{s}}{\ell_\perp})$. This is natural since the charge does not change at the scattering gluon's abelian vertex -- the only charge acceleration is that due to the $\ell_\perp$ of this exchanged gluon. Thus the radiation will be confined to a finite region of rapidity in the vicinity of the original projectile rapidity.

Intuitively the non-abelian situation is different. The color of the quark is instantaneously altered at the vertex of the L-N exchange gluon. Radiation filling the allowed rapidity range will in general occur for any fixed $\ell_\perp$. Taking $C_{(a)}-C_{(b)} = C_{(c)}$, as appropriate for color averaging or as will be appropriate in the hadron context, we obtain in the region $y\ell_\perp << q_\perp$ (which includes $\eta \gtrsim 0$)

$$\frac{dn}{dy\,d^2q_\perp} = \frac{\langle C_{(c)}^{\,2}\rangle\alpha_s}{\pi^2}\,\frac{1}{y}\,\frac{\ell_\perp^2}{q_\perp^2(q_\perp-\ell_\perp)^2} \tag{24}$$

or

$$\frac{dn}{d\eta\,d^2q_\perp} = \frac{\langle C_{(c)}^{\,2}\rangle\alpha_s}{\pi^2}\,\frac{\ell_\perp^2}{q_\perp^2(q_\perp-\ell_\perp)^2}\,. \tag{24}$$

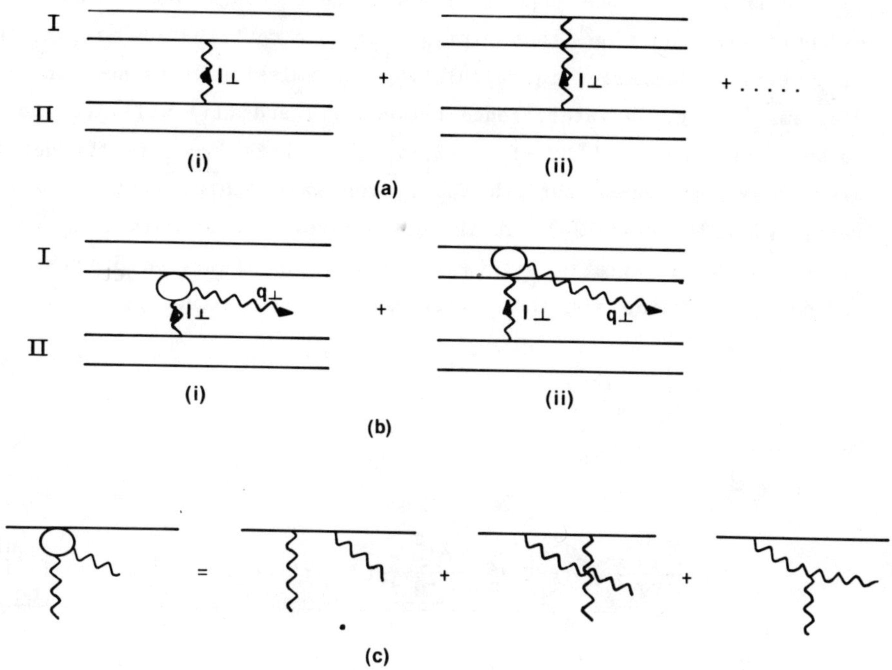

Figure 6. a) Low-Nussinov gluon exchange diagrams for π-π scattering.
b) Associated gluon radiation diagrams.
c) The subprocess amplitude for gluon + quark → gluon + quark appearing in b).

Thus we have a uniformly populated central plateau with a limited $q_\perp^2$ spectrum. At small $q_\perp < \ell_\perp$ one obtains a $d^2q_\perp/q_\perp^2$ spectrum closely analogous to the e^+e^- result (14), but at large $q_\perp$ there is cancellation among the amplitudes contributing to (23) and the spectrum behaves as $d^2q_\perp/q_\perp^4$. Thus, unlike e^+e^- annihilation, the gluon plateau height does not increase with energy in this order of perturbation theory.

It is straight forward to place this calculation in the hadron context. At the amplitude level we consider for the moment the two π-π cross section diagrams of Fig. 6a. (There are two others with the exchange gluon connected to the other line of state II). The two corresponding gluon emission diagrams are shown in Fig. 6b; these contain as a "subprocess" the three diagrams, Fig. 6c, for gluon+quark → gluon+quark which appeared in Fig. 5. In the cross section calculation of Fig. 6a we recall from the first section of these lectures that (i) and (ii) enter with opposite signs and that their interference requires routing the exchanged momentum $\ell_\perp$ through a hadronic wave function, thus giving rise to a contribution proportional to $-f_{q\bar{q}}^{I}(\ell_\perp^2)$. Similarly, in calculating the emission cross section from the diagrams of Fig. 6b interference between (i) and (ii) will lead to a contribution proportional to $-f_{q\bar{q}}^{I}((\ell_\perp - q_\perp)^2)$ -- here $\ell_\perp - q_\perp$ is the net transverse momentum routed through the hadron wave function of state I. The result of (23), (24)-(24)' at the quark level thus appears as a subprocess cross section integrated within a form factor framework provided by the hadrons. Including all four diagrams at the amplitude level yields (for $y\ell_\perp \ll q_\perp$)

$$\frac{dn}{dyd^2q_\perp} = \frac{\dfrac{d\sigma^{1\text{-gluon}}}{dy\, d^2 q_\perp}}{\sigma^{0\text{-gluon}}} \tag{25}$$

$$= \frac{\dfrac{C_A \alpha_s}{\pi^2 y q_\perp^2} \displaystyle\int \frac{d^2\ell_\perp}{\ell_\perp^4}\, \alpha_s^2 \frac{\ell_\perp^2}{(\ell_\perp - q_\perp)^2} (2)[1-f^{I}_{q\bar{q}}((\ell_\perp - q_\perp)^2)](2)[1-f^{II}_{q\bar{q}}(\ell_\perp^2)]}{\displaystyle\int \frac{d^2\ell_\perp}{\ell_\perp^4}\, \alpha_s^2\, (2)\, [1-f^{I}_{q\bar{q}}(\ell_\perp^2)](2)\, [1-f^{II}_{q\bar{q}}(\ell_\perp^2)]} .$$

If $q_\perp^2 < \langle \ell_\perp^2 \rangle$ ($\langle \ell_\perp^2 \rangle$ is determined by the bound state form factors) one obtains an e^+e^--like result

$$\frac{dn}{dyd^2q_\perp} = \frac{C_A \alpha_s}{\pi^2 y q_\perp^2} \tag{26}$$

with the obvious difference that the color factor is sensitive to the total charge of the radiating octet jets and is thus $C_A/C_F = 9/4$ larger than that obtained in e^+e^- annihilation.[11] At large $q_\perp^2 > \ell_\perp^2$, as in (24)-(24)', the $d^2q_\perp/q_\perp^2$ perturbative gluon tail disappears via cancellation among the full gauge invariant set of subprocess diagrams of Fig. 6c. Instead a $d^2q_\perp/q_\perp^4$ spectrum emerges at high $q_\perp$. Thus, as stated previously the central gluon plateau height in a hadronic collision does not increase with s in this order of perturbation theory. Using Eq. (2) with $m_V^2 = m_\rho^2$ one obtains roughly

$$\frac{\left.\dfrac{dn}{d\eta}\right|_{\eta=0}^{\text{hadron}}}{\left.\dfrac{dn}{d\eta}\right|_{\eta=0}^{e^+e^-}} \sim \frac{C_A}{C_F} \frac{\ln m_\rho^2/\Lambda^2}{\ln s/\Lambda^2} . \tag{27}$$

If we take $\Lambda \approx .2$ GeV from our fit to the absolute e^+e^- plateau height then at $\sqrt{s} = 30$ GeV the above ratio is .6. The plateaus have equal heights at lower energy, roughly $s \sim 18\ \text{GeV}^2$.

The experimental situation regarding hadronic collision plateau heights is far from clear. The data of Thome et al.[16] shows a slow rise with s of the pp plateau height; the plateau height is, however, well below the e^+e^- plateau height at the same energy. Other presentations[17] at this conference have suggested that there may be a much greater similarity between e^+e^- and hadronic collision plateaus. This would be natural if much of the e^+e^- plateau height rise was indeed of kinematic origin, as discussed earlier. Presumably similar kinematic effects would be at work in the hadron collision case. Only systematics of the type apparent in the Thome et al. data are consistent with a large "perturbative" component to the final particle multiplicity.

Several other features of the calculation are noteworthy.

1) On an event by event basis $\ell_\perp$ varies. Larger $\ell_\perp$'s will allow a longer perturbative $d^2q_\perp/q_\perp^2$ tail and a higher multiplicity. Thus there should be a correlation in which a big value for $\langle n\rangle_{\eta>0}$ would imply, on average, a bigger value for $\langle n\rangle_{\eta<0}$. Such a correlation is observed in the data.[18]

2) Smaller hadrons with bigger quark masses and bigger m_V^2 values have a bigger $\langle\ell_\perp\rangle$ than the standard light hadrons. This in turn implies a higher hadronic pleateau according to the perturbative calculation just presented. Of course, in a collision involving a heavy hadron, the final state will, in general, be required to contain at least one related heavy hadron. This means less phase space and a possible decrease in the non-perturbative contribution to the central plateau height. At high enough s the perturbative prediction should hold. Unfortunately we have no ϕ, ψ, or Υ beams; and for the K^+ and other hadrons which are only partly composed of strange quarks the effect is likely to be too small to be observed.

3) The present calculation may be placed in the context of the work of Pokorski and Wolfram[19] by noting that the prediction (25) corresponds to a predicted distribution in the longitudinal momentum $\frac{1}{2}(\ell_+ - \ell_-) = \ell_3$ transferred between the two colliding

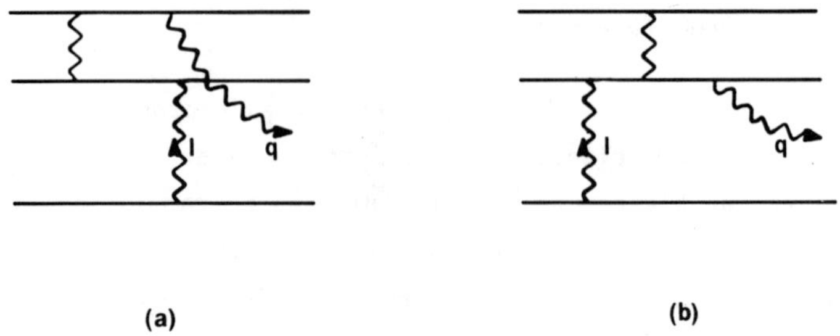

Figure 7. Radiation diagrams in which the radiated gluon is separated from the L-N exchange gluon by a bound state interaction. These are suppressed in the central region.

hadrons. (See Eq. (22) for the relationship between $\ell_\perp$, $q_\perp$, y, $\sqrt{s}$, and ℓ_+, ℓ_-.) The consequences of this distribution in ℓ_3 are as stated.

4) The final state radiated gluon is necessarily contiguous to the exchanged gluon as in Fig. 6c. The alert reader might have asked if graphs like those in Fig. 7 could not be important. There the radiation and exchange interactions occur on different bound state quark lines (Fig. 7a) or are separated by a bound state gluon interaction on their common quark line (Fig. 7b). In fact, both Fig. 7a and Fig. 7b are suppressed in the central region under consideration, where $y\ell_\perp \ll q_\perp$. This is because the exchange gluon must bring in a "large" $\ell_- \gtrsim q_\perp^2/y\sqrt{s}$ (Eq. (22)) in order to provide the necessary off-shellness for one of the bound state quarks to radiate a soft gluon. Amplitudes like those of Fig. 7 in which this off-shellness must be routed through more than one quark propagator will have extra factors of $y/q_\perp^2$ and will not contribute in the central plateau.

5) Higher Fock states of the incoming hadrons will yield very similar results. For instance, a hard gluon in the Fock state will radiate as shown in Fig. 8. The gluon+gluon → gluon+gluon subprocess which enters is more complicated than the quark-gluon subprocess but exhibits exactly the same cancellations and central plateau limits.

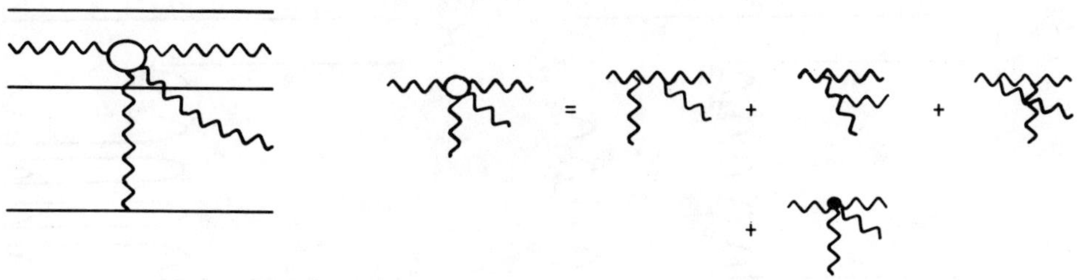

Figure 8. Radiation diagram involving a hard gluon in the initial Fock state of the incident hadron.

6) Soft gluons in the Fock state cannot materialize on-shell without a contiguous exchange/interaction gluon attachment. For instance Figures 5b and 5c correspond to soft virtual gluons in the initial particle Fock state which are put on-shell by an exchange interaction. However, these diagrams must be considered together with Fig. 5a as we have done.

Finally it is illuminating to make a few comparisons to other models of hadronic collisions. We will single out the Multiperipheral Model (MPM)[20] and Dual Topological Unitarization (DTU) approach.[21] In the MPM all particles including soft "central" ones which appear in the final state are present virtually in the initial states of the colliding hadrons; the interaction of the MPM particle chain of the projectile hadron with the MPM particle chain of the target hadron allows these virtual particles to materialize as on-shell central plateau hadrons. We have already stressed, under item 6), that in a vector gluon theory (unlike the ϕ^3 model on which the MPM approach is based) diagrams in which the final gluon is a virtual part of the initial state must be considered together with diagrams in which the final gluon is radiated subsequently to the collision. All these diagrams interfere and it is the interference which provides the distinguishing features of a hadronic collision plateau compared to an e^+e^- collision plateau. The MPM model does not contain this type of interference.

In the DTU approach the Pomeron and the final state particles are closely related, through unitarity. The Pomeron is a reflection of the

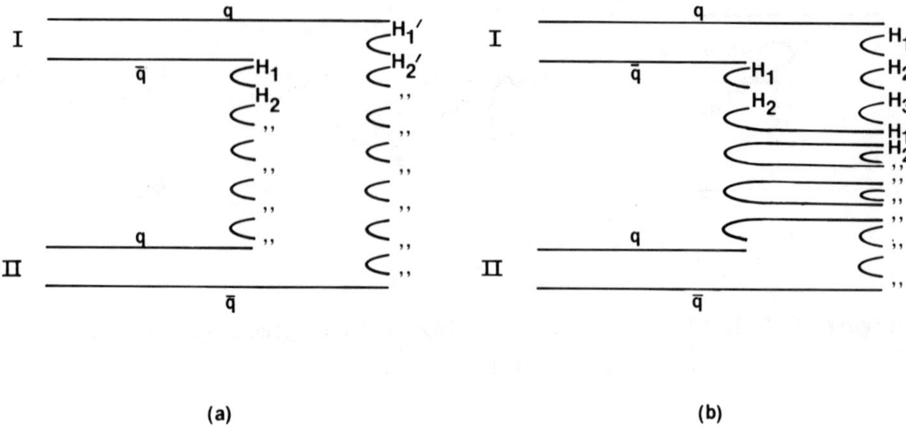

Figure 9. a) The DTU diagram for particle production in a π-π collision, Ref. 21.

b) A DTU-like particle production diagram in which the two chains interfere by co-mingling with one another over part of the chain rapidity intervals.

production of two non-interfering "chains" of particles as illustrated in the amplitude of Fig. 9a. It is stated that the multiplicity of each chain should be identified with the e^+e^-, q-$\bar{q}$ separation multiplicity at an appropriate reduced energy. To me it is not clear why the q and $\bar{q}$ of state I would choose to radiate in response to color separation (as in e^+e^-) in concert with their $\bar{q}$ and q counterparts in state II unless some interaction involving color had previously established the necessity for radiation. The hadrons are color singlets and have no need to radiate unless color is exchanged prior to chain development. It is possible that these chains should be regarded like the MPM chains of virtual soft clouds waiting to be put on shell, but then any close analogue to the e^+e^- radiation is hard to justify. It is also hard to understand why the two chains cannot interfere (co-mingle) with one another as illustrated in Fig. 9b. Once they have co-mingled with one another the color structure becomes much more complicated and the relation to e^+e^- annihilation less clear. A related point is that the DTU diagrams may accurately reflect the flavor structure of the quark-level amplitude underlying the Pomeron but it is not clear that the color structure is correct. Hard virtual gluons could be present in either initial state (probably they are necessary to "slow down" the interacting quarks as assumed in the DTU model.); the flavor structure would be exactly that assumed by the DTU models but the color structure completely different.[22]

In summary, it is probably fair to say that all approaches have their weaknesses. The perturbative approach relies (at the moment) on a low order calculation of the hadronic collision amplitude via gluon exchange. The radiation produced in response can then be computed in close analogy with e^+e^- annihilation at lowest order and a comparison made. The phenomenology is very reasonable. The DTU approach would claim that a low order calculation of the cross section is not necessarily relevant. There is an existing dual topological picture of the Pomeron which specifies the routings of flavor at the quark level and one need make only a few additional assumptions involving color to make a connection to e^+e^- annihilation. The assumptions are not strongly motivated, in my opinion, but the phenomenology is very good. In the future one could hope to understand the QCD basis of the DTU flavor structure and thus make specific statements about the all important color structure.

4. Fragmentation in QCD

Our ultimate goal in this section is to calculate hadron production in the fragmentation region of a Low-Nussinov initiated hadronic collision. In order to understand the framework for such a calculation and because it is interesting in itself we first discuss quark and diquark fragmentation in e^+e^- annihilation and deep inelastic scattering. The relationship between the approach I shall advocate[23] here and the recombination model[24] will be apparent already at this level. We begin by considering quark fragmentation in e^+e^- annihilation, Fig. 10. We consider the frame

$$\begin{aligned} p_M &= (z\sqrt{s},\ \frac{p_\perp^2}{z\sqrt{s}},\ p_\perp, 0) \\ p' &= (0,\ \sqrt{s} - \frac{p_\perp^2}{z(1-z)\sqrt{s}},\ 0,\ 0) \\ k &= ((1-z)\sqrt{s},\ \frac{p_\perp^2}{(1-z)\sqrt{s}},\ -p_\perp,\ 0)\ . \end{aligned} \tag{28}$$

The variable z is the "+" momentum of the observed meson relative to the "+" momentum, $\sqrt{s}$, of the underlying source quark. In the limit of $z\to 1$ this fragmentation can be computed rigorously. This is because the momentum-squared of the source quark, $(p_M+k)^2$, becomes large as $z\to 1$,

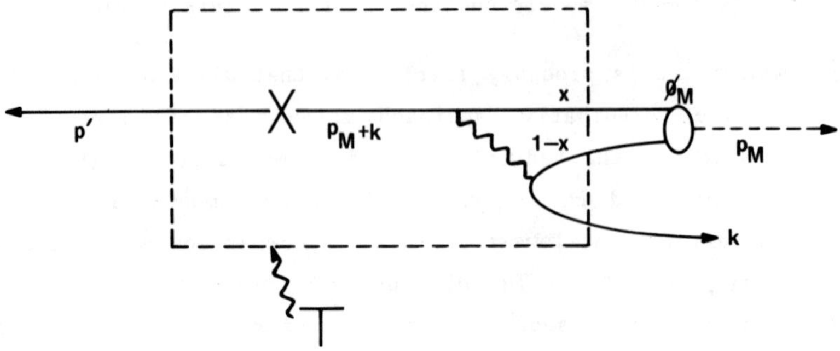

(a)

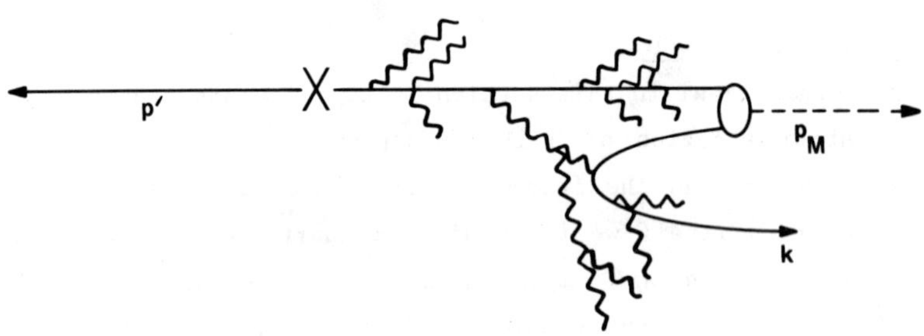

(b)

Figure 10. a) Lowest order "exclusive" tree-graph level amplitude for quark fragmentation to a meson, M, in e^+e^- annihilation.
b) A higher order "inclusive" amplitude for the same fragmentation.

$$(p_M+k)^2 \sim \frac{p_\perp^{\,2}}{z(1-z)} \tag{29}$$

The situation is closely analogous to the form factor calculation considered by Brodsky and Lepage.[25] They show that the form factor can be treated by short distance techniques since one of the quarks of a typical

QCD form factor diagram is forced far off-shell. In the form factor calculation, only the most elementary (i.e., lowest order) tree graph survives in leading order as the Q^2 of the photon probe goes to infinity. This tree graph or subprocess amplitude is convoluted with short distance valence Fock state wave functions (for the hadron involved) which are obtained from a solvable QCD-axial-gauge ladder integral equation.

In quark fragmentation a similar division occurs, see the sample diagram of Fig. 10a. There is a short distance $O(\alpha_s)$ amplitude for the tree graph $e^+e^- \to \bar{q}+(q\bar{q})_M q$ and this is convoluted with a wave function $\phi_M(x,"Q^2")$, where $"Q^2" = \frac{p_\perp^2}{z(1-z)}$ in the present instance relative to $"Q^2"=Q^2$ in a form factor calculation. The result for the amplitude is proportional to

$$\int dx\, T_{e^+e^- \to \bar{q}+(q\bar{q})_M q}(x,z,s,p_\perp^2)\ \phi_M(x, "Q^2") \ . \tag{30}$$

The form of T is easily computed. For spinless quarks it would simply be proportional to $[(p_M+k)^2]^{-1} \overset{z\to 1}{\sim} (1-z)/p_\perp^2$; spin leads to additional structure. We square the amplitude and divide by the bare $e^+e^- \to q\bar{q}$ cross section, with the q and $\bar{q}$ in the analogous momentum configuration, to obtain the fragmentation function. As a simple example consider just the graph of Fig. 10a calculated in Feynman gauge for the gluon. We define

$$I_M("Q^2") = \int \frac{dx\ \phi_M\ (x,"Q^2")}{x(1-x)} \ , \tag{31}$$

related in leading order to the charged meson form factor through

$$F_M(Q^2) = \frac{4\pi\, C_F\, \alpha_s(Q^2)}{Q^2}\, I_M^2(Q^2); \tag{32}$$

we obtain for this one graph[26]

$$\frac{dN}{dzd^2p_\perp} \equiv D_{M/q}\ (z,p_\perp) \overset{z\to 1}{\sim} \frac{1}{6\pi}\ \frac{(1-z)^2}{p_\perp^4}\ C_F^2\ \alpha_s^2\ ("Q^2")\ I_M^2("Q^2"). \tag{33}$$

For spinless quarks $D_{M/q}$ would have been proportional to (1-z). The normalization of the $p_\perp$-integrated z distribution is sensitive to the quark and meson masses that we have neglected. For masses of the order of the experimental $\langle p_\perp \rangle$ the result for dN/dz from Eq. (33) is somewhat small but of the same general magnitude as that observed experimentally.[27] A complete calculation[26] must include another contributing graph which (in Feynman gauge) has the same $(1-z)^2$ power law and will thus add to the overall normalization. Also, at moderate z values the $\langle "Q^2" \rangle$ is not large and the leading log approximation of (33) and Fig. 10a is not obviously valid. (A sample higher order graph of importance has the general structure illustrated in Fig. 10b.) Nonetheless, the leading-log result will provide a good approximation to the shape and, very roughly, to the normalization of the observed distribution. In particular, the basic power law, $(1-z)^2$ in the present case, deriving from off-shell propagators in the "Born" or tree-graph amplitudes for the fragmentation, will not be substantially altered by including non-leading effects. Thus we will determine the basic power law expected for a given fragmentation by finding and calculating the exclusive tree graph amplitude(s) with leading $z\to1$ behavior which can give rise to that fragmentation. Further work on normalization is in progress.[26]

At this point a direct comparison to the recombination/valon model of Hwa and collaborators[24,28] is possible. Basically they claim that quark fragmentation is dominated by higher order graphs of the type drawn in Fig. 10b. In their approach the off-shellness of the initial quark is not set by $\frac{\langle p_\perp^2\rangle}{1-z}$ (though kinematically it must be as $z\to1$) -- it is chosen at a value Q_o^2 which is adjusted to fit the data. This initial quark radiates, one of the radiated gluons creates a q-$\bar{q}$ pair, more radiation occurs and finally the meson is made via recombination. The meson fragmentation function is computed as a convolution of several QCD evolved Alterelli-Parisi $q\to q$, $q\to g$, etc., distribution functions which describe the sequence of radiations shown in Fig. 10b. The $(1-z)$ power law is a result of this convolution and not of any off-shell behavior proportional $\langle p_\perp^2\rangle\,(1-z)^{-1}$ within the amplitude drawn. We would claim that higher order "inclusive" graphs of the type drawn may well be important at moderate z values but should be computed using off-shell techniques of the type discussed for the "exclusive" (no radiation) amplitude of Fig. 10a. The power laws predicted by the "exclusive" and "inclusive" amplitudes would then be very similar.

Let us illustrate the exclusive leading tree-graph results for a variety of other fragmentations. In what follows we will quote predictions which neglect possible quark spin effects. Results for specific graphs will refer to calculations done in a Feynman gauge for the gluon(s). Consider first unfavored quark fragmentation $d\to\pi^+$. The minimal tree graph is illustrated in Fig. 11a. The spinless calculation yields[1]

$$D_{\pi^+/d}(z) \propto (1-z)^2. \tag{34}$$

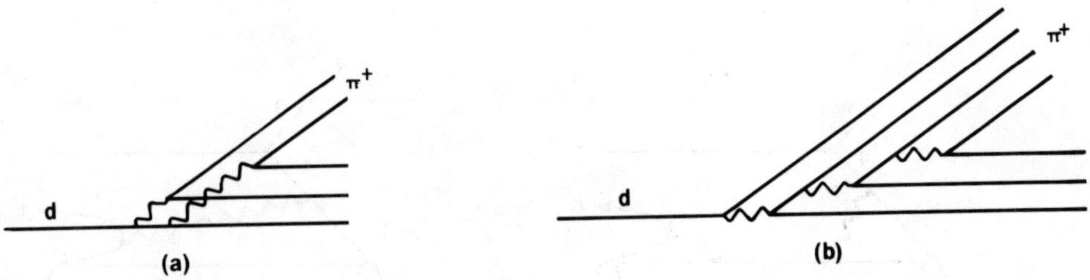

Figure 11. "Unfavored" quark fragmentation:
a) via a "point-like" lowest order QCD tree graph.
b) via a "counting-rule" or "hadronic" fragmentation graph in which all quarks appearing in the final state are connnected to a higher Fock state of the meson fragment.

It is interesting to compare this to the old "counting rule" result[29] based on the diagram of Fig. 11b which yields

$$D^{\text{counting rule}}_{\pi^+/d}(z) \propto (1-z)^5; \tag{35}$$

the higher power compared to (34) is basically due to the greater number of quarks which must be "slowed down", as the z of the π^+ approaches 1, in Fig. 11b compared to Fig. 11a. Experiment is much closer to (34).

We next consider deep inelastic spectator jet fragmentation. The struck quark will, of course, fragment as in e^+e^- annihilation. But what about the spectator jet? To illustrate an important subtlety consider a pion target. Two fragmentation diagrams of interest are shown in Fig. 12. We consider the Bjorken deep inelastic limit followed by the limit where

$$z = \frac{p^+_{M'}}{(p_{M'}+k)^+} = \frac{p^+_{M'}}{(1-x_{Bj})p^+_M} \to 1.$$

First we note that we are not restricted to valence states of the incoming meson M and outgoing meson M'. This is because there is no growing transverse momentum (unlike a form factor calculation) involved in the fragmentation as z→1 and thus, in the sample higher Fock state cases drawn, quarks 2 and 3 in Fig. 12a and quarks 1, 2, and 3 in Fig. 12b can be freely transmitted from the initial M to the final M' without being forced off-shell. The "intuitive" diagram in which the spectator jet is first created by the deep inelastic scattering and then fragments to the observed meson is

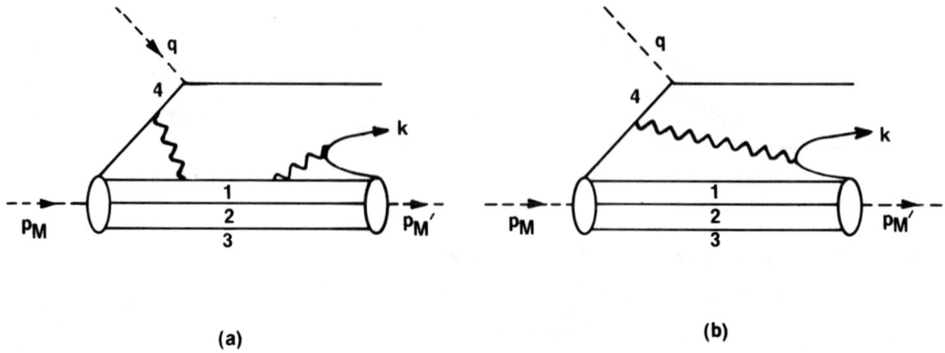

Figure 12. Meson fragmentation of the spectator jet in a deep inelastic collision on a meson target.

a) The intuitive diagram in which the fragmentation originates via a quark fragmentation within the spectator jet.

b) The dominant diagram in which the quark struck by the deep inelastic probe first fragments to a $q\bar{q}$ pair, one of which then "recombines" with the spectator jet to form the observed M' meson.

Fig. 12a. There both quark lines 1 and 4 are forced far off-shell as $z\to 1$; the off-shellness of quark line 1 must be "supplied" by the deep inelastic probe through line 4. The result for $D_{M'/SJ}(z)$ (SJ = spectator jet) is

$$D^{12a}_{M'/SJ}(z) \sim (1-z)^3. \quad (36)$$

A less intuitive diagram is Fig. 12b in which the quark to be struck by the deep inelastic probe first radiates a gluon which supplies a q and $\bar{q}$, one of which enters the final meson, M'. In Fig. 12b the only quark line forced off-shell is line 4, neighboring the deep inelastic photon. The result is

$$D^{12b}_{M'/SJ} \sim (1-z), \quad (37)$$

which is the answer one might have guessed by the incorrect reasoning that "the spectator jet is quark like and will fragment as in e^+e^- annihilation". The relative normalization of Figs. 12a and 12b has not yet been computed. However, the competition between Fig. 12a and Fig. 12b is an important subtlety of a type which will reappear in the hadronic collision situation.

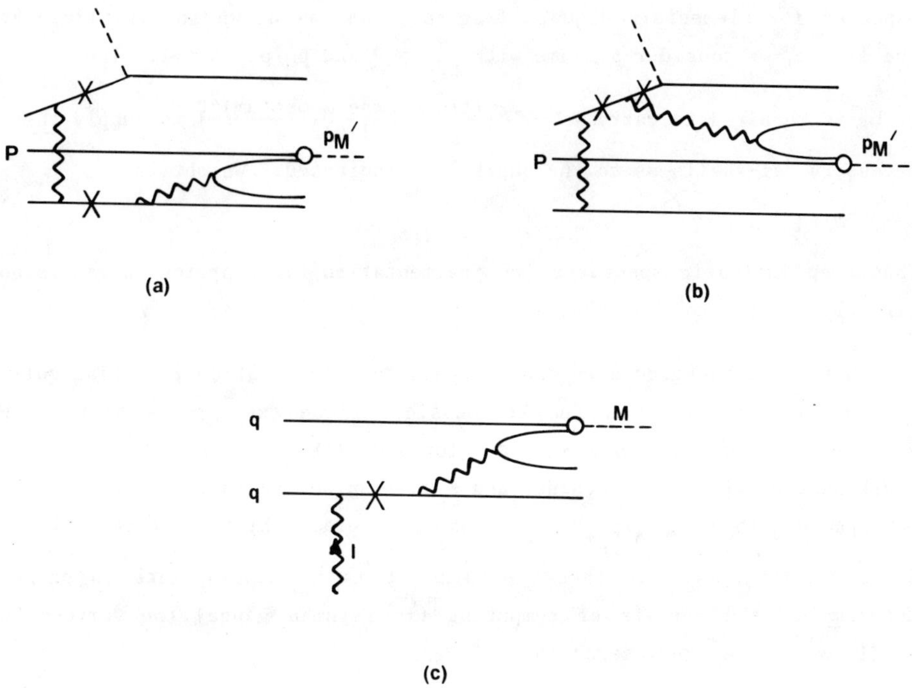

Figure 13. a) and b) Meson fragmentation of the spectator (diquark) jet in deep inelastic collisions with a proton target: a) is the intuitive diagram where the extra $q\bar{q}$ pair originates within the spectator jet; in b) the $q\bar{q}$ pair originates from the quark to be struck by the deep inelastic probe.
c) Diquark jet fragmentation in its most elementary form -- 2q collision with a probe.

One would naively anticipate dominance by (37).

Analogous results for a proton target are easily computed. At the valence Fock state level two competing diagrams are those shown in Figs. 13a and 13b. In this case both diagrams require two far off-shell quark propagators and both predict

$$D_{M'/SJ}(z) \sim (1-z)^3 \tag{38}$$

as observed experimentally. Neither diagram yields the behavior actually expected for elementary diquark fragmentation, as drawn in Fig. 13c. For the latter, we consider a frame with $p^+_{2q} = P$ and $p^+_M/p^+_{2q} = z\to 1$. The probe is required to carry $\ell_+ \sim \mathcal{O}(\frac{(\text{transverse momentum})^2}{(1-z)P})$ to supply the necessary off-shellness to the quark line indicated. We obtain

$$D_{M/2q}(z) \sim (1-z). \tag{39}$$

Thus deep inelastic spectator jet fragmentation for a proton target is not directly related to diquark fragmentation.

With this background we can now turn to fragmentation following collision of two hadrons via the Low-Nussinov gluon exchange mechanism. We illustrate by considering $\pi^+ + q \to \pi^o$ (forward) + X where the π^+ and quark scatter via gluon exchange and a π^o is produced in the forward π^+ hemisphere with $z = p^+_{\pi^o}/p^+_{\pi^+} \to 1$. Sample diagrams which could contribute appear in Fig. 14. In these lectures it is not appropriate to present details but the results of computing (in Feynman gauge) the various QCD amplitudes shown are interesting.

1) In Fig. 14a-c an extra $u\bar{u}$ pair is made by the scattering ($\bar{d}$) quark. One of these ($\bar{u}$) recombines with the other (u) quark of the incoming π^+ state to form the observed π^o. This will be termed "recombination". The amplitudes behave as $\sim(1-z)$; the indicated quark line must be far off-shell. The off-shellness is supplied from the target quark via the exchanged L-N gluon of momentum ℓ; ℓ^2 itself is, however, not far off-shell. (The off-shellness of the pair-creating gluon is compensated by a trace factor.)

2) In Fig. 14d the extra $u\bar{u}$ pair is made from a target quark. This might be termed "recombination with the target sea". However, the momentum of the $\bar{d}$ must be routed through the Low-Nussinov exchange gluon which is thus forced off-shell. The two uncompensated off-shell denominators (as indicated by x's) lead to an amplitude behaving as $\sim (1-z)^2$.

3) In Fig. 14e, where the interaction/exchange gluon creates the required $u\bar{u}$ pair, there is only one uncompensated off-shell (quark) line but there is an extra (1-z) factor from the trace associated with this gluon. Thus we obtain $\sim (1-z)^2$ again.

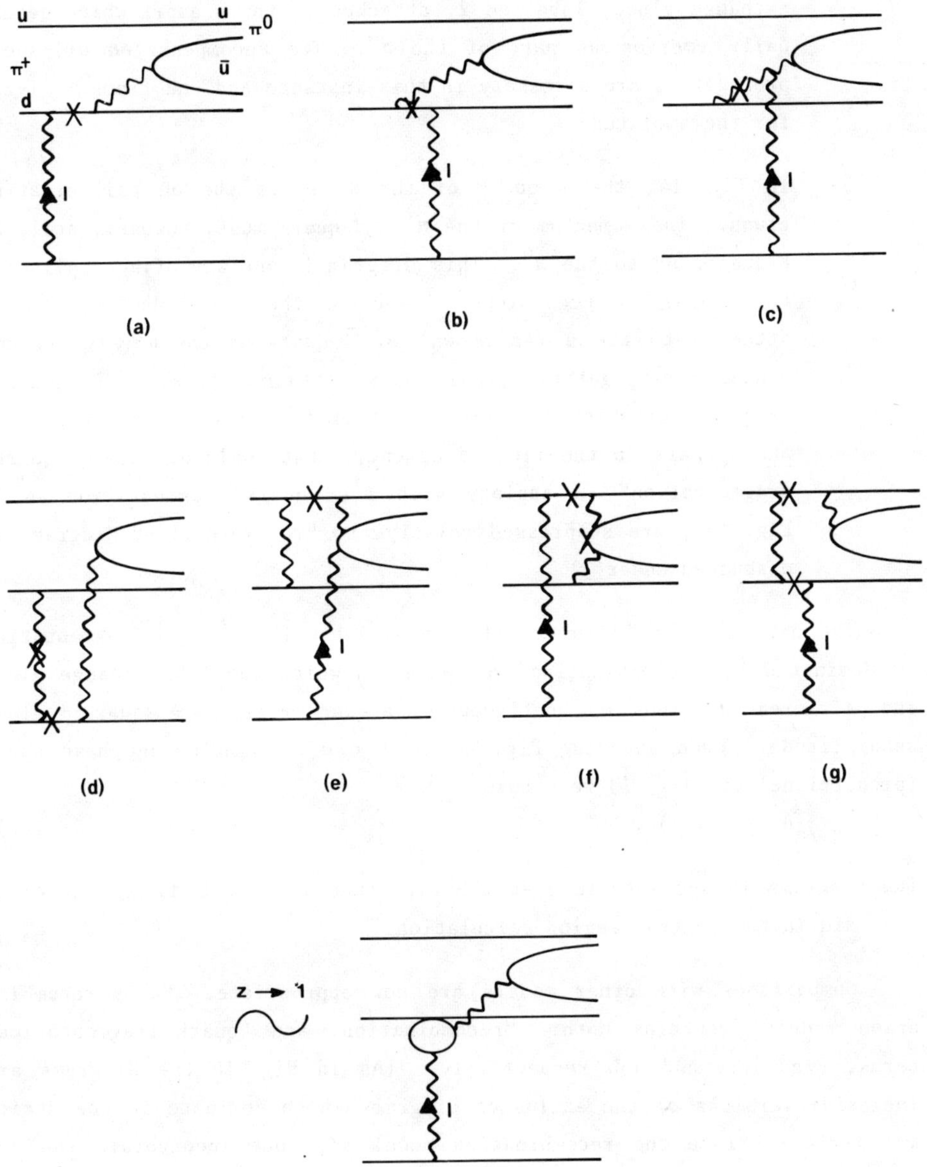

Figure 14. Fragmentation of $\pi^{+} \rightarrow \pi^{0}$. For discussion see text.

4) In Fig. 14f we draw a diagram very like 14a-c except that the L-N exchange gluon, labelled ℓ, attaches to the u quark which eventually emerges as part of the π^o. Two uncompensated off-shell propagators are necessary in this instance and one finds $\sim (1-z)^2$ for the amplitude.

5) In Fig. 14g the u quark of the π^+ emits the $u\bar{u}$ pair-creating gluon. The momentum of the π^+'s $\bar{d}$ quark must, however, still be transferred to the π^o. This diagram is one way of accomplishing the transfer. Regardless of whether the L-N exchange gluon is attached to the u (as shown) or $\bar{d}$ quark of the π^+, two uncompensated propagators appear and we obtain $\sim (1-z)^2$. In general diagrams in which the same quark emits the pair creating gluon <u>and</u> appears in the final fragment, (what could be termed "quark-fragmentation" in analogy with e^+e^- quark fragmentation as in Fig. 10), are suppressed relative to "recombination" diagrams as discussed under 1).

The net result is illustrated in Fig. 14h. The $\pi^+ \to \pi^o$ fragmentation is dominated by "recombination" diagrams in which the L-N exchange gluon and pair creating gluon are contiguous in a quark + gluon $\to$ quark + gluon subamplitude. Upon squaring Fig. 14a + b + c and including phase space (proportional to $(1-z)^{-1}$) we obtain

$$D_{\pi^+/\pi^o} \sim (1-z) \ . \tag{40}$$

The diagrams 14 a-c interfere at the amplitude level but do not cancel as they did in the central region calculation.

Comparisons with other models are now appropriate. Hwa's recombination model[24] contains both: "recombination" and "quark fragmentation" terms, Figs. 15a and 15b respectively. (As in Fig. 10 the diagrams are inclusive versions of the exclusive diagrams which dominate in the strict $z\to 1$ limit.) Since the recombination model does not incorporate the $z\to 1$ off-shell behaviors, these two types of diagram yield similar (1-z) power laws. In particular, the diagram of Fig. 15b, in which an isolated u quark gives rise to a secondary quark and antiquark, which then rejoin to give the π^o, is not posible in the $z\to 1$ limit; the u quark has to have interacted in order to acquire the necessary off-shellness. This yields extra suppression. In general, no interaction/collision mechanism is specified; it is like erasing the L-N gluon line labelled by ℓ in Fig. 14 a-c and g. As

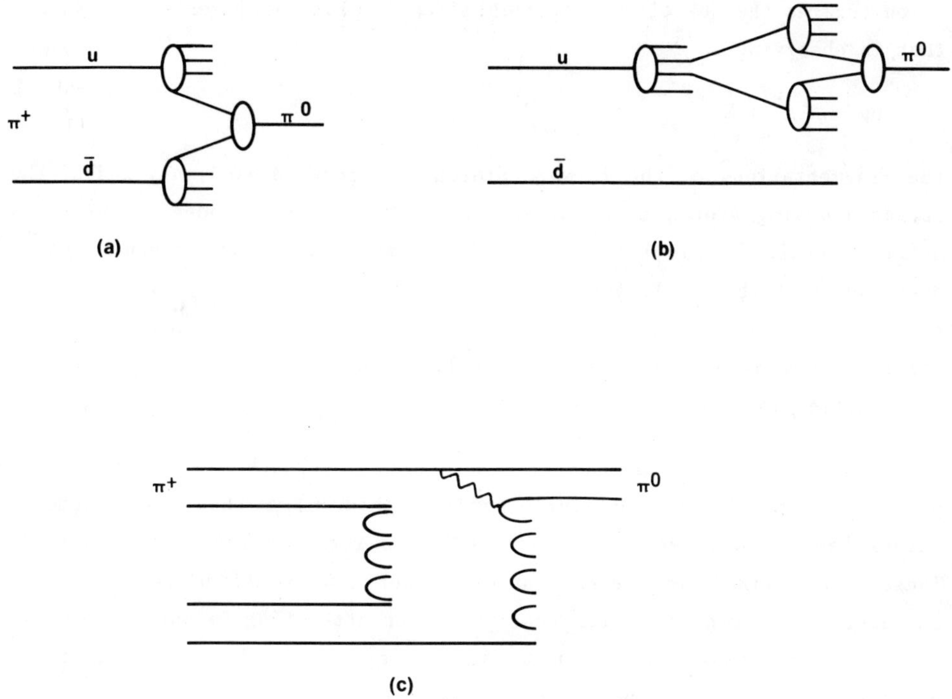

Figure 15. a) and b) Recombination model diagrams of the "recombination" type (a) and "quark fragmentation" type (b).
c) "Quark fragmentation" in the DTU model.

a result there is also no interference between different diagrams at the amplitude level. Even Figs. 15a and 15b are treated incoherently. In contrast, the QCD exchange interaction introduces a variety of interfering diagrams (each with a different color factor) and, if correct, would imply that the recombination model normalization procedure (putting aside the off-shellness problem) could not be reliable. In the DTU approach[21] the visualization of interaction followed by fragmentation, Fig. 15c, is closest to the "quark-fragmentation" diagram of Fig. 14g. We have argued that this is suppressed as $z \to 1$ relative to Fig. 14a-c, the "recombination" diagrams. Last year there was optimism[1] that the various fragmentation approaches would converge towards a common picture, if not a common calculation; instead it seems that the gap is widening.

A further phenomenological point concerning the L-N gluon exchange interaction together with "recombination" as the mechanism for fragmentation regards the question of factorization. It is experimentally apparent[31] that in observing

$$pp \to \pi^F \pi^B + X \tag{41}$$

the fragmentations of the forward proton to a forward pion, π^F, and of the backward moving proton to a backward pion, π^B, are independent. This is a natural result in the L-N gluon exchange model since the gluon exchange does not disturb the longitudinal "+" ("-") momentum distribution of the colliding forward-moving (backward-moving) hadron or the quark content. Other hadron interaction models involving quark exchange are prone to difficulties in this regard (see Ref. 1).

The implications of the "recombination" approach for fragmentation power laws of all types of single particle $(\text{hadron})_1 \to (\text{hadron})_2$ fragmentations have been given elsewhere[23] and were recently reviewed in Ref. 1. These "point-like" or "recombination" power laws continue be remarkably successful. Future work will concentrate on attempting to obtain approximate normalizations for the underlying diagrams (including interference effects) and on extending the power law predictions to processes such as $p \to (\pi^+\pi^-)$ and $p \to (\pi^-\pi^-)$, i.e., bi-hadron fragmentation.

5. Summary and Conclusion:

These lectures have attempted to show that pertubative QCD provides a simple, phenomenologically successful, and yet calculationally well-defined description of low-p_T phenomena. Starting with the lowest order Low-Nussinov gluon exchange model for hadron interactions we have obtained a qualitatively and on occasion quantitatively successful description of total cross sections, central region multiplicity, and fragmentation distributions. The multiplicity and fragmentation distributions are computed using the lowest possible order diagrams: one gluon radiation for central multiplicity and leading order exclusive diagrams for fragmentation. Ample room for qualitative disagreement with data existed. Instead the perturbative diagrams appear to give a good first order description of the data, and in addition provide natural dynamical explanations for some features of experiment which have previously been treated on a largely phenomenological basis (e.g., total cross section counting; the relation between hadron size and total cross section size; the possible plateau height difference between hadron-hadron and e^+e^- collisions; and the power counting rules for quark, deep inelastic spectator jet and hadron fragmentation). In all cases the subtle gauge invariance/color features of QCD play a significant role. The obvious question is: "Why should one believe that perturbative calculations provide a reliable guide to the dynamics underlying processes which are not obviously short distance dominated"? We have not provided an answer to this question. It may be pursued through attempting to make contact between the perturbative and dual model approaches and by pursuing the perturbative calculations to higher order. The perturbative calculations are sufficiently successful that effort in these directions may prove worthwhile.

Acknowledgments

I would like to thank A. Capella, R. Hwa, S.J. Brodsky, W. Iley, and G. Bertsch for helpful conversations. I would like to thank the Institute for Theoretical Physics for its hospitality during preparation of these lectures and, of course, the Erice School and R. Van de Walle for the opportunity to present them. The work was also supported by DOE (grant #DE AS03-76SF00034Pa101) and the A.P. Sloan foundation.

References

1. J.F. Gunion, "Quarks and Gluons in Low-p_T Physics". Proceedings of the XI International Symposium on Multi-Particle Dynamics, Bruges, Belgium, June 22-27, 1980. Edited by E. De Wolf and F. Verbeure.

2. F.E. Low, Phys. Rev. D12,163 (1975); S. Nussinov, Phys. Rev. Lett. 34, 1286 (1975).

3. J.F. Gunion and D.E. Soper, Phys. Rev. D15,2617 (1977).

4. G. Bertsch, S.J. Brodsky, F. Goldhaber, J.F. Gunion, Institute for Theoretical Physics Preprint, May 1980.

5. J. Pumplin and E. Lehman, Zeit. f. Phys. C, to be published.

6. G.P. Lepage and S.J. Brodsky, Phys. Rev. D22,2157 (1980).

7. S.J. Brodsky, P. Hoyer, C. Peterson, and N. Sakai, Phys. Lett. 93B, 451 (1980); and S.J. Brodsky, C. Peterson, and N. Sakai, SLAC-PUB-2660 (1980).

8. J.F. Donoghue and E. Golowich, Phys. Rev. D15,3425 (1977).

9. H. Wachsmuth, Proceedings of the Photon-Lepton Conference, Fermilab, 1979.

10. See the review by K. Konishi, Proceedings of the XI International Symposium on Multi-Particle Dynamics, Bruges, Belgium (1980).

11. Early discussion of such comparisons appears in: S.J. Brodsky and J.F. Gunion, Phys. Rev. Lett. 37,402 (1976) and Proceedings of 7th International Colloquium on Multiparticle Reactions, Munich 1976; J.F. Gunion, Proceedings of the 13th Rencontre de Moriond, Vol. I, (1978).

12. G. Bertsch and J.F. Gunion in preparation.

13. A. Bassetto, M. Ciafaloni and G. Marchesini, Nuclear Phys. B163,477 (1980), and Phys. Lett. 83B,207 (1979). W. Furmanski, R. Petronzio and S. Pokorski, Nuclear Phys. B155,253 (1979).

14. See H.V. Martyn, Proceedings of the 1981 Moriond Conference, Les Arcs, France.

15. R. Feynman and R. Field, Nucl. Phys. B136,1 (1978). The calculations discussed by Ref. 14 include all modifications as incorporated by the experimental groups at DESY.

16. W. Thome et al., Nucl. Phys. B129,365 (1977).

17. See presentations by H. Lubatti and W. Kittel. See also M. Basile et al., Phys. Lett. 92B,367 (1980).

18. S. Uhlig et al. Nucl. Phys. B132,15 (1978); T. Kafka et al., Phys. Rev. Lett. 34,687 (1975).

19. S. Pokorski and S. Wolfram, Zeit Phys. C4,2 (1980).

20. For a parton level model of such an approach see S.J. Brodsky, J.F. Gunion, and J.H. Kühn, Phys. Rev. Lett. 39,1120 (1977). The basic model is due to R.P. Feynman; Photon-Hadron Interactions, (Benjamin, 1972) and references therein.

21. See the talks by A. Capella and B. Andersson at this meeting. See also B. Andersson, G. Gustafson, and C. Peterson, Phys. Lett. 71B,337 (1977); R. Peschanski, Proceedings of XI Int. Symposium on Multiparticle Dynamics, Bruges, Belgium (1980); and A. Capella, ibid. and A. Capella, U. Sukhatme, and J. Tran Thanh Van, Zeit. f. Phys. C,329 (1980). The particular comparison made is based mostly on the work of Capella, et al.

22. This point is discussed at greater length in Ref. 1.

23. J.F. Gunion, Phys. Lett. 88B,150 (1979).

24. See R. Hwa's lectures at this conference, and references given therein.

25. G.P. Lepage and S.J. Brodsky SLAC-PUB-2478 (1980), Phys. Rev. D22,2157 (1980).

26. J.F. Gunion and W. Iley, in preparation; the other contributing graph (in Feynman gauge) is more complex in nature and will be dealt with in this paper.

27. See for instance: N. Schmitz, Proceedings of 1978 Lepton Photon Symposium (Fermilab), for the experimental distribution.

28. For quark fragmentation see, in particular, Van Chang and R. Hwa, Phys. Rev. Lett. 44,439 (1980).

29. R. Blankenbecler and S.J. Brodsky, Phys. Rev. D10,2973 (1974); J.F. Gunion, ibid. 242; G.R. Farrar, Nucl. Phys. B77,429 (1974).

30. This is contrary to the more naive survey given in Ref. 1.

31. G.J. Bobink et al., Phys. Rev. Lett. 44,118 (1980).

[illegible] (1980).

[illegible] Gutiérrez, and [illegible] Phys. Rev. Letters 49, [illegible] [illegible] (1972) [illegible]

[illegible] B. Andersen, [illegible] (1977) [illegible] (1960) [illegible]

[illegible]

[illegible] (1967).

[illegible] J.R. [illegible]

[illegible] See, for instance, [illegible] Stanley, [illegible] (1971) [illegible] experimental situation.

[illegible] Phys. Rev. Lett. [illegible]

[illegible] Blackman [illegible] Phys. Rev. [illegible]

[illegible]

[illegible] Smith et al, Phys. Rev. [illegible] (1968).

Chapter IX: CONCLUDING REMARKS

H.I. MIETTINEN
Research Institute for Theoretical Physics
University of Helsinki
Helsinki, Finland

Abstract:

Some remarks on the dynamics of soft multiparticle collisions are presented. The observed similarity between jets produced in small-p_t hadron-hadron collisions and in hard collisions such as e^+e^- annihilation is emphasized. Several models for particle production and the relationships between them are discussed.

In this talk I shall consider the problem of multiparticle dynamics from a very global point of view. During the past week, we have heard many lectures on specific phenomenological approaches to multiparticle phenomena. As we have seen, some of these approaches very succesfully describe the experimental data. I am quite sure that this is not an accident. These approaches employ much good physics, and probably provide a fairly correct overall picture of the evolution of a multiparticle production process. On the other hand, Quantum Chromodynamics provides a theory for strong interaction phenomena, and we do not yet know how to calculate the properties of soft collisions from this theory. Nevertheless, we may speculate that the general picture of soft multi-body phenomena which follows from QCD will turn out to be very similar to the phenomenological pictures provided by our current models.

I do not have any deep insights to offer, and I shall therefore limit myself to some remarks on the data and ideas presented at this Conference. I shall first discuss the experimentally observed similarity between multiparticle jets produced in soft hadron-hadron collisions and in hard collisions such as electron-positron annihilation. I believe that this similarity deserves serious attention, both from experimentalists and from theorists. Next I shall discuss some models. I feel that some of the difference of thought between the "Recombination School" and the "Fragmentation School" has been overemphasized. In fact, it seems to me that these two camps have much in common and that it may well turn out that they both present an approximately correct point of view. The extremely interesting field of hadron-nucleus collisions was reviewed by A. Dar(1). Since I have nothing to add to (or substract from) his excellent review, I shall not discuss this topic.

Universal Hadronization?

The data on multiparticle jets were reviewed by W. Kittel(2) and by H. Lubatti(3). The comparisons between the jets produced in soft and in hard collisions are ad hoc comparisons, since we have no good theoretical reasons to believe the two kinds of jets to be similar (or different, for that matter). The most important jet properties to be compared are, in order of increasing com[lexity:

a) multiplicity
b) shape i.e. differential particle density
c) flavour content
d) correlation structure

Experiment shows that the jets produced in soft and in hard collisions have a) the same multiplicity and b) the same event shape. This statement holds on about ten percent level. On points c) and d), not much is yet known. What little information is available (on the kaon-to-pion ratio, on the vector meson production rates and on the charge distributions within the jets) seems not to disagree with the hyphothesis of jet universality.

It seems to me fair to conclude that the data provide clear support for the idea that the jets produced in soft and in hard collisions have the same overall structure. Further experimental studies are needed to learn if the microscopic correlations in the two types of jets are similar.

The dynamics of particle production in hard collisions, such as electron-positron annihilation, is qualitatively, although not quantitatively, understood. The underlying reason for particle production is separation of colour sources in space, and the mechanism is hadronization of the colour flux tube connecting these sources. The properties of the produced jets, such as the particle density, should be the same in all processes in which similar colour charges are separated. The jets develop from inside to outside, the fastest particles being produced last.

Let us accept the above picture of jet development in e^+e^- collisions and also - tentatively - the similarity between the e^+e^- jets and the soft jets produced in hadron-hadron collisions. Then it is very natural - almost necessary, I would say - to conclude that particle production in hadron-hadron collisions must also be due to separation of colour charges. But which colour charges? How and why are they separating? How is such a picture related to Feynman's parton-model picture of soft-hadron collisions?(4) I believe that these are the key questions, and by providing good answers to them we will male real progress in understanding multiparticle dynamics.

Fragmentation Spectra

Another important experimental conclusion which has been determined is the similarity between the longitudinal momentum spectra of fast particles produced in hadron-hadron collisions and the hadrons' quark momentum distributions(5). Some people take the point of view that this similarity is accidental. I do not believe that this is so. It seems to me that the observed regularities are very clear indeed, and that the simple explanation the parton model approach provides for them, namely, that fast particles are produced by fast quarks and thus their spectra should be similar(6), is likely to be a correct one.

Models

Many authors have tried to understand soft multiple production within the framework of QCD. Some of these attempts were discussed at length by J. Gunion(7). Although I find these studies extremely interesting, I do not believe that they bear relation to what is seen experimentally. My point is, simply, that the QCD calculation apply low-order perturbation theory, whereas the low-p_t phenomena are entirely dominated by nonperturbative effects due to the confinement of colour.

Consider a two-parton interaction due to gluon exchange. The cross-section will behave as

$$\sigma \sim \exp[(J-1)y_{12}]$$

where y_{12} is the rapidity difference of the two partons and J is the gluon spin. For a vector gluon J=1 and the interaction probability is independent of the rapidity difference y_{12}. Such a behaviour directly contradicts the hypothesis of short-range-order of the hadrons' parton wave-functions, which corresponds to $\sigma \sim \exp(-y_{12}/L)$ with the correlation length $L \approx 1 - 2$ units of rapidity.

The short-range-order parton model and QCD may well turn out to be compatible. If so, the solution must lay in multi-gluon exchanges and in gluon-gluon interactions, which may build up an approximately exponentially decreasing short-range rapidity interaction. I do not know how to prove this results in QCD.

Several models relate soft multiparticle production directly to separation of colour charges in space-time.[7,8,9] In the model of Low and Nussinov, the separating colour systems are two octets. The colour flux tube connecting two 8's should be much "thicker" than that connecting a 3 and a $\bar{3}$, and the multiplicity resulting from its break-up should consequently be larger. Naive QCD arguments suggest a multiplicity ratio of 9 to 4. This prediction disagrees with experiment which suggests a ratio of 1:1. Both in the Brodsky-Gunion model and in the Lund model the separating coloured systems are assumed to be a 3 and a $\bar{3}$. The equality of the hadron-hadron and the e^+e^- multiplicities then follows. These models have several nice features but also some weak points. In particular, no natural explanation is given why the separating coloured systems should be a 3 and a $\bar{3}$. Furthermore, the qualitative picture of the collision dynamics is different from the short-range-order Feynman picture.

The quark recombination model[6] in its various versions (the Kuti-Weisskopf approach, the valon approach, the quark-diquark approach) is a simple phenomenological realization of the Feynman picture of soft hadron collisions. It emphasizes short-range-order and colour confinement. As far as I understand, the quark-antiquark recombination mechanism for the transformation of partons into hadrons should not be interpreted literally as a model for the microscopic dynamics but rather as an average description which emphasizes the semi-local nature (in rapidity) of this transformation.

The recombination model provides a good phenomenological description for the spectra of fast particles produced in hadron-hadron collisions.[2] Baryon production has not yet been treated very satisfactorily. The problem with baryons is in part related to that of describing the Regge behaviour of the inclusive spectra (the baryon spectra at large x_F exhibit clear Regge behaviour). It is easy to see qualitatively how Regge behaviour should appear through coherent many-parton recombination in this model. A quantitative treatment is outside the scope of the present versions of the model, since Regge behaviour follows from coherence of amplitudes, whereas this model deals with probabilities.

The Dual Topological Model[9] described by A. Capella in his lectures has many appealing features. It is a short-range-order model, it provides a correct description for the space-time evolution of the hadronization process, and it connects nicely to the two-component duality picture of Harari and Freund. Furthermore, the connection to the Regge model approach is very straightforward.

A somewhat disturbing property of this model is that the asymptotic ratio of the hadron-hadron to the e^+e^- jet multiplicity is predicted to be 2:1. In actual fits to data, the experimentally observed ratio 1:1 is obtained. This result, however, is a numerical accident and valid only in a limited energy range.

The weakest point of this model is, in my opinion, the assumption that the two dual "sheets" of quark lines should evolve independently. The production of these sheets is interpreted as being due to two $3\bar{3}$ systems ($3_{beam}\bar{3}_{target}$ and $\bar{3}_{beam}3_{target}$) separating independently. To me such an assumption appears rather unjustified. The 3 and the $\bar{3}$ of the beam particle together form a colour neutral bound state. Thus it seems strange that the 3 of the beam particle would interact with the $\bar{3}$ of the target particle with no interference from the $\bar{3}$ it is bound to.

Fragmentation = Recombination?

My last remark concerns the relationship between the fragmentation model and the recombination model. During the past week's discussions, the supporters of these two approaches have emphasized the differences between these points of view. So let me emphasize their similarity. In particular, the space-time evolution of the hadronization process is the same in both approaches: slow particles materialize first, and faster particles are produced later, until finally (after a long time ~ p_{lab}) the very fast beam fragments materialize. In the recombination approach (which, as I emphasized, follows closely Feynman's original ideas) the partons are present in the beam particle's wavefunction before the initial interaction with the target. In the fragmentation model approach, on the other hand, the sea partons are seen appearing only <u>after</u> the interaction. How important is this difference?

My own point of view is that this difference is probably not very important. The two approaches may well be just (somewhat) different ways of looking at the same phenomenon: soft materialization of a short-range-ordered distribution of almost-on-mass-shell partons. The important question to be answered is: How is this distribution produced? Either as a result of boosting a bound system of coloured quarks from the rest frame to the large-momentum frame (the recombination point of view) or by the constituent collisions (the fragmentation point of view)?

Conclusions

In conclusion, I repeat three points:

i) Particle production in soft hadron-hadron collisions shows features very similar to those of electron-positron annihilation and of deep-inelastic lepton-hadron collisions. Further experimental studies are needed to clarify if this similarity also remains true on the level of multiparticle correlations.

ii) The "fragmentation" and the "recombination" approaches are not as dissimilar as it might seem. Important questions remain to be answered within both of these approaches. In particular, the colour dynamics underlying these approaches needs clarification.

iii) A long step forward would be to understand - at least qualitatively, if not quantitatively - how these phenomenological approaches relate to QCD. My quess is that QCD will yield Feynman's short-range-order parton picture of soft collisions.

Acknowledgements

It is a pleasure to thank S. Ratti and R.T. Van de Walle for the invitation to speak and for making the conference pleasant and stimulating. Discussions with many people, but especially with Tom DeGrand and Rudy Hwa, have helped me to understand the parton model approach to soft collisions. I am grateful to Larry McLerran for comments on the manuscript.

References

1. A. Dar, "The Quark-Parton Model and Production of Low p_t Particles in Nuclei", Lectures at this Conference, p.

2. W. Kittel, "Soft Hadronic Interactions - Review of Three Basic Observations", Lectures at this Conference, p.

3. H. Lubatti, "Lepton-Hadron Interactions and Nucleon Structure", Lectures at this Conference, p.

4. R.P. Feynman, Photon-Hadron Interactions, W.A. Benjamin, Inc., 1972; J. Kogut and L. Susskind, Phys. Reports 8, 75 (1973).

5. W. Ochs, Nucl. Physics B118, 397 (1977); Data are reviewed by Kittel (Ref. 2).

6. K.P. Das and R.C. Hwa, Phys. Letters 68B, 459 (1977); R.C. Hwa, "Hadron Structure and Soft Processes", Lectures at this Conference, p.

7. J. Gunion, "QCD and Low-p_t Physics", Lectures at this Conference, p.

8. O. Sukhatme, "Models for Color Jets and their Fragmentation Functions", Lectures at this Conference, p.

9. A. Capella, "Dual Fragmentation Models of Soft Hadron-Hadron and Hadron-Nucleus Interactions", Lectures at this Conference, p.